A Guide to Six Sigma and Process Improvement for Practitioners and Students

Second Edition

A Guide to Six Sigma and Process Improvement for Practitioners and Students

Foundations, DMAIC, Tools, Cases, and Certification
Second Edition

Howard S. Gitlow
Richard J. Melnyck
David M. Levine

Publisher: Paul Boger
Editor-in-Chief: Amy Neidlinger
Executive Editor: Jeanne Glasser Levine
Operations Specialist: Jodi Kemper
Cover Designer: Alan Clements
Managing Editor: Kristy Hart
Senior Project Editor: Betsy Gratner
Copy Editor: Geneil Breeze
Proofreader: Laura Hernandez
Indexer: Erika Millen
Senior Compositor: Gloria Schurick
Manufacturing Buyer: Dan Uhrig

© 2015 by Pearson Education, Inc.
Old Tappan, New Jersey 07675

For information about buying this title in bulk quantities, or for special sales opportunities (which may include electronic versions; custom cover designs; and content particular to your business, training goals, marketing focus, or branding interests), please contact our corporate sales department at corpsales@pearsoned.com or (800) 382-3419.

For government sales inquiries, please contact governmentsales@pearsoned.com.

For questions about sales outside the U.S., please contact international@pearsoned.com.

Company and product names mentioned herein are the trademarks or registered trademarks of their respective owners.

All rights reserved. No part of this book may be reproduced, in any form or by any means, without permission in writing from the publisher.

Printed in the United States of America

Second Printing May 2017

ISBN-10: 0-13-392536-6
ISBN-13: 978-0-13-392536-4

Pearson Education LTD.
Pearson Education Australia PTY, Limited
Pearson Education Singapore, Pte. Ltd.
Pearson Education Asia, Ltd.
Pearson Education Canada, Ltd.
Pearson Educación de Mexico, S.A. de C.V.
Pearson Education—Japan
Pearson Education Malaysia, Pte. Ltd.

Library of Congress Control Number: 2015932281

This book is dedicated to:
Shelly Gitlow
Ali Gitlow
Abraham Gitlow
Beatrice Gitlow

Jack Melnyck
Eileen Melnyck

Lee Levine
Reuben Levine

Contents

Section I   **Building a Foundation of Process Improvement Fundamentals**

Chapter 1   **You Don't Have to Suffer from the Sunday Night Blues! 1**
- What Is the Objective of This Chapter? 1
- Sarah's Story 2
- Nine Principles of Process Improvement to Get the Most Out of This Book 3
- Structure of the Book 14
- Let's Go! 16
- References 16

Chapter 2   **Process and Quality Fundamentals 17**
- What Is the Objective of This Chapter? 17
- Process Fundamentals 18
  - What Is a Process? 18
  - Where Do Processes Exist? 18
  - Why Does Understanding Processes Matter? 19
  - What Is a Feedback Loop and How Does It Fit into the Idea of a Process? 19
  - Some Process Examples to Bring It All Together! 19
- Variation Fundamentals 24
  - What Is Variation in a Process? 24
  - Why Does Variation Matter? 25
  - What Are the Two Types of Variation? 25
  - How to Demonstrate the Two Types of Variation 27
  - Red Bead Experiment 30
- Quality Fundamentals 31
  - Goal Post View of Quality 31
  - Continuous Improvement Definition of Quality—Taguchi Loss Function 32
  - More Quality Examples 33
- Takeaways from This Chapter 33
- References 34

Chapter 3   **Defining and Documenting a Process**..........................35
    What Is the Objective of This Chapter?..35
    A Story to Illustrate the Importance of Defining and Documenting a Process......35
    Fundamentals of Defining a Process..36
        Who Owns the Process? Who Is Responsible for the Improvement of the Process?..36
        What Are the Boundaries of the Process?..................................37
        What Are the Process's Objectives? What Measurements Are Being Taken on the Process with Respect to Its Objectives?..................38
    Fundamentals of Documenting a Process..39
        How Do We Document the Flow of a Process?..............................39
        Why and When Do We Use a Flowchart to Document a Process?............39
        What Are the Different Types of Flowcharts and When Do We Use Each?....40
        What Method Do We Use to Create Flowcharts?............................43
    Fundamentals of Analyzing a Process...44
        How Do We Analyze Flowcharts?..44
        Things to Remember When Creating and Analyzing Flowcharts............45
    Takeaways from This Chapter..46
    References..46

Section II   Creating Your Toolbox for Process Improvement

Chapter 4   **Understanding Data: Tools and Methods**......................47
    What Is the Objective of This Chapter?..47
    What Is Data?..47
        Types of Numeric Data..47
    Graphing Attribute Data...50
        Bar Chart..50
        Pareto Diagrams..51
        Line Graphs..52
    Graphing Measurement Data..54
        Histogram..54
        Dot Plot...55
        Run Chart..56
    Measures of Central Tendency for Measurement Data...............................59
        Mean...59
        Median...60
        Mode...60

> Measures of Central Tendency for Attribute Data 61
>> Proportion 61
> Measures of Variation 62
>> Range 62
>> Sample Variance and Standard Deviation. 63
>> Understanding the Range, Variance, and Standard Deviation 64
> Measures of Shape. 66
>> Skewness 66
> More on Interpreting the Standard Deviation 68
> How-To Guide for Understanding Data: Minitab 17 User Guide 70
>> Using Minitab Worksheets 70
>> Opening and Saving Worksheets and Other Components 71
>> Obtaining a Bar Chart 74
>> Obtaining a Pareto Diagram 76
>> Obtaining a Line Graph (Time Series Plot). 78
>> Obtaining a Histogram 79
>> Obtaining a Dot Plot. 82
>> Obtaining a Run Chart. 84
>> Obtaining Descriptive Statistics 85
> Takeaways from This Chapter. 87
> References 88
> Additional Readings 88

Chapter 5 Understanding Variation: Tools and Methods 89
> What Are the Objectives of This Chapter?. 89
> What Is Variation? 89
>> Common Cause Variation. 89
>> Special Cause Variation 90
> Using Control Charts to Understand Variation 90
>> Attribute Control Charts 90
>> Variables Control Charts 91
>> Understanding Control Charts 91
>> Rules for Determining Out of Control Points. 93
> Control Charts for Attribute Data. 98
>> P Charts 98
>> C Charts 104
>> U Charts 106

         Control Charts for Measurement Data..108
              Individuals and Moving Range (I-MR) Charts........................109
              X Bar and R Charts...112
              X Bar and S Charts...115
         Which Control Chart Should I Use?..119
              Control Chart Case Study..119
         Measurement Systems Analysis..126
              Measurement System Analysis Checklist..............................126
              Gage R&R Study...127
         How-To Guide for Understanding Variation: Minitab User Guide
         (Minitab Version 17, 2013)...131
              Using Minitab to Obtain Zone Limits................................131
              Using Minitab for the P Chart......................................131
              Using Minitab for the C Chart......................................134
              Using Minitab for the U Chart......................................134
              Using Minitab for the Individual Value and Moving Range Charts.........136
              Using Minitab for the X Bar and R Charts............................137
              Using Minitab for the X Bar and S Charts............................139
         Takeaways from This Chapter...142
         References...143
         Additional Readings..143

Chapter 6   **Non-Quantitative Techniques: Tools and Methods**............**145**
         What Is the Objective of This Chapter?....................................145
         High Level Overview and Examples of Non-Quantitative Tools and Methods....145
              Flowcharting..146
              Voice of the Customer (VoC).......................................146
              Supplier-Input-Process-Output-Customer (SIPOC) Analysis..............149
              Operational Definitions...151
              Failure Modes and Effects Analysis (FMEA)..........................153
              Check Sheets..153
              Brainstorming...155
              Affinity Diagrams..156
              Cause and Effect (Fishbone) Diagrams...............................157
              Pareto Diagrams...159
              Gantt Charts..159
              Change Concepts...160
              Communication Plans...163

        How-To Guide for Using Non-Quantitative Tools and Methods 165
            How to Do Flowcharting 165
            How to Do a Voice of the Customer (VoC) Analysis 166
            How to Do a SIPOC Analysis 172
            How to Create Operational Definitions 173
            How to Do a Failure Modes and Effects Analysis (FMEA) 174
            How to Do Check Sheets 177
            Brainstorming 179
            How to Do Affinity Diagrams 181
            How to Do Cause and Effect Diagrams (C&E Diagrams) 182
            How to Do Pareto Diagrams 182
            How to Do Gantt Charts 185
            How to Use Change Concepts 185
            How to Do Communication Plans 198
        Takeaways from This Chapter 200
        References 201
        Additional Readings 201

Chapter 7   Overview of Process Improvement Methodologies 203
        What Is the Objective of This Chapter? 203
        SDSA Cycle 203
            SDSA Example 204
        PDSA Cycle 206
            PDSA Example 207
        Kaizen/Rapid Improvement Events 209
            Kaizen/Rapid Improvement Events Example 210
        DMAIC Model: Overview 212
            Define Phase 213
            Measure Phase 213
            Analyze Phase 214
            Improve Phase 216
            Control Phase 216
            DMAIC Model Example 216
        DMADV Model: Overview 218
            Define Phase 218
            Measure Phase 218
            Analyze Phase 218
            Design Phase 219

      Verify/Validate Phase...219
      DMADV Model Example...219
   Lean Thinking: Overview...221
      The 5S Methods 221
      Total Productive Maintenance (TPM)................................223
      Quick Changeover (Single Minute Exchange of Dies—SMED)............224
      Poka-Yoke 225
      Value Streams 226
   Takeaways from This Chapter..227
   References 228

Chapter 8 **Project Identification and Prioritization:**
      Building a Project Pipeline..................................231
   What Is the Objective of This Chapter?......................................231
   Project Identification 231
      Internal Proactive...232
      Internal Reactive..234
      External Proactive...235
      External Reactive 236
   Using a Dashboard for Finding Projects 238
      Structure of a Managerial Dashboard 238
      Example of a Managerial Dashboard.................................239
      Managing with a Dashboard...240
   Project Screening and Scoping 240
      Questions to Ask to Ensure Project Is Viable 241
      Estimating Project Benefits...242
      Project Methodology Selection—Which Methodology Should I Use?.......243
      Estimating Time to Complete Project 245
      Creating a High Level Project Charter................................246
      Problem Statement...247
   Prioritizing and Selecting Projects 247
      Prioritizing Projects Using a Project Prioritization Matrix 248
      Final Project Selection 250
   Executing and Tracking Projects 250
      Allocating Resources to Execute the Projects..........................250
      Monthly Steering Committee (Presidential) Reviews 251
   Takeaways from This Chapter..251
   References 252

Section III Putting It All Together—Six Sigma Projects

Chapter 9 Overview of Six Sigma Management 253

What Is the Objective of This Chapter? 253
Non-Technical Definition of Six Sigma Management 253
Technical Definition of Six Sigma. 253
Where Did Six Sigma Come From? 253
Benefits of Six Sigma Management 254
Key Ingredient for Success with Six Sigma Management 255
Six Sigma Roles and Responsibilities 255
    Senior Executive 255
    Executive Steering Committee 256
    Project Champion. 256
    Process Owner. 257
    Master Black Belt 257
    Black Belt. 258
    Green Belt 259
    Green Belt Versus Black Belt Projects 260
Six Sigma Management Terminology 260
Next Steps: Understanding the DMAIC Model. 264
Takeaways from This Chapter. 264
References 265
Additional Readings 265
Appendix 9.1 Technical Definition of Six Sigma Management 266

Chapter 10 DMAIC Model: "D" Is for Define. 273

What Is the Objective of This Chapter? 273
Purpose of the Define Phase 273
The Steps of the Define Phase 274
    Activate the Six Sigma Team. 274
    Project Charter 276
    SIPOC Analysis. 283
    Voice of the Customer Analysis 286
    Definition of CTQ(s) 288
    Create an Initial Draft of the Project Objective. 289
    Tollgate Review: Go-No Go Decision Point 290
Keys to Success and Pitfalls to Avoid in the Define Phase. 291

	Case Study of the Define Phase: Reducing Patient No Shows in an Outpatient Psychiatric Clinic—Define Phase 292
	Activate the Six Sigma Team. 292
	Project Charter 293
	SIPOC Analysis 299
	Voice of the Customer Analysis 299
	Definition of CTQ(s) 308
	Initial Draft Project Objective. 308
	Tollgate Review: Go-No Go Decision Point 308
	Takeaways from This Chapter. 309
	References 310
	Additional Readings 310

Chapter 11 DMAIC Model: "M" Is for Measure 311

What Is the Objective of This Chapter? 311
Purpose of the Measure Phase. 311
The Steps of the Measure Phase 312
    Operational Definitions of the CTQ(s) 312
    Data Collection Plan for CTQ(s). 312
    Validate Measurement System for CTQ(s) 313
    Collect and Analyze Baseline Data for the CTQ(s). 317
    Estimate Process Capability for CTQ(s) 321
Keys to Success and Pitfalls to Avoid 324
Case Study: Reducing Patient No Shows in an Outpatient Psychiatric
Clinic—Measure Phase 324
    Operational Definition of the CTQ(s) and Data Collection
    Plan for the CTQ(s) 325
    Validate Measurement System for CTQ(s) 326
    Collect and Analyze Baseline Data. 328
    Tollgate Review: Go-No Go Decision Point 330
Takeaways from This Chapter. 330
References 331
Additional Readings 331

Chapter 12 DMAIC Model: "A" Is for Analyze. 333

What Is the Objective of This Chapter? 333
Purpose of the Analyze Phase 333
The Steps of the Analyze Phase. 334
    Detailed Flowchart of Current State Process 334
    Identification of Potential Xs for CTQ(s) 335

    Failure Modes and Effects Analysis (FMEA) to Reduce the Number of Xs ...338
    Operational Definitions for the Xs..338
    Data Collection Plan for Xs..339
    Validate Measurement System for X(s)...................................340
    Test of Theories to Determine Critical Xs................................340
    Develop Hypotheses/Takeaways about the Relationships between the
    Critical Xs and CTQ(s)..342
    Go-No Go Decision Point 342
  Keys to Success and Pitfalls to Avoid..343
  Case Study: Reducing Patient No Shows in an Outpatient Psychiatric Clinic—
  Analyze Phase 344
    Detailed Flowchart of Current State Process 344
    Identification of Xs for CTQ(s)..344
    Failure Modes and Effects Analysis (FMEA) to Reduce the Number of Xs ...346
    Operational Definitions of the Xs.......................................346
    Data Collection Plan for Xs...348
    Validate Measurement System for Xs 348
    Test of Theories to Determine Critical Xs................................348
    Develop Hypotheses/Takeaways about the Relationships between the
    Critical Xs and CTQ(s)..354
    Tollgate Review—Go-No Go Decision Point 354
  Takeaways from This Chapter...355
  References 356
  Additional Readings 356

Chapter 13 DMAIC Model: "I" Is for Improve........................357
  What Is the Objective of This Chapter? 357
  Purpose of the Improve Phase..357
  The Steps of the Improve Phase 358
    Generate Alternative Methods for Performing Each Step in the Process.....358
    Select the Best Alternative Method (Change Concepts) for All of the CTQs ..360
    Create a Flowchart for the Future State Process..........................361
    Identify and Mitigate the Risk Elements for New Process 362
    Run a Pilot Test of the New Process......................................362
    Collect and Analyze the Pilot Test Data..................................362
    Go-No Go Decision Point 364
  Keys to Success and Pitfalls to Avoid..365

        Case Study: Reducing Patient No Shows in an Outpatient Psychiatric Clinic—
        Improve Phase..366
            Generate Alternative Methods for Performing Each Step in the Process.....366
            Select the Best Alternative Method (Change Concept) for All the CTQs.....367
            Create a Flowchart of the New Improved Process......................368
            Identify and Mitigate the Risk Elements for the New Process.............369
            Run a Pilot Test of the New Process..................................369
            Collect and Analyze the Pilot Test Data...............................372
            Tollgate Review—Go-No Go Decision Point 373
        Takeaways from This Chapter..373
        References 374
        Additional Readings...374

Chapter 14   **DMAIC Model: "C" Is for Control**.........................375
        What Is the Objective of This Chapter?.....................................375
        Purpose of the Control Phase 375
        The Steps of the Control Phase 376
            Reduce the Effects of Collateral Damage to Related Processes 376
            Standardize Improvements (International Standards Organization [ISO])...379
            Develop a Control Plan for the Process Owner 380
            Identify and Document the Benefits and Costs of the Project 383
            Input the Project into the Six Sigma Database.........................383
            Diffuse the Improvements throughout the Organization..................383
            Conduct a Tollgate Review of the Project 384
        Keys to Success and Pitfalls to Avoid 385
        Case Study: Reducing Patient No Shows in an Outpatient Psychiatric Clinic—
        Control Phase..386
            Reduce the Effects of Collateral Damage to Related Processes 386
            Standardize Improvements (International Standards Organization [ISO])...386
            Develop a Control Plan for the Process Owner 387
            Financial Impact...387
            Input the Project into the Six Sigma Database.........................390
            Diffuse the Improvements throughout the Organization..................390
            Champion, Process Owner, and Black Belt Review the Project 390
        Takeaways from This Chapter..391
        References 392
        Additional Readings...392

Chapter 15　Maintaining Improvements in Processes, Products-Services, Policies, and Management Style.........................393

What Is the Objective of This Chapter?......................................393
Improving Processes, Products-Services, and Processes: Revisited.............393
Case Study 1: Failure in the Act Phase of the PDSA Cycle in Manufacturing......393
Case Study 2: Failure in the Act Phase of the PDSA Cycle in Accounts Receivable..394
A Method for Promoting Improvement and Maintainability..................396
    Dashboards...396
    Presidential Review of Maintainability Indicators......................397
The Funnel Experiment and Successful Management Style....................398
    Rule 1 Revisited...398
    Rule 4 Revisited...398
Succession Planning for the Maintainability of Management Style.............399
    Succession Planning by Incumbent Model............................399
    Succession Planning by Creating Talent Pools Model....................400
    Succession Planning Using the Top-Down/Bottom-Up Model............400
    Process Oriented Top-Down/Bottom-Up Succession Planning Model......401
Egotism of Top Management as a Threat to the Maintainability of Management Style...403
    Six Indicators of Egotism That Threaten the Maintainability of Management Style..403
    Summary..404
The Board of Directors Fails to Understand the Need for Maintainability in the Organization's Culture and Management Style............................404
    Definition of Culture/Management Style.............................404
    Components of Board Culture......................................405
    Shared Mission and Shared Values/Beliefs............................405
    Allocation of Work...405
    Reducing Variability..406
    Engagement..406
    Trust..406
Takeaways from This Chapter...406
References..407

Section IV The Culture Required for Six Sigma Management

Chapter 16 W. Edwards Deming's Theory of Management: A Model for Cultural Transformation of an Organization 409

Background on W. Edwards Deming....................................... 409
    Deming's System of Profound Knowledge 409
    Purpose of Deming's Theory of Management 410
    Paradigms of Deming's Theory of Management 410
    Components of Deming's Theory of Management 411
Deming's 14 Points for Management....................................... 413
Deming's 14 Points and the Reduction of Variation 430
Transformation or Paradigm Shift.. 433
    The Prevailing Paradigm of Leadership 433
    The New Paradigm of Leadership 434
    Transformation... 434
Quality in Service, Government, and Education............................ 434
Quotes from Deming 434
Summary 435
References and Additional Readings 436

Index **439**

Section V Six Sigma Certification

Chapter 17 Six Sigma Champion Certification .... Online

Chapter 18 Six Sigma Green Belt Certification .... Online

ACCESS TO DATA FILES AND CHAPTERS 17 AND 18

Go to www.ftpress.com/sixsigma and click the Downloads tab to access Minitab practice data files and Chapters 17 and 18.

Acknowledgments

First, we thank the late W. Edwards Deming for his philosophy and guidance. Second, we thank everyone at the University of Miami and the University of Miami Miller School of Medicine for collaborating with us in all of our process improvement efforts. We have learned something from every single one of you! Third, we thank all the people who provided life lessons to us to make this book a reality. Finally, we thank Jeanne Glasser Levine for giving us the opportunity to write this second edition. Thank you one and all.

About the Authors

Dr. Howard S. Gitlow is Executive Director of the Institute for the Study of Quality, Director of the Master of Science degree in Management Science, and a Professor of Management Science, School of Business Administration, University of Miami, Coral Gables, Florida. He was a visiting professor at the Stern School of Business at New York University from 2007 through 2013, and a visiting professor at the Science University of Tokyo in 1990 where he studied with Dr. Noriaki Kano. He received his PhD in Statistics (1974), MBA (1972), and BS in Statistics (1969) from New York University. His areas of specialization are Six Sigma Management, Dr. Deming's theory of management, Japanese Total Quality Control, and statistical quality control.

Dr. Gitlow is a Six Sigma Master Black Belt, a fellow of the American Society for Quality, and a member of the American Statistical Association. He has consulted on quality, productivity, and related matters with many organizations, including several Fortune 500 companies.

Dr. Gitlow has authored or coauthored 16 books, including *America's Research Universities: The Challenges Ahead,* University Press of America (2011); *A Guide to Lean Six Sigma,* CRC Press (2009); *Design for Six Sigma for Green Belts and Champions,* Prentice-Hall, (2006); *Six Sigma for Green Belts and Champions,* Prentice-Hall, (2004); *Quality Management: Tools and Methods for Improvement,* 3rd edition, Richard. D. Irwin (2004); *Quality Management Systems,* CRC Press (2000), *Total Quality Management in Action,* Prentice-Hall, (1994); *The Deming Guide to Quality and Competitive Position,* Prentice-Hall (1987); *Planning for Quality, Productivity, and Competitive Position,* Dow Jones-Irwin (1990); and *Stat City: Understanding Statistics Through Realistic Applications,* 2nd edition, Richard D. Irwin (1987). He has published more than 60 academic articles in the areas of quality, statistics, management, and marketing.

While at the University of Miami, Dr. Gitlow has received awards for outstanding teaching, outstanding writing, and outstanding published research articles.

Richard J. Melnyck is Assistant Vice President for Medical Affairs and Executive Director of Process Improvement at the University of Miami Miller School of Medicine and Health System. He is a Six Sigma Master Black Belt, the University of Miami faculty advisor for the American Society for Quality, the University of Miami Miller School of Medicine faculty advisor for the Institute for Healthcare Improvement, and a member of the Beta Gamma Sigma International Honor Society. Melnyck has taught process improvement in both the School of Business and the Miller School of Medicine at the University of Miami. He has consulted on quality, productivity, and related matters with many organizations. He received his MS in Management Science (2008), MBA (2002), and MS in Computer Information Systems (2002) from the University of Miami.

David M. Levine is Professor Emeritus of Statistics and Computer Information Systems at Baruch College (City University of New York). He received B.B.A. and M.B.A. degrees in Statistics from City College of New York and a PhD from New York University in industrial engineering and operations research. He is nationally recognized as a leading innovator in statistics education and is the coauthor of 14 books, including such bestselling statistics textbooks as *Statistics for Managers Using Microsoft Excel*, *Basic Business Statistics: Concepts and Applications*, *Business Statistics: A First Course*, and *Applied Statistics for Engineers and Scientists Using Microsoft Excel and Minitab*. He also is the coauthor of *Even You Can Learn Statistics & Analytics: A Guide for Everyone Who Has Ever Been Afraid of Statistics*, currently in its third edition, and *Design for Six Sigma for Green Belts and Champions*, and the author of *Statistics for Six Sigma Green Belts*, all published by Pearson, and *Quality Management*, third edition, McGraw-Hill/Irwin. He is also the author of *Video Review of Statistics* and *Video Review of Probability*, both published by Video Aided Instruction, and the statistics module of the MBA primer published by Cengage Learning. He has published articles in various journals, including *Psychometrika*, *The American Statistician*, *Communications in Statistics*, *Decision Sciences Journal of Innovative Education*, *Multivariate Behavioral Research*, *Journal of Systems Management*, *Quality Progress*, and *The American Anthropologist*, and he has given numerous talks at the Decision Sciences Institute (DSI), American Statistical Association (ASA), and Making Statistics More Effective in Schools and Business (MSMESB) conferences. Levine has also received several awards for outstanding teaching and curriculum development from Baruch College.

1

You Don't Have to Suffer from the Sunday Night Blues!

What Is the Objective of This Chapter?

We all know someone who dreads Sunday night because he or she isn't looking forward to going to work the next day. In fact, many of us know that person very well because that person is us!

Many employees are highly respected and well paid, and you may believe that they are happy with their jobs, but do not be fooled by their smiles. Many of them dislike their jobs. Many people are "burned out" at work. So, if you are an employee just trying to do your job and you think your job is boring, draining, and depressing, just think—you may have to do it for the rest of your work life! How's that for something to look forward to?

Well, we are here to tell you that you don't have to suffer from the Sunday night blues!

Before we tell you what you can do to make that happen we need to first tell you a little bit about intrinsic motivation. Intrinsic motivation comes from the sheer joy or pleasure of performing an act, in this case such as improving a process or making your job better. It releases human energy that can be focused into improvement and innovation of a system. As amazing as it may seem, work does not have to be a drain on your energy. If you can release the intrinsic motivation that lies within all of us it can actually fill you with energy so you can enjoy what you do and look forward to doing it, day after day and year after year. Many artists, athletes, musicians, and professors enjoy their work over the course of their lives. You can enjoy your work also, or at least you can enjoy it much more than you currently do. It just requires a redefinition of work and a management team that promotes the redefined view of work to release the intrinsic motivation within each of us.

In today's world, many of us are asked to self-manage to a great extent, meaning we are given the autonomy and opportunity to direct our work to accomplish important organizational objectives. However, many of us do not take advantage of that opportunity. Why? The reason is that we do not have the tools to release that intrinsic motivation to make our jobs, our organizations, and most importantly our lives better. Now we do!

This book not only explains how it is possible for you to make both your work life and your personal life better using process improvement and Six Sigma, but it gives you the tools and methods to make it happen.

Sarah's Story

Most people go into work every day and are confronted with a long list of crises that require immediate attention. Consider the story of Sarah who is an administrative assistant in a department in a large, urban, private university. Please note that Sarah has not read this book—yet. So she comes to work every day only to be greeted by a long to-do list of mini crises that are boring and repetitive. Sound familiar?

The mini crises include answering the same old questions from faculty and students, week after week after week:

- What room is my class in?
- Does the computer in room 312 work?
- What are my professor's office hours?
- Are the copies I need for class (and requested only 5 minutes ago) ready? Blah, blah, blah.

These crises prevent Sarah from doing her "real" work, which keeps piling up. It is frustrating and depressing. If you ask Sarah what her job is, she will say: "I do whatever has to be done to get through the day without a major disaster."

No one is telling Sarah she cannot improve her processes so that she doesn't have to answer the same questions over and over again. In fact, her bosses would rather her not focus on answering the same old questions and instead prefer her to work on projects that actually add value. The problem is not that she doesn't want to improve her processes; the problem is that she doesn't know how.

Then one day somehow the stars align and Sarah finds a copy of our book on her desk, so she reads it. She starts to apply some of the principles of the book to her job and to her life, and guess what? Things begin to change for the better.

For example, instead of having people call her to see what room their class is in she employs something that she learns in the book called *change concepts,* which are approaches to change that have been found to be useful in developing solutions that lead to improvements in processes. In this case, she uses a change concept related to automation and sends out a daily autogenerated email to all students and staff to let them know what room their classes are located in. Utilizing the change concept eliminates the annoying calls she used to receive to see what room classes are in.

Can you identify with Sarah? Do you want to learn tools and methods that will help you transform your job, your organization, and your life? The upcoming chapters take you on that journey, the journey of process improvement.

Before we go through the structure of the book, it is important for you to understand some key fundamental principles. These are principles that you need to understand as a prerequisite to reading this book and are principles you need to keep referring back to if you want to transform your job (to the extent management allows you to do it), your organization (if it is under your control), and your life through process improvement.

A young violinist in New York City asks a stranger on the street how to get to Carnegie Hall; the stranger's reply is, "Practice, practice, practice." The same thing applies to process improvement. The only way you get better at it is through practice, practice, practice, and it starts with the nine principles outlined in this chapter.

Nine Principles of Process Improvement to Get the Most Out of This Book

Process improvement and Six Sigma embrace many principles, the most important of which in our opinion are discussed in this section. When understood, these principles may cause a transformation in how you view life in general and work in particular (Gitlow, 2001; Gitlow, 2009).

The principles are as follows:

- **Principle 1**—Life is a process (a process orientation).
- **Principle 2**—All processes exhibit variation.
- **Principle 3**—Two causes of variation exist in all processes.
- **Principle 4**—Life in stable and unstable processes is different.
- **Principle 5**—Continuous improvement is always economical, absent capital investment.
- **Principle 6**—Many processes exhibit waste.
- **Principle 7**—Effective communication requires operational definitions.
- **Principle 8**—Expansion of knowledge requires theory.
- **Principle 9**—Planning requires stability. Plans are built on assumptions.

These principles are outlined in the following sections and appear numerous times throughout the book. Illustrated from the point of view of everyday life, it is your challenge to apply them to yourself, your job, and your organization.

Principle 1: Life is a process. A process is a collection of interacting components that transform inputs into outputs toward a common aim called a mission statement. Processes exist in all facets of life in general, and organizations in particular, and an understanding of them is crucial.

The transformation accomplished by a process is illustrated in Figure 1.1. It involves the addition or creation of time, place, or form value. An output of a process has *time value* if it is available when needed by a user. For example, you have food when you are hungry, or equipment and tools available when you need them. An output has *place value* if it is available where needed by a user. For example, gas is in your tank (not in an oil field), or wood chips are in a paper mill. An output has *form value* if it is available in the form needed by a user. For example, bread is sliced so it can fit in a toaster, or paper has three holes so it can be placed in a binder.

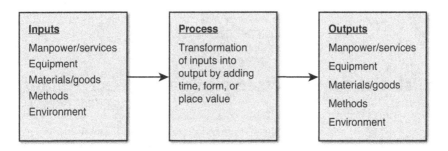

Figure 1.1 Basic process

An example of a personal process is Ralph's "relationship with women he dates" process. Ralph is 55 years old. He is healthy, financially stable, humorous, good looking (at least he thinks so!), and pleasant. At age 45 he was not happy because he had never had a long-term relationship with a woman. He wanted to be married and have children. Ralph realized that he had been looking for a wife for 20 years, with a predictable pattern of four to six month relationships—that is, two relationships per year on average; see Figure 1.2. That meant he had about 40 relationships over the 20 years.

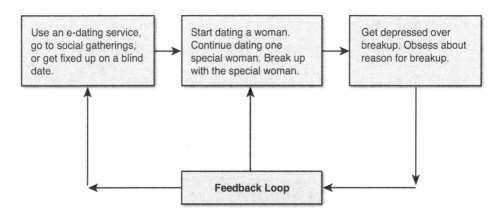

Figure 1.2 Ralph's relationship with women process

Ralph continued living the process shown in Figure 1.2 for more than 20 years. It depressed and frustrated him, but he did not know what to do about it. Read on to the next principles to find out more about Ralph's situation.

Principle 2: All processes exhibit variation. Variation exists between people, outputs, services, products, and processes. It is natural and should be expected, but it must be reduced. The type of variation being discussed here is the unit-to-unit variation in the outputs of a process (products or services) that cause problems down the production or service line and for customers. It is *not* diversity, for example, racial, ethnic, or religious, to name a few sources of diversity. Diversity makes an organization stronger due to the multiple points of view it brings to the decision making process.

Let's go back to our discussion of unit-to-unit variation in the outputs of a process. The critical question to be addressed is: "What can be learned from the unit-to-unit variation in the outputs of a process (products or services) to reduce it?" Less variability in outputs creates a situation in which it is easier to plan, forecast, and budget resources. This makes everyone's life easier.

Let's get back to Ralph's love life or lack thereof. Ralph remembered the reasons for about 30 of his 40 breakups with women. He made a list with the reason for each one. Then he drew a line graph of the number of breakups by year; see Figure 1.3.

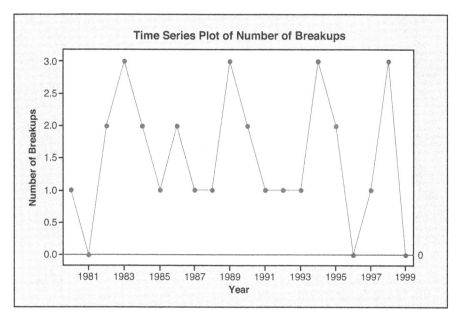

Figure 1.3 Number of breakups by year

As you can see, the actual number of breakups varies from year to year. Ralph's ideal number of breakups per year is zero; this assumes he is happy and in a long-term relationship with a woman whom he has children with. The difference between the actual number of breakups and the ideal number of breakups is unwanted variation. Process improvement and Six Sigma management help you understand the causes of unwanted waste and variation, thereby giving you the insight you need to bring the actual output of a process and the ideal output of a process closer to each other.

Another example: Your weight varies from day to day. Your *ideal* daily weight would be some medically determined optimum level; see the black dots on Figure 1.4. Your *actual* daily weights may be something entirely different. You may have an unacceptably high average weight with great fluctuation around the average; see the fluctuating squares on Figure 1.4. Unwanted variation is the difference between your ideal weight and your actual weights. Process improvement and Six Sigma management help you understand the causes of this variation, thereby giving you the insight you need to bring your actual weight closer to your ideal weight.

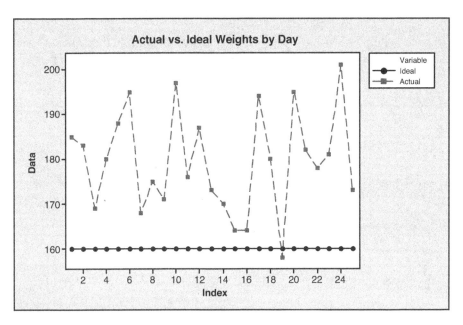

Figure 1.4 Actual versus ideal weights by day

Principle 3: Two causes of variation exist in all processes; they are special causes and common causes of variation. Special causes of variation are due to assignable causes external to the process. Common causes of variation are due to the process itself—that is, variation caused by the structure of the process. Examples of common causes of variation could be stress, values and beliefs, or the level of communication between the members of a family. Usually, most of the variation in a process is due to common causes. A process that exhibits special and common causes of variation is unstable; its output is not predictable in the future. A process that exhibits only common causes of variation is stable (although possibly unacceptable); its output is predictable in the near future.

Let's visit Ralph again. Ralph learned about common and special causes of variation and began to use some basic statistical thinking and tools to determine whether his pattern of breakups with women was a predictable system of common causes of variation. Ralph constructed a control chart (see Figure 1.5) of the number of breakups with women by year. After thinking about himself from a statistical point of view using a control chart, he realized his relationships with women were not unique events (special causes); rather, they were a common cause process (his relationship with women process).

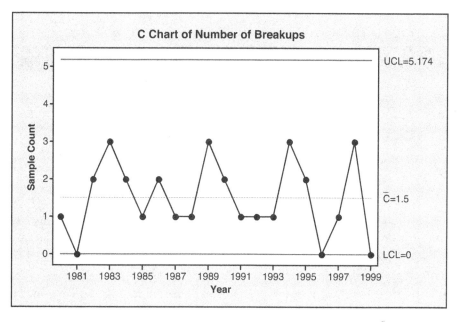

Figure 1.5 Number of breakups with women by year

Control charts are statistical tools used to distinguish special from common causes of variation. All control charts have a common structure. As Figure 1.5 shows, they have a center line, representing the process average, and upper and lower control limits that provide information on the process variation. Control charts are usually constructed by drawing samples from a process and taking measurements of a process characteristic, usually over time. Each set of measurements is called a *subgroup,* for example, a day or month. In general, the center line of a control chart is taken to be the estimated mean of the process; the upper control limit (UCL) is a statistical signal that indicates any point(s) above it are likely due to special causes of variation, and the lower control limit (LCL) is a statistical signal that indicates any point(s) below it are likely due to special causes of variation. Additional signals of special causes of variation are not discussed in this chapter, but are discussed later in the book.

Back to Ralph's love life; Figure 1.5 shows that the number of breakups by year are all between the UCL = 5.174 and the LCL = 0.0. So, Ralph's breakup process with women only exhibits common causes of variation; it is a stable and predictable process, at least into the near future. This tells Ralph that he should analyze all 30 data points for all 20 years as being part of his "relationship with women" process; he should not view any year or any relationship as special.

Ralph was surprised to see that the reasons he listed for the 30 breakups collapsed down to five basic categories, with one category containing 24 (80%) of the relationships. The categories (including repetitions) are grouped into the frequency distribution shown in Table 1.1.

Table 1.1 Frequency Distribution of Reasons for Breakups with Women for 20 Years

Reason	Frequency	Percentage
Failure to commit	24	80.00
Physical	03	10.00
Sexual	01	3.33
Common interests	01	3.33
Other relationships	01	3.33
Total	30	100.00

Ralph realized that there were not 30 unique reasons (special causes) that moved him to break up with women. He saw that there were only five basic reasons (common causes of variation in his process) that contributed to his breaking up with women, and that "failure to commit" is by far the most repetitive common cause category.

Principle 4: Life in stable and unstable processes is different. This is a *big* principle. If a process is stable, understanding this principle allows you to realize that most of the crises that bombard you on a daily basis are nothing more than the random noise (common causes of variation) in your life. Reacting to a crisis like it is a special cause of variation (when it is in fact a common cause of variation) will double or explode the variability of the process that generated it. All common causes of variation (formerly viewed as crises) should be categorized to identify *80-20 rule* categories, which can be eliminated from the process. Eliminating an 80-20 rule category eliminates all, or most, future repetition of the common causes (repetitive crises) of variation generated by the problematic component of the process.

Let's return to the example of Ralph. Ralph realized that the 30 women were not individually to blame (special causes) for the unsuccessful relationships, but rather, he was to blame because he had not tended to his emotional well-being (common causes in his stable emotional process); refer to Figure 1.5. Ralph realized he was the process owner of his emotional process. Armed with this insight, he entered therapy and worked on resolving the biggest common cause category (80-20 rule category) for his breaking up with women, failure to commit.

The root cause issue for this category was that Ralph was not getting his needs met by the women. This translated into the realization that his expectations were too high because he had a needy personality. In therapy he resolved the issues in his life that caused him to be needy and thereby made a fundamental change to himself (common causes in his emotional process). He is now a happily married man with two lovely children. Ralph studied and resolved the common causes of variation between his ideal and real self, and moved himself to his ideal; see the right side of Figure 1.6. He did this by recognizing that he was the process owner of his emotional process and that his emotional process was stable, and required a common cause type fix, not a special cause type fix. Ralph is the manager of his life; only he can change how he interacts with the women he forms relationships with.

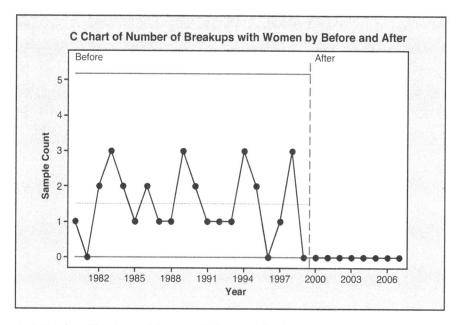

Figure 1.6 Number of breakups with women before and after therapy

Principle 5: Continuous improvement is always economical, absent capital investment. Continuous improvement is possible through the rigorous and relentless reduction of common causes of variation and waste around a desired level of performance in a stable process. It is always economical to reduce variation around a desired level of performance, *without capital investment*, even when a process is stable and operating within specification limits. For example, elementary school policy states that students are to be dropped off at 7:30 a.m. If a child arrives before 7:25 a.m., the teacher is not present and it is dangerous because it is an unsupervised environment. If a child arrives between 7:25 a.m. and 7:35 a.m., the child is on time. If a child arrives after 7:35 a.m., the entire class is disrupted. Consequently, parents think that if their child arrives anytime between 7:25 a.m. and 7:35 a.m. it is acceptable (within specification limits). However, principle 5 promotes the belief that for every minute a child is earlier or later than 7:30 a.m., even between 7:25 am and 7:35 am, a loss is incurred by the class. The further from 7:30 a.m. a child arrives to school, the greater the loss. Please note that the loss may not be symmetric around 7:30 a.m. Under this view, it is each parent's job to continuously reduce the variation in the child's arrival time to school. This minimizes the total loss to all stakeholders of the child's classroom experience (the child, classmates, teacher, and so on). Table 1.2 shows the loss incurred by the class of children in respect to accidents from early arrivals of children and the disruptions by late arrivals of children for a one year period.

Table 1.2 Loss from Minutes Early or Late

Arrival Times (a.m.)	# Minutes Early or Late	Loss to the Classroom
7:26	4	2 accidents
7:27	3	2 accidents
7:28	2	1 accident
7:29	1	1 accident
7:30	0	0 accidents
7:31	1	1 minor disruption
7:32	2	1 minor disruption
7:33	3	1 medium disruption
7:34	4	1 major disruption
Total		6 accidents 2 minor disruptions 1 medium disruption 1 major disruption

If parents can reduce the variation in their arrival time processes from the distribution in Table 1.2 to the distribution in Table 1.3, they can reduce the loss from early or late arrival to school. Reduction in the arrival time process requires a fundamental change to parents' arrival time behavior, for example, laying out their child's clothes the night before to eliminate time. As you can see, Table 1.2 shows 6 accidents, 2 minor disruptions, 1 medium disruption, and 1 major disruption, while Table 1.3 shows 4 accidents, 2 minor disruptions, and 1 medium disruption. This clearly demonstrates the benefit of continuous reduction of variation, even if all units conform to specifications.

Table 1.3 Improved Loss from Minutes Early or Late

Arrival Times (a.m.)	# Minutes Early or Late	Loss to the Classroom
7:26	4	0 accidents
7:27	3	2 accidents
7:28	2	1 accident
7:29	1	1 accident

Arrival Times (a.m.)	# Minutes Early or Late	Loss to the Classroom
7:30	0	0 accidents
7:31	1	1 minor disruption
7:32	2	1 minor disruption
7:33	3	1 medium disruption
7:34	4	0 disruptions
Total		4 accidents
		2 minor disruptions
		1 medium disruption

Principle 6: Many processes exhibit waste. Processes contain both value added activities and non-value added activities. Non-value added activities in a process include any wasteful step that

- Customers are not willing to pay for
- Does not change the product or service
- Contains errors, defects, or omissions
- Requires preparation or setup
- Involves control or inspection
- Involves overproduction, special processing, and inventory
- Involves waiting and delays

Value added activities include steps that customers are willing to pay for because they positively change the product or service in the view of the customer. Process improvement and Six Sigma management promote reducing waste through the elimination of non-value added activities (streamlining operations), eliminating work in process and inventory, and increasing productive flexibility and speed of employees and equipment.

Recall Ralph and his love life dilemma. If you consider Ralph's failure to commit as part of his relationship with women process, you can clearly see that it is a non-value added activity. This non-value added activity involves some wasteful elements. First, the women Ralph dates do not want to spend their valuable time dating a man who cannot commit to a long-term relationship. Second, the women ultimately feel tricked or lied to because Ralph failed to discuss his commitment issues early in the relationship. Third, the women resent the emotional baggage (unwanted inventory) that Ralph brings to the prospective relationship. Clearly, Ralph needed to eliminate these forms of waste from his love life.

Principle 7: Effective communication requires operational definitions. An operational definition promotes effective communication between people by putting communicable meaning into a word or term. Problems can arise from the lack of an operational definition such as endless bickering and ill will. A definition is operational if all relevant users of the definition agree on the definition. It is useful to illustrate the confusion that can be caused by the absence of operational definitions. The label on a shirt reads "75% cotton." What does this mean? Three quarters cotton on average over this shirt, or three quarters cotton over a month's production? What is three quarters cotton? Three quarters by weight? Three quarters at what humidity? Three quarters by what method of chemical analysis? How many analyses? Does 75% cotton mean that there must be some cotton in any random cross-section the size of a silver dollar? If so, how many cuts should be tested? How do you select them? What criterion must the average satisfy? And how much variation between cuts is permissible? Obviously, the meaning of 75% cotton must be stated in operational terms; otherwise confusion results.

An operational definition consists of

- A criterion to be applied to an object or a group
- A test of the object or group in respect to the criterion
- A decision as to whether the object or group did or did not meet the criterion

The three components of an operational definition are best understood through an example.

Susan lends Mary her coat for a vacation. Susan requests that it be returned clean. Mary returns it dirty. Is there a problem? Yes! What is it? Susan and Mary failed to operationally define clean. They have different definitions of clean. Failing to operationally define terms can lead to problems. A possible operational definition of clean is that Mary will get the coat dry-cleaned before returning it to Susan. This is an acceptable definition if both parties agree. This operational definition is shown here:

> Criteria: The coat is dry-cleaned and returned to Susan.
>
> Test: Susan determines if the coat was dry-cleaned.
>
> Decision: If the coat was dry-cleaned, Susan accepts the coat. If the coat was not dry-cleaned, Susan does not accept the coat.

From past experience, Susan knows that coats get stained on vacation and that dry cleaning may not be able to remove a stain. Consequently, the preceding operational definition is not acceptable to Susan. Mary thinks dry cleaning is sufficient to clean a coat and feels the preceding operational definition is acceptable. Since Susan and Mary cannot agree on the meaning of clean, Susan should not lend Mary the coat.

An operational definition of clean that is acceptable to Susan follows:

> Criteria: The coat is returned. The dry-cleaned coat is clean to Susan's satisfaction or Mary must replace the coat, no questions asked.
>
> Test: Susan examines the dry-cleaned coat.
>
> Decision: Susan states the coat is clean and accepts the coat. Or, Susan states the coat is not clean and Mary must replace the coat, no questions asked.

Mary doesn't find this definition of clean acceptable. The moral is: Don't do business with people without operationally defining critical quality characteristics.

Operational definitions are not trivial. Statistical methods become useless tools in the absence of operational definitions because data does not mean the same thing to all its users.

Principle 8: Expansion of knowledge requires theory. Knowledge is expanded through revision and extension of theory based on systematic comparisons of predictions with observations. If predictions and observations agree, the theory gains credibility. If predictions and observations disagree, the variations (special and/or common) between the two are studied, and the theory is modified or abandoned. Expansion of knowledge (learning) continues forever.

Let's visit Ralph again. He had a theory that each breakup had its own and unique special cause. He thought deeply about each breakup and made changes to his behavior based on his conclusions. Over time, Ralph saw no improvement in his relationships with women; that is, the difference between the actual number of breakups by year was not getting any closer to zero; that is a long-term relationship. Coincidently, he studied process improvement and Six Sigma management and learned that there are two types of variation in a process, special and common causes. He used a control chart to study the number of breakups with women by year; refer to the left side of Figure 1.6. Ralph developed a new theory for his relationship with women process based on his process improvement and Six Sigma studies. The new theory recognized that all Ralph's breakups were due to common causes of variation. He categorized them, went into therapy to deal with the biggest common cause problem, and subsequently, the actual number of breakups with women by year equaled the ideal number of breakups with women by year; refer to the right side of Figure 1.6. Ralph tested his new theory by comparing actual and ideal numbers, and found his new theory to be helpful in improving his relationship with women process.

Principle 9: Planning requires stability. Plans are built on assumptions. Assumptions are predictions concerning the future conditions, behavior, and performance of people, procedures, equipment, or materials of the processes required by the plan. The predictions have a higher likelihood of being realized if the processes are stable with low degrees of variation. If you can stabilize and reduce the variation in the processes involved with the plan, you can affect the assumptions required for the plan. Hence, you can increase the likelihood of a successful plan.

Example: Jan was turning 40 years old. Her husband wanted to make her birthday special. He recalled that when Jan was a little girl she dreamed of being a princess. So, he looked for a castle that resembled the castle in her childhood dreams. After much searching, he found a castle in the middle of France that met all the required specifications. It had a moat, parapets, and six bedrooms; perfect. Next, he invited Jan's closest friends, three couples and two single friends, filling all six bedrooms. After much discussion with the people involved, he settled on a particular three day period in July and signed a contract with the count and countess who owned the castle. Finally, he had a plan and he was happy.

As the date for the party drew near, he realized that his plan was based on two assumptions. The first assumption was that the castle would be available. This was not a problem because he had a contract. The second assumption was that all the guests would be able to go to the party. Essentially, each guest's life is a process. The question is: Is each "guest's life process" stable with a low enough degree of variation to be able to predict attendance at the party. This turned out to be a substantial problem. Due to various situations, several of the guests were not able to attend the party. One couple began to have severe marital problems. One member of another couple lost his job. Jan's husband should have realized that the likelihood of his second assumption being realized was problematic and subject to chance; that is, he would be lucky if all the guests were okay at the time of the party. He found out too late that the second assumption was not met at the time of the party. If he had he realized this, he could have saved money and heartache by renting rooms that could be cancelled in a small castle-type hotel. As a postscript, the party was a great success!

Structure of the Book

We structured the book strategically into five main sections, each building upon each other and each expanding your knowledge so that eventually you can complete a process improvement project on your own.

We use the analogy of building a house in how we structured this book.

Section I—Building a Foundation of Process Improvement Fundamentals

- **Chapter 1**—You Don't Have to Suffer from the Sunday Night Blues!
- **Chapter 2**—Process and Quality Fundamentals
- **Chapter 3**—Defining and Documenting a Process

One of the first steps to building a house is to lay down a foundation. The first section creates your foundation in process improvement by taking you through the process and quality fundamentals you need as you build up your knowledge base. It goes into further detail on many of our nine principles for process improvement, principles critical to your understanding of this material.

Section II—Creating Your Toolbox for Process Improvement

- **Chapter 4**—Understanding Data: Tools and Methods
- **Chapter 5**—Understanding Variation: Tools and Methods
- **Chapter 6**—Non-Quantitative Techniques: Tools and Methods
- **Chapter 7**—Overview of Process Improvement Methodologies
- **Chapter 8**—Project Identification and Prioritization: Building a Project Pipeline

You cannot build a house without tools and without understanding how and when to use them, right? The second section creates your toolbox for process improvement by not only teaching you the tools and methods you need to improve your processes but teaching you when and how to use them.

Section III—Putting It All Together—Six Sigma Projects

- Chapter 9—Overview of Six Sigma Management
- Chapter 10—DMAIC Model: "D" Is for Define
- Chapter 11—DMAIC Model: "M" Is for Measure
- Chapter 12—DMAIC Model: "A" Is for Analyze
- Chapter 13—DMAIC Model: "I" Is for Improve
- Chapter 14—DMAIC Model: "C" Is for Control
- Chapter 15—Maintaining Improvements in Processes, Products-Services, Policies, and Management Style

When you build a house you need a framework or guide to follow to make sure you build the house correctly; it's called a blueprint! Once that beautiful house is built you need to maintain it so it stays beautiful, right?

The third section is analogous to the blueprint of a house, and it is where we put everything you have learned together to complete a project. We use a specific set of steps—kind of like a blueprint—to keep us focused and make sure we do the project correctly. Those steps are called the Six Sigma management style. Like the maintenance of a new house, once we improve the process, the last thing we want is for the process to backslide to its former problematic state. We show you how to maintain and sustain those improvements.

Section IV—The Culture Required for Six Sigma Management

- Chapter 16—W. Edwards Deming's Theory of Management: A Model for Cultural Transformation of an Organization

The fourth section of this book discusses an appropriate culture for a successful Six Sigma management style. We can use the house building analogy because a house has to be built on a piece of property that can support all its engineering, social, psychological, and so on needs and wants. Without a proper piece of property, the house could fall into a sinkhole.

Section V—Six Sigma Certification

- Chapter 17—Six Sigma Champion Certification (online-only chapter)
- Chapter 18—Six Sigma Green Belt Certification (online-only chapter)

The fifth section discusses how you can become Six Sigma certified at the Champion and Green Belt levels of certification. Certification is like getting your house a final inspection and receiving a Certificate of Occupancy so you can move in. (This section can be found online at www.ftpress.com/sixsigma.)

We hope you enjoy this book. Feel free to contact the authors concerning any mistakes you have found, or any ideas for improvement. Thank you for reading our book. We hope you find it an invaluable asset on your journey toward a Six Sigma management culture.

Howard S. Gitlow, PhD
Professor
Six Sigma Master Black Belt
Department of Management Science
University of Miami
hgitlow@miami.edu

Richard J. Melnyck, MBA, MS in MAS, and MAS in CIS
Six Sigma Master Black Belt
Assistant Vice President for Medical Affairs
Executive Director of Process Improvement
Office of the Senior Vice President for Medical Affairs and Dean
University of Miami Miller School of Medicine
rmelnyck2@med.miami.edu

David M. Levine, PhD
Professor Emeritus
Department of Statistics and Computer Information Systems
Baruch College
City University of New York
DavidMLevine@msn.com

Let's Go!

We are excited to begin this journey with you—the journey of process improvement that we hope transforms your job and more importantly your life! While this is a technical book, we want to make it fun and interesting so that you will remember more of what we are teaching you. We tried to add humor and stories to make the journey a fun one. So what are we waiting for? Let's go!

References

Gitlow, H. (2009), *A Guide to Lean Six Sigma Management Skills* (New York: CRC Press).

Gitlow, H., "Viewing Statistics from a Quality Control Perspective," *International Journal of Quality and Reliability Management,* vol.18, issue 2, 2001.

2

Process and Quality Fundamentals

What Is the Objective of This Chapter?

The objective of this chapter is to teach you the meaning of (1) a process or system, (2) variation in a process or system, and (3) quality of a process, product, service, policy, procedure, and so on. These concepts prepare you for what is to come in the rest of the book. Before you can improve a process, it is crucial that you understand the building blocks of process, variation in a process, and the definitions of quality and their significance.

Now That's an Interesting Process!

A man walks into a bar and engages in an interesting process. He orders three beers, takes a drink out of the first one and sets it down, takes a drink out of the second one and sets it down, takes a drink out of the third one and sets it down. He comes in multiple times per week and performs the same process for a month before the bartender gets curious and asks him, "I can't help but notice your odd process of ordering three beers, taking a drink out of each, and then leaving. What's the deal?" The man responds, "Growing up, my two brothers and I were very close and we used to drink together all the time. When we all got married and moved far away from each other we made a pact that we would each do this to remember the good ol' times."

The man repeated the same process for another month and then all of a sudden he changed his process and ordered only two beers. Again the bartender got curious and asked, "I don't mean to pry, but I cannot help but notice you have changed your process and are only ordering two beers. Did one of your brothers die?" The man responded with a smile, "No, no, they are both alive and well. The reason I changed my process was because my drinking was affecting my marriage, and my wife said I couldn't drink anymore. But she didn't say anything about my brothers!"

Moral of the story: When all else fails, just blame it on the process!

Process Fundamentals

Process fundamentals include the following topics: (1) What is a process? (2) Where do processes exist? (3) Why does understanding processes matter? (4) What is a feedback loop, and how does it fit into the idea of a process? These questions need to be answered to begin to perform process improvement activities.

What Is a Process?

A *process* is a collection of interacting components that transform *inputs* (elements that the process needs to execute) into *outputs* (the results of the process) toward a common *aim*, called a mission statement (Gitlow et al., 2015). A mission statement for a hospital could be "to be a world class provider of medical services to the community." A mission statement for an individual, in this case for me, is "to generate positive energy into the universe." For example, this is helpful when I am driving. I am an impatient driver and get frustrated when someone is in my way. My mission statement directs me to just relax and take it easy; getting to my destination 2 minutes later isn't worth the frustration. Figure 2.1 shows how a process transforms inputs into outputs.

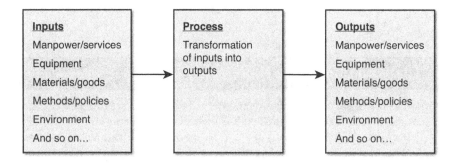

Figure 2.1 Transforming inputs into outputs

Where Do Processes Exist?

Now that we know what a process is, the next question is where do processes exist (Gitlow et al., 2015). Processes exist in all facets of organizations, as well as in everyday life. An understanding of processes is crucial if you want to improve the quality of your organization and/or the quality of your life. Remember, work and life processes are not mutually exclusive, so feel free to improve them both at the same time. Many of us mistakenly think only of production processes. However, if you think about it processes are everywhere; administration, sales, service, human resources, training, maintenance, paper flows, interdepartmental communication, and vendor relations are all processes you see at work. Importantly, relationships between people are also processes. How about your daily life? From when you wake up in the morning to when you leave for school or work is a process, cleaning your house is a process (not exactly a fun one), and making plans with your friends is a process (depending on your friends it can be a complicated one!).

It is critical for top management of an organization to have a process-oriented view of their organization and life if they are to be successful with the style of management presented in this book.

Why Does Understanding Processes Matter?

Most processes can be studied, documented, defined, standardized, improved, and innovated. Any situation in your life or work, where an input is turned into an output, involves a process. Consequently, acquiring the tools to improve processes makes your life and work a whole lot better!

What Is a Feedback Loop and How Does It Fit into the Idea of a Process?

A feedback loop is the part of a system that takes the actual outputs (data in one form or another) of a step in a process and feeds them back/forward/sideways to another step in the process to determine whether the desired outputs were generated by the process step (Gitlow et al., 2015). The purpose of a feedback loop is to use the data from it to close the gap between what was desired of the process output and what outputs were actually delivered by the process. In layman's terms, did you get what you wanted out of the process? If not, the feedback loop provides data to help you improve the process. As a result of the process improvement, the next time there is a better chance you will get the desired process output.

Some Process Examples to Bring It All Together!

Let's review some examples that will help you understand the process concepts covered so far, namely, inputs, outputs, feedback loops, and aims/missions.

Process Example #1—The Background Check Process

An example of an important human resources process is the background check that is usually the last major process completed before hiring a new employee. The hiring manager gives human resources the name of the person she wants to hire and then human resources begins the background check process. The *aim* of this process is to make sure the new employee does not have any showstopper skeletons in his closet that puts the organization at risk. Figure 2.2 shows the inputs, process, and outputs of performing a background check on an employee. The inputs include the candidate's hiring application, the candidate's consent form for the background check, and the recommendation to hire by the hiring manager. The process involved in this transformation of inputs into outputs includes reviewing criminal records, reviewing educational records, reviewing driving records, and so on. The output in this example is the completed background check report. An important aspect of this process is the *feedback loop* that enables the human resources manager to report back to the hiring manager on the employee's appropriateness for a given job.

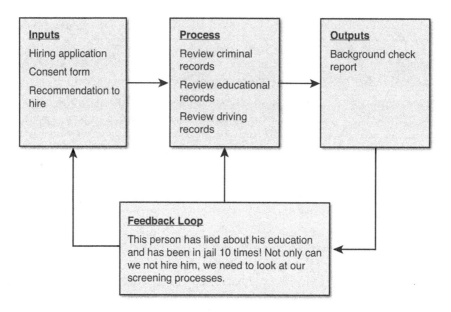

Figure 2.2 Background check process

Process Example #2—Leonard's Head Searching Process

The execution of different types of diagnostic testing can also be thought of as processes. Consider the case of Leonard who goes to the doctor. The doctor senses something is really "off" with Leonard, so his *aim* is to figure out what it is. The *inputs* to the process are the doctor checking Leonard, looking at his past medical records, and ordering an MRI of Leonard's head. The *process* involves Leonard being escorted to the Radiology department, the technologist prepping and positioning Leonard, the technologist telling him to hold his breath and then taking his picture. Unfortunately, the technologist does not tell Leonard to stop holding his breath, so his face starts to turn blue and he almost passes out. Luckily, the technologist is on her game so she immediately tells him to breathe and he quickly recovers. The radiologist then reads the report and the *output* involves the report being sent to the doctor. As for the *feedback loop*, it seems like the doctor's hunch was correct as to what is going on in Leonard's head—nothing! Since there is nothing going on in Leonard's head, the doctor sends Leonard to see a psychiatrist.

Figure 2.3 depicts the process of trying to figure out what is going on in Leonard's head.

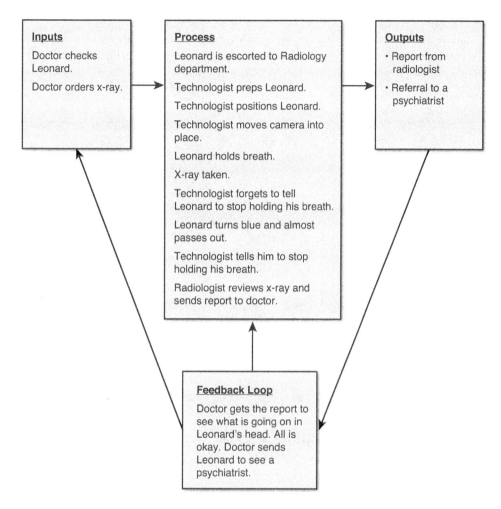

Figure 2.3 Leonard's head searching process

Process Example #3—Patient's Day of Surgery Process

Another example of a process is the presurgery process that a patient goes through the day of surgery. The *aim* is to ensure that the patient is ready to proceed with the surgery as planned. Figure 2.4 shows the inputs, process, outputs, and feedback loop of the patient's day of surgery process. The *inputs* include the patient's surgery information packet, the blood bank band (if required), medications currently taken and pre-op testing clearance form. The *process* involved in this transformation of inputs into outputs includes being driven to the hospital by a responsible adult, waiting, filling out necessary paperwork, waiting, filling out more paperwork, waiting, receiving ID bracelet with your name misspelled, waiting, receiving new ID bracelet with your name now spelled correctly, waiting, getting in awkward looking gown, waiting, giving your belongings to a nurse, praying you will see your belongings again, waiting, speaking to anesthesiologist, waiting, going to pre-op area and confirming with staff the procedure you are having with them, waiting. The *output* in this example would be the patient ready for surgery. An important aspect of this process is the *feedback loop* that enables the pre-op area to give feedback to staff upstream in the process if the patient is not ready for surgery.

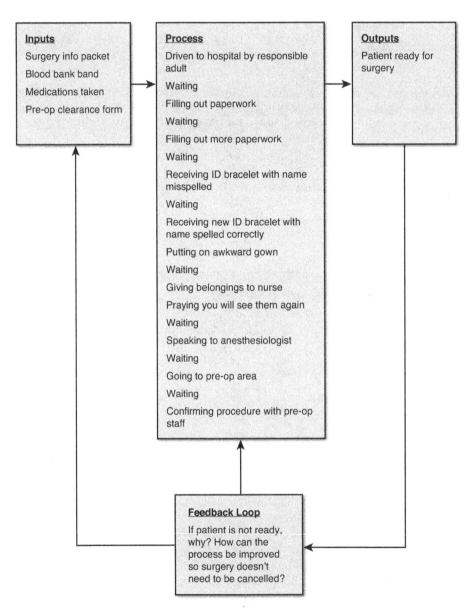

Figure 2.4 Patient's day of surgery process

An Anecdote about No Feedback Loops

John and Mary have a troubled marriage, and they have had it for 40 years. They never talk to each other; in other words, they have no feedback loops in their marriage to use to possibly make their lousy situation better. Over time, their relationship continues to deteriorate; now they have a loveless marriage highlighted by angry silence.

An Anecdote about Only Common Cause Feedback Loops

I grew up in a two-story house. My bedroom was on the second floor. When I was about five years old, one night I snuck downstairs to see what my parents were doing. Lucky for me they weren't making me a baby sister, so I missed that trauma. Well, they were having a conversation about my father's upcoming trip to China. He said to my mother: "I have to get a Gama globulin shot and the needle is a foot long." At that point two things stuck in my mind. First, I heard a really big word and wanted to brag to my friends that I knew a big word like Gama globulin, even though I had no idea what it meant. Second, I learned that needles existed that were a foot long, and I hoped I would never see one.

Fast forward 50 years, and I am booked to go on a trip to China. My physician tells me that I need a Gama globulin shot. As amazing as it may seem, I actually remembered that the needle was a foot long. Dear reader, please understand that this foot long fact was hard-wired into my brain. I didn't think about it rationally; I just unconsciously accepted it.

Needless to say, I was sweating bullets when I went to get the shot. I assumed the position for the shot: arms on the examining table with my pants and underwear around my ankles, butt pushed out and ready for the railroad spike to be inserted into my butt cheek. In my opinion, the nurse was taking too long to give me the shot, so I angrily said to her: "Give me the damn shot." She said, "I already did." I called her a liar because if someone was going to stick a railroad spike in my butt I would be the first to know about it. She got offended and came over to me and flicked my butt where she had given me the shot, and sure enough she wasn't lying. Well, I went from hating her to thinking she was Florence Nightingale; the best nurse and shot giver ever born.

I left the office and sent an email to my physician, who also is the Chief of the Clinic, saying that the nurse was the best shot giver in the world and they should all congratulate themselves on being world class!

> The next part of the story is my fantasy of what happened after the email was sent. In reality I have no idea what happened, but here is my fictional account of the transpiring events. Everyone in the office knew that I was an expert in process improvement, so my email may have carried some weight. My doctor went out into the office and read the email out loud to the entire office staff, congratulating them on jobs well done.
>
> The next bit of the story takes a dark turn; I imagined that another patient came in and got a Gama globulin shot and moved while it was being inserted into her butt, causing great pain. She yelled at the nurse and complained to my doctor, the Chief of the Clinic, about the horrible and painful medical service delivered by the clinic. At this point he assembled all the office staff and berated them for sloppy work. They went from elated to depressed. The moral to this story is that if you always respond to common causes of variation as if they were special causes of variation you are micromanaging people, and it will drive them crazy and increase the variation in their performance of their jobs.

Variation Fundamentals

Variation fundamentals are critical to understanding how to improve a process. Understanding that there are two types of variation in a process, not one as is usually assumed, allows process improvers to stabilize and improve a process, as opposed to over-reacting to process variation and making the process worse that it was before. These concepts are explained in the following sections.

What Is Variation in a Process?

As we saw in the previous examples, each process has one or many outputs, and each of these outputs may be measured. The distribution of these measurements varies, and the differences between these measurements are what we call *process variation*.

> **An Anecdote about Your Daily Moods**
>
> Each day you wake up and your mood is slightly different, or maybe a lot different, than it was the day before, and the day before that, and so on. Your daily moods form a distribution of daily moods. Most of the time your mood is around your average mood. The further your daily mood is from your average daily mood, the less frequently this particular mood happens. Your daily moods may well form a bell shaped distribution. If the distribution is skinny, your daily mood may well be predictable into the near future with a high degree of belief. If the distribution is fat, your daily mood may not be so predictable into the near future with any degree of

> belief. That's life! Process improvement is largely about making your distribution of daily moods skinny around a mean of high happiness. Remember, that 10% of the time you will be in your bottom 10% of your moods; this is a mathematical fact. The only way you can do anything about this is to shift your mood distribution further in the direction of your happiness metric. Process improvement can help.

Why Does Variation Matter?

Variation matters because it is a fact of life. All processes exhibit variation, even if it is too small to measure. When making a plan, you must consider whether the assumptions of your plan will be in place when you execute the plan. Unfortunately, each assumption of your plan is the output of some process, which exhibits variation.

For example, I plan to go to a 7:00 p.m. movie with my friend Neal, but he shows up at 7:25 p.m. Variation has ruined my movie plans. Variation matters; it needs to be reduced so prediction into the near future is possible with some degree of comfort.

What Are the Two Types of Variation?

The two types of variation (Deming, 1982, 1986, and 1994; Gitlow and Gitlow, 1987) are common causes and special causes of variation.

Common variation is due to the design, management, policies, and procedures of the system itself; this type of variation is the responsibility of management. An employee cannot change the system he works in; only management can do that.

Special variation is external to the system; it disrupts the system from its routine generation of common variation. Special variation is the responsibility of front line employees; however, front line employees may need management's help sometimes to deal with a special cause of variation.

The outputs from all processes and their component parts vary over time. Your body has a process that does a good job of generating and removing heat. The measurement used to measure your body's ability to do this is your body temperature. Despite large variation in temperatures outside of you, your body does a great job of keeping your body temperature within a safe and stable range. If your body is too hot your blood vessels expand to carry the excess heat to the surface of your skin in the form of sweat, and when the sweat evaporates it cools your body. If your body is too cold your blood vessels contract so that the blood flow to your skin slows to conserve heat; your body may shiver in response to create more heat. Under normal conditions both sweating and shivering help keep your body temperature within safe and stable levels.

Normal body temperature is said to be 98.6 degrees Fahrenheit, so it can go as high as 99.6 degrees Fahrenheit and as low as 97.6 degrees Fahrenheit throughout the day depending on the time of day and how active you are. This slight variation in your body temperature occurs all the time and is an example of common variation, or variation due to the system.

Now, let's say you get sick and your body temperature jumps up to 101 degrees Fahrenheit. This is a special cause of variation because it is caused by a change external to your body's heat management process—that is, getting sick.

If you had not gotten sick your body's heat management process would have continued on its former path of common variation. Note that if you get sick a lot, it might be a common cause of variation. Later in the book we give you specific tools to understand how to distinguish between special and common causes of variation.

The capability of a process is determined by inherent common causes of variation such as poor hiring or training processes, inadequate work environments, poor information technology systems, or a lack of policies and procedures to name just a few. Front line employees cannot change those types of processes, so they should not be held accountable for their outcomes. Managers need to realize that unless changes are made to those processes, the capability of the outcomes will remain the same. And only they can change the processes!

Special causes of variation on the other hand are due to events outside the system, such as a natural disaster, problematic new supplies or equipment, or problems with a new IT system to name just a few. Please note that special variation is *always* detected by statistical signal, not human logic.

Due to the fact that variation causes the customer to lose confidence in the reliability of the dependability and uniformity of outputs, managers need to understand how to identify and reduce variation. Using statistical methods, employees can identify different types of variation so that special causes of variation can be resolved (they can be good or bad) and common variation can be reduced via process improvement projects resulting in more uniform and reliable outputs.

Anecdote on Special Versus Common Causes of Variation

To demonstrate the difference between special and common causes of variation let's give another example. A friend of the authors was sitting at the kitchen table on a Sunday morning practicing his penmanship, more specifically his little letter "a." He kept writing the little a over and over hoping he could perfect it when all of a sudden his wife walked by and accidently hit the arm he was writing with, causing a much larger tail on one of the a's (see Figure 2.5). This example helps explain the difference between special and common causes of variation. Most of the variation in the size and shape of the a's is due to common causes of variation as they are all pretty close in size and shape, but when something external happens to the process (his wife hitting his arm) you see a big difference in the size and shape of that letter a. You are probably wondering why someone would spend time actually practicing one letter; the only explanation we can come up with is that he is a Type A personality!

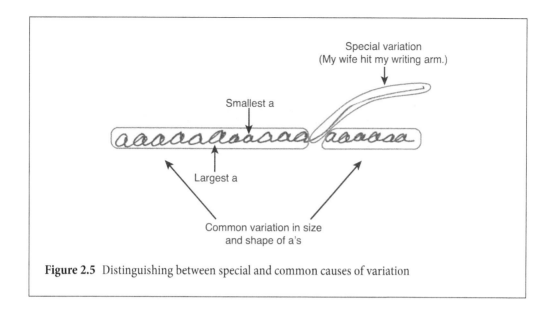

Figure 2.5 Distinguishing between special and common causes of variation

How to Demonstrate the Two Types of Variation

Dr. Deming used to conduct workshops to demonstrate special and common causes of variation; the workshops are called the Funnel Experiment and the Red Bead experiment.

The Funnel Experiment

Purpose and Introduction: The Funnel Experiment (Boardman and Iyer, 1986; Gitlow et al., 2015) shows that adjusting (or tampering with) a stable process that is exhibiting only common causes of variation will increase the variation in the output of the process, depending on how the process is adjusted.

In the experiment a marble is dropped through a funnel onto a piece of paper that has a point that serves as the designated target. The objective is to get the marble as close to the target as possible, the experiment uses various "rules" to attempt to minimize the distance between the spread of the marble drops (results) and the position of the target. For each rule, the marble is dropped 50 times, and its landing spot is marked on the piece of paper.

Rule #1:

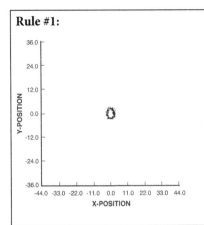

Description: The marble is dropped, but the position of the funnel does not change after each drop. This rule serves as our baseline.

Results: A small circle of points emerges.

Takeaway: The size of the circle is due entirely to common causes of variation in the process. This is analogous to management understanding common and special variation and knowing how to manage each without tampering with the process.

Rule #2:

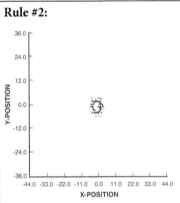

Description: We take the distance and direction away from the target where the previous marble landed and move the funnel to the equal and opposite direction for the next drop.

Results: A circle of points emerges with double the variation of the results in Rule #1.

Takeaway: The size of the circle is twice as large as Rule #1, which is analogous to management tampering with the process.

Real life example: The level of overtime at your hospital was 12% over budget last month, so this month you make sure to reduce overtime to be less than 12% of budget.

Rule #3

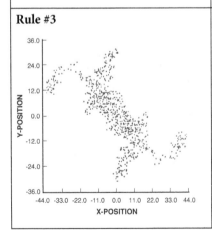

Description: We take the distance and direction of the spot where the previous marble landed away from the target and move the funnel first back to the target, and second in an equal and opposite direction from the target for the next drop.

Results: This rule produces an unstable, explosive pattern of resting points on the surface.

Takeaway: This is analogous to management tampering with the process causing variation to explode.

Real life example: Making up the previous month's miss to budget on inpatient admissions during the current period.

Rule #4:

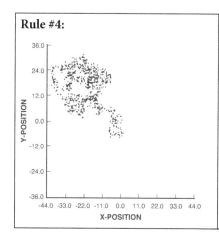

Description: We take the spot where the previous marble landed and move the funnel to that spot for the next drop.

Results: This rule produces an unstable, explosive pattern of resting points on the surface and it eventually moves farther and farther away from the target in one direction

Takeaway: Again, this is analogous to management tampering with the process causing variation to explode.

Real life example: On the job training where worker trains worker and the job skills deteriorate without bounds.

Conclusion and takeaways: The Funnel Experiment illustrates how a system behaves when it is tampered with.

- Rules 2, 3, and 4 are analogous to management tampering with a process without profound knowledge of how to improve the process through statistical thinking. This increases the process's variation and reduces the ability to manage that process.

- Instead of using rules 2, 3, and 4, the better approach would be to create a control chart of the output of rule 1. We would see that the process is stable and in control, and instead of tampering the best course of action would be to improve the process. Some ways to improve the process could be lowering the height of the funnel or using a funnel with a smaller diameter.

An Anecdote about Rule 4 of the Funnel Experiment

On one occasion I was consulting for a factory and met an engineer who was working 12- to 14-hour days, seven days a week, for many years. He was still married, which surprised me. One day he asked me a statistics question that didn't seem to have anything to do with his job. I answered his question, but started to observe him so I could understand his job better. I discovered that every time a manager senior to him changed positions, which was relatively frequently given the number of years he was on the job, the new manager would give Doug some new tasks to do. Doug was a "get things done" kind of guy. As the years progressed his list of things to do increased to the point of absurdity. Doug's job is a perfect example of rule 4 of the Funnel Experiment; it kept expanding without bounds. He never asked his new manager if all the data he was collecting and the reports he was writing were still desired by the his new manager. Most likely, the new manager would have eliminated many of the tasks he was doing for previous bosses, but he never asked and the manager never questioned Doug about his workload. Doug was a wheel that didn't squeak. This is a perfect example of rule 4 in operation.

Red Bead Experiment

Purpose and Introduction: W. Edwards Deming used the Red Bead Experiment to show the negative effects of treating common variation as special variation. It is discussed in this section to further enhance your understanding of common causes and special causes of variation (Deming, 1994; Gitlow et al., 2015).

The experiment involves using a paddle to select beads from a box that contains 4,000 beads. The box contains 3,200 white beads and 800 red beads. This fact is unknown to the participants in the experiment. Components of the experiment include a paddle with 50 bead size depression, another box for mixing the beads, as well as a foreman in the Quality Bead Company who hires, trains, and supervises four "willing workers" to produce white beads, two inspectors of the willing workers' output, one chief inspector to inspect the findings of the inspectors, and one recorder of the chief inspector's findings.

The job of the workers is to produce white beads, since red beads are unacceptable to customers. Strict procedures are to be followed. Work standards call for the production of 50 beads by each worker (a strict quota system): no more and no less than 50. Management has established a standard that no more than three red beads (6%) per worker are permitted on any given day. The paddle is dipped into the box of beads so that when it is removed, each of the 50 holes contains a bead. Once this is done, the paddle is carried to each of the two inspectors, who independently record the count of red beads. The chief inspector compares their counts and announces the results to the recorder who writes down the number and percentage of red beads next to the name of the worker.

The Results: On each day, some of the workers were above the average and some below the average. On day 1, Sharyn did best with 7 red beads, but on day 2, Peter (who had the worst record on day 1) was best with 6 red beads, and on day 3, Alyson was best with 6 red beads. Table 2.1 displays the results for the four workers over three days.

Table 2.1 Red Bead Experiment Results for Four Workers Over Three Days

| | Day | | | |
Name	1	2	3	All 3 Days
Alyson	9 (18%)	11 (22%)	6 (12%)	26 (17.33%)
David	12 (24%)	12 (24%)	8 (16%)	32 (21.33%)
Peter	13 (26%)	6 (12%)	12 (24%)	31 (20.67%)
Sharyn	7 (14%)	9 (18%)	8 (16%)	24 (16.00%)
All 4 workers	41	38	34	113
Average ($\bar{X}$)	10.25	9.5	8.5	9.42
Proportion	20.5%	19%	17%	18.83%

Some Takeaways from the Red Bead Experiment:

1. Common variation is an inherent part of any process. The variation between the number of red beads by worker by day is only due to common variation.

2. Managers are responsible for the common variation in a system; they set the policies and procedures. If managers are unhappy with the number of red beads, they should take actions; for example, get a new supplier of white beads with less red bead per load.

3. Workers are not responsible for the problems of the system—that is, common causes of variation. The system primarily determines the performance of workers. The quota of no more than three red beads per day per worker is insane. This is the case because it is beyond the capability of this company to achieve that quota; 20% of the beads are red, so in a load of 50 beads you would expect an average of 10 red beads per worker, per day.

4. Some workers will always be above the average, and some workers will always be below the average.

5. Some workers will always be in the bottom 10% because all distributions have a bottom 10% of units.

Quality Fundamentals

Quality is a term we hear frequently: That is a quality automobile, she is a quality person, and this is a quality stock. Most people equate high quality with a big price tag, and low quality with a small price tag. The purpose of this section is to debunk this outdated notion of quality and to explain what it really means.

Goal Post View of Quality

When watching a game of American football, as long as the ball is kicked between the goalposts, no matter how close to either goalpost, it is considered "good" (Gitlow et al., 2015).

In quality the same used to be true as quality meant "conformance to valid customer requirements." That is, as long as an output fell within acceptable limits (the goal posts), called *specification limits*, around a desired value, called the *nominal value* (denoted by "m"), or *target value*, it was deemed conforming, good, or acceptable. The nominal value and specification limits are set based on the perceived needs and wants of customers.

Figure 2.6 shows the goal post view of losses arising from deviations from the nominal value. That is, losses are zero until the *lower specification limit (LSL)* or *upper specification limit (USL)* is reached. Then suddenly they become positive and constant, regardless of the magnitude of the deviation from the nominal value.

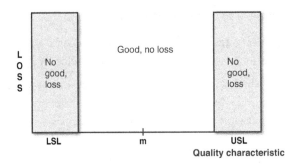

Figure 2.6 Goal posts view of losses arising from deviations from the nominal value

As an example of the goal post view, the desired diameter of a prescription container in a hospital pharmacy is 25 mm (the nominal value). A tolerance of 2 mm above or below 25 mm is acceptable to pharmacists. Thus, if the diameter of a prescription container measures between 23 mm and 27 mm (inclusive), it is deemed conforming to specifications. If the diameter of a prescription container measures less than 23 mm or greater than 27 mm, it is deemed not conforming to specifications as the lids will hardly fit and is scrapped at a cost of $1.00 per container, which is a stiff pill to swallow. In this example, 23 mm is the lower specification limit and 27 mm is the upper specification limit.

Takeaway: Output merely has to be between specification limits; that is, the diameter of the prescription container must be anywhere from 23 to 27 inclusive.

Continuous Improvement Definition of Quality— Taguchi Loss Function

As the definition of quality has evolved, its meaning has shifted. A modern definition of quality states that "quality is a predictable degree of uniformity and dependability, at low cost and suited to the market" (Deming, 1982). Figure 2.7 shows a more realistic loss curve developed by Dr. Genichi Taguchi in 1950 (Taguchi and Wu, 1980) called the Taguchi Loss Function (TLF). Using the TLF view of quality, losses begin to accrue as soon as a quality characteristic of a product or service deviates from the nominal value. As with the goal post view of quality, once the specification limits are reached the loss suddenly becomes positive and constant, regardless of the deviation from the nominal value beyond the specification limits.

Curve A represents the distribution of output of the process before process improvement. Curve B represents the distribution of output of the process after process improvement. The shaded area under the loss curve framed by the distribution of output represents the cost of poor quality. As you can see, the area under the Taguchi Loss Function for curve A is shaded both gray and black; however, the area under the Taguchi Loss Function for curve B is only shaded black. So, decreasing the variation in the distribution of output (from curve A to curve B) decreases cost, absent capital investment.

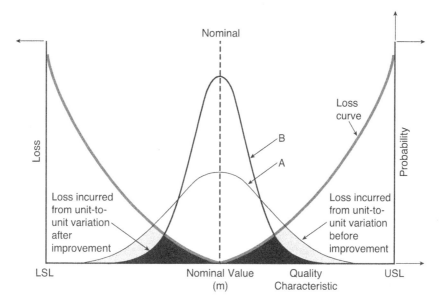

Figure 2.7 Continuous improvement view of losses from deviations from the nominal value

Takeaway: Reducing unit-to-unit variation around the nominal value always makes sense, absent capital investment. At that point it becomes a cost-benefit decision for management.

More Quality Examples

An individual visiting an outpatient clinic expects to wait an hour to see the physician, but if she has to wait five hours she will perceive the quality to be poor and probably won't come back.

If a materials management worker receives orders from a supplier without any missing supplies, her needs will be met and she will perceive the quality of that supplier as good, that is, unless the supplies are bad.

If a patient in the hospital finds a clean, comfortable room on a quiet floor, he will feel that his expectations were met. But if the room is not clean or there is constant noise that affects his ability to sleep, the patient will perceive that the quality is poor and seek revenge by giving the hospital crappy patient satisfaction scores.

Takeaways from This Chapter

- Processes transform inputs into outputs.
- Processes are everywhere in our lives and our jobs; improving our lives and/or jobs is accomplished by improving our processes.

- Feedback loops utilize data to help us improve our processes.
- Variation consists of the differences in the distribution of the measurements of outputs from a process.
- Variation limits your ability to predict the future outcomes of a process.
- There are two types of variation: special causes and common causes.
- Common causes are inherent to the system and are the responsibility of management.
- Special causes are external to the system and the responsibility of front line workers.
- Dr. Deming used the Funnel Experiment to show that tampering with a process makes things worse and the Red Bead Experiment to show that common causes of variation are inherent in any system and that you cannot expect employees to produce beyond the capability of the system.
- The goal post view of quality states that as long as the output is within specification limits it is considered good or conforming no matter how close it is to either specification limit.
- The Taguchi Loss Function shows that it is always advised to reduce variation around nominal, absent capital investment.

References

Boardman, T. J. and H. Iyer (1986), *The Funnel Experiment* (Fort Collins, CO: Colorado State University).

Deming, W. E. (1982), *Quality, Productivity, and Competitive Position* (Cambridge, MA: Massachusetts Institute of Technology).

Deming, W. E. (1986), *Out of the Crisis* (Cambridge, MA: Massachusetts Institute for Technology Center for Advanced Engineering Study).

Deming, W. E. (1994), *The New Economics for Industry, Government, Education,* 2nd ed. (Cambridge, MA: Massachusetts Institute for Technology Center for Advanced Engineering Study).

Gitlow, H., and S. Gitlow (1987), *The Deming Guide to Quality and Competitive Position* (Englewood Cliffs, NJ: Prentice-Hall).

Gitlow, H., A. Oppenheim, R. Oppenheim, and D. Levine (2015), *Quality Management: Tools and Methods for Improvement*, 4th ed. (Naperville, IL: Hercher Publishing Company). This book is free online at hercherpublishing.com.

Taguchi, G. and Y. Wu (1980), *Introduction to Off-Line Quality Control* (Nagoya, Japan: Central Japan Quality Control Association).

3
Defining and Documenting a Process

What Is the Objective of This Chapter?

To improve a process you must define it, then you must document it, and finally you must analyze it.

The objective of this chapter is to teach you how to do the following:

- Define a process by looking at who owns and is responsible for the process, by understanding the boundaries of the process, and by understanding the objectives and metrics of the process used to measure its success.

- Document a process by understanding flowcharts.

- Analyze a process by understanding how to analyze a flowchart to begin improving a process.

A Story to Illustrate the Importance of Defining and Documenting a Process

Defining and documenting a process is a critical step toward improvement and/or innovation of a process. The following example demonstrates this point. In a study of a hospital laundry office, an analyst began to diagram the flow of paperwork using a flowchart. While walking the green copy of an invoice through each step of its life cycle, he came upon an administrative assistant transcribing information from the green invoice copy into large black loose-leaf books. The analyst asked what she was doing so that he could record it on his flow diagram. She responded, "I'm recording the numbers from the green papers into the black books." He asked, "What are the black books?" She said, "I don't know." He asked, "What are the green papers?" She said, "I have no idea." He asked, "Who looks at the black books?" She said, "Well, Mr. Johnson used to look at the books, but he died 5 years ago." The analyst asked, "Who has looked at them for the last 5 years?" She said, "Nobody." He asked, "How long have you been doing this job?" She said, "Seven and one half years." He asked, "What percentage of your time do you spend working on the black books?" She said, "100 percent."

Next, the analyst did two things. First, he examined the black books. From examining the black books he realized that they were inventory registers. Every day all laundry pulled from central supply was recorded on the green papers (invoices) by item onto the appropriate page in the black books. At the end of each month, page totals were calculated, yielding monthly used laundry inventory by item. Second, he asked the administrative assistant how long ago the person who hired and trained her had left the hospital, to which she responded, "Five years ago." At this point, the consultant realized he had solved a problem the hospital did not know it had. Nobody was looking at the books because nobody knew what the administrative assistant was doing. The current manager assumed the administrative assistant was doing something important. Surprisingly, the current manager had computerized the inventory registers as soon as he came on the job five years ago! So, the administrative assistant had been doing a redundant job for five years. Nobody bothered to document the process. The administrative assistant was a wheel that didn't squeak; so why study her? As an epilogue, the administrative assistant was reassigned to other needed duties because she was a good employee.

This problem is an example of a failure to define and document a process to make sure that it is logical, complete, and efficient.

Fundamentals of Defining a Process

Defining a process requires answers to the following questions: (1) Who own the process? (2) What are the boundaries of the process? (3) What are the process's objectives? (4) What measurements are being taken on the process with respect to its objectives? A process definition is created by answering those questions.

Who Owns the Process? Who Is Responsible for the Improvement of the Process?

Every process must have an owner; that is, an individual responsible for the process (Gitlow and Levine, 2004). Process owners can be identified because they can change the flowchart of a process using only their signature. Process owners may have responsibilities extending beyond their departments, called cross-functional responsibilities; their positions must be high enough in the organization to influence the resources necessary to take action on a cross-functional process. In such cases, a process owner is the focal point for the process, but each function of the process is controlled by the line management within that function. The process owner requires representation from each function; these representatives are assigned by the line managers. They provide functional expertise to the process owner and are the promoters of change within their functions. A process owner is the coach and counsel of her process in an organization.

The identification and participation of a process owner are critical in defining a process. It is usually a waste of time to be involved in defining and documenting a process, as part of process improvement activities, without the complete commitment of the process owner.

What Are the Boundaries of the Process?

Next, boundaries must be established for processes; in other words, before a flowchart of the process can be created, the process owner must help you identify where the process starts and stops (Gitlow and Levine, 2004). These boundaries make it easier to establish process ownership and highlight the process's key interfaces with other (customer/vendor) processes. Process interfaces frequently are the source of process problems, which result from a failure to understand downstream requirements; they can cause turf wars. Process interfaces must be carefully managed to prevent communication problems. Construction of operational definitions for critical process characteristics agreed upon by process owners on both sides of a process boundary goes a long way toward eliminating process interface problems. Operational definitions are discussed later in this book.

An Anecdote about Process Boundaries

One time I was consulting at a paper mill, and I noticed that the entire mill was surrounded by a nine-foot-tall chain link fence. I realized that since a paper mill is a dangerous place, even potential intruders have to be protected.

As was my custom, I started my tour from the beginning of the process, in this case the wood procurement area. This is the where trees enter the mill on flatbed trucks and are cut into 40-foot lengths by large saws. Everyone I met was nice and helpful.

The next part of the process was the wood yard. This is the area where the 40-foot logs are turned into wood chips for making paper. I noticed that there was a chain link fence between these two areas. That was to prevent truckers from wandering into the wood yard. However, the chain link fence also had a door that was padlocked, and on top of the chain link fence was concertina barbed wire, coils of wire with razor blades attached. It would slice to pieces anyone trying to get over it.

I wondered why the wood procurement area would be more concerned about the wood yard employees than they would be about outsiders. The wood procurement folks opened the padlocked door between the two areas and let me enter. I thought I was about to be attacked by wild lions. In the distance I saw a man walking toward me, and as he got close, I stuck out my hand to shake the wood yard manager's hand.

The wood yard manager did not reciprocate and called me an ethnic slur that was relevant only to a small area in Europe that my ancestors are from. How this man would have known this term was a mystery to me. Apparently he was a savant of bigotry.

Now I knew why there was concertina wire between the two mill areas; it was a statement of mutual hatred. When I asked the wood yard manager why there was so much hatred between the two mill areas, he responded: "They are morons! They

> cut the logs into 40-foot lengths and we need 20-foot lengths for our equipment to operate at its best." When I asked why they didn't know to cut the logs into 20-foot lengths the manager said, "They should know."
>
> At that point, I realized that the wood procurement area was so put off by the manager of the wood yard that they had no communication whatsoever, even to the extent of which size saws to purchase. This is a classic example of a clear dysfunctional process boundary. Most disagreements occur at process boundaries.

What Are the Process's Objectives? What Measurements Are Being Taken on the Process with Respect to Its Objectives?

A key responsibility of a process owner is to clearly state the objectives of the process and indicators that are consistent with organizational objectives (Gitlow and Levine, 2004). An example of an organizational objective is "Provide our customers with higher-quality products/services at an attractive price that will meet their needs." Each process owner can find meaning and a starting point in the adaptation of this organizational objective to his process's objectives. For example, a process owner in the purchasing department of a health system could translate the preceding organizational objective into the following subset of objectives and metrics:

Objective: Decrease the number of days from purchase request to item/service delivery.

Metric: Number of days from purchase request to item/service delivery by delivery overall, and by type of item purchased, by purchase.

Objective: Increase ease of filling out purchasing forms.

Metric: Number of questions received about how to fill out forms by form by month.

Objective: Increase employee satisfaction with purchased material

Metric: Number of employee complaints about purchased material by month.

Objective: Continuously train and develop purchasing personnel with respect to job requirements.

Metric: Number of errors per purchase order by purchase order.

Metric: Number of minutes to complete a purchase order by purchase order.

Whatever the objectives of a process are, all involved persons must understand them and devote their efforts toward those objectives. A major benefit of clearly stating the objectives of a process is that everybody works toward the same aim/mission.

Fundamentals of Documenting a Process

Now that we have identified the process owner, know where the process starts and stops, and understand the objectives/metrics to measure success, it is time to document the process. Documenting a process requires input from all stakeholders of the process, as they may have different points of view on the flow of the process.

How Do We Document the Flow of a Process?

To document a process we use a flowchart, which is a pictorial summary of the steps, flows, and decisions that comprise a process (Fitzgerald and Fitzgerald, 1973; Silver and Silver, 1976). Figure 3.1 shows a simple generic flowchart.

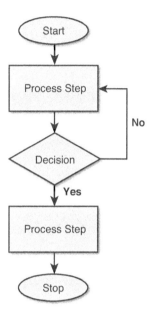

Figure 3.1 A simple generic flowchart

Why and When Do We Use a Flowchart to Document a Process?

Documenting a process using a flowchart, as opposed to using written or verbal descriptions has several advantages:

- A flowchart makes it easier for people who are unfamiliar with a process to understand it.

- A flowchart allows for employees to visualize what actually happens in a process, as opposed to what is supposed to happen.

- A flowchart functions as a communications tool. It provides an easy way to convey ideas between engineers, managers, hourly personnel, vendors, and others in the interdependent system of stakeholders for the organization. It is a concrete, visual way of representing a complex system.

- A flowchart functions as a planning tool. Designers of processes are greatly aided by flowcharts. They enable a visualization of the elements of new or modified processes and their interactions.

- A flowchart removes unnecessary details and breaks down the system so designers and others get a clear, unencumbered look at what they're creating.

- A flowchart defines roles. It demonstrates the functions of personnel, workstations, and subprocesses in a system. It also shows the personnel, operations, and locations involved in the process.

- Flowcharts can be used in the training of new and current employees.

- A flowchart helps you understand what data needs to be collected when trying to measure and improve a process.

- A flowchart can also be used to be compliant with regulatory agencies, i.e., JCAHO (Joint Commission on Accreditation of Healthcare Organizations)

- A flowchart can be used to compare the current state of the process (how the process is), the desired state of the process (how the process should be with standardization), and the future state of the process (how the process could be with improvement).

- And last, but not least, a flowchart helps to identify the weak points in a process; that is, the points in the process that are causing problems for the stakeholders of the process.

Flowcharts can be applied in any type of process to aid in defining and documenting it, and ultimately to improve and innovate the process.

What Are the Different Types of Flowcharts and When Do We Use Each?

This chapter covers two types of flowcharts used in process improvement activities: process flowcharts and deployment flowcharts. Each type of flowchart has different features discussed in the following sections.

Process Flowchart

What is it?

- A flowchart that lays out process steps and decision points in a downward direction from the starting point to the stopping point of the process.

When to use it?

- When you want to depict a process at a high level or when you want to drill down into a detailed portion of a process.

What does it look like?

- An example of a process flowchart for a typical inpatient cardiology consult process is shown in Figure 3.2.

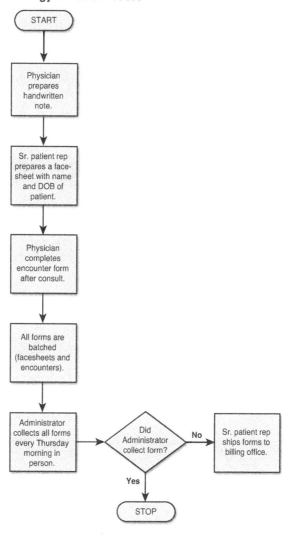

Figure 3.2 Process flowchart example

Deployment Flowchart (Also Known As Cross-Functional or Swim Lane Flowcharts)

What is it?

- A flowchart organized into "lanes" that show processes that involve various departments, people, stages, or other categories.

When to use it?

- When you want to show who is responsible for different parts of a process as well as track the number and location of handoffs within the process.

What does it look like?

- An example of a deployment flowchart for a surgical biopsy sign-out process is shown in Figure 3.3.

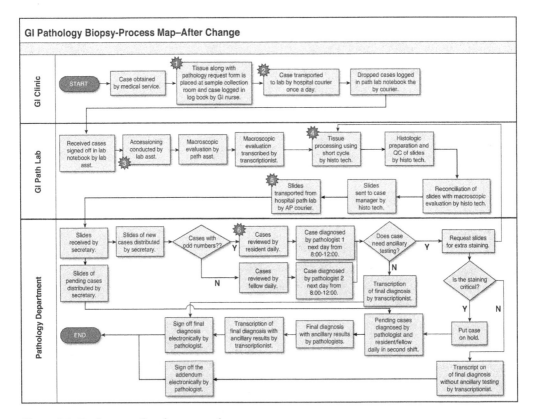

Figure 3.3 Deployment flowchart example

The starbursts in Figure 3.3 show areas of the process that are suspected of causing problems in the process's outputs. The starbursts can come from process experts, reviews of the available literature, benchmarking with similar processes in other organizations, or many other possible sources. All the possible sources of the starbursts are discussed later in this book.

What Method Do We Use to Create Flowcharts?

The American National Standards Institute, Inc. (ANSI), developed a standard set of flowchart symbols used for defining and documenting a process, some of which are shown in Table 3.1. The shape of the symbol and the information written within the symbol provide information about that particular step or decision in a process.

Table 3.1 American National Standards Institute Approved Standard Flowchart Symbols

Symbol	Function
Start/stop symbol	The general symbol used to indicate the beginning and end of a process is an oval.
Basic processing symbol	The general symbol used to depict a processing operation is a rectangle.
Decision symbol	A diamond is the symbol that denotes a decision point in the process. This includes decisions such as pass-fail or yes-no, which creates branches on the flowchart.
Flowline symbol	A line with an arrowhead is the symbol that shows the direction of the stages in a process. The flowline connects the elements of the system.

- Flowcharts provide a common language for the stakeholders of a process to discuss the process, for example, where it begins and ends, who is responsible for each step in the process, how does the process flow, how should the process flow, to name a few benefits of flowcharts.

- Flowcharts help managers see that they are responsible for the outputs of the process, as opposed to the employees who work in the process, because the managers design and manage the process.

- Flowcharts provide stakeholders a tool to walk the process from front to back (process owner's point of view) and back to front (customer's point of view).

- Flowcharts provide an opportunity for the stakeholders of a process to identify the problematic steps in the process (starbursts).

Fundamentals of Analyzing a Process

One of the best methods to define and analyze a process is a flowchart. We discuss flowcharts in detail in this section.

How Do We Analyze Flowcharts?

Process improvers can use a flowchart to change a process by paying attention to the following five points:

1. Process improvers find the steps of the process that are weak (for example, parts of the process that generate a high defect rate).
2. Process improvers improve the steps of the process that are within the process owner's control; that is, the steps of the process that can be changed without higher approval than the process owner.
3. Process improvers isolate the elements in the process that affect customers.
4. Process improvers find solutions that don't require additional resources.
5. Process improvers don't have to deal with political issues.

If these five conditions exist simultaneously, an excellent opportunity to constructively modify a process has been found. Again, process improvements have a greater chance of success if they are either nonpolitical or have the appropriate political support, and either do not require capital investment or have the necessary financial resources.

Other questions that the process improver can ask are the following:

- Why are steps done? How else could they be done?
- Is each step necessary?
    - Is it value added and necessary? Is it repetitive?
    - Would a customer pay for this step specifically? Would the customer notice if it's gone?
    - Is it necessary for regulatory compliance?
- Does the step cause waste or delay?
- Does the step create rework?
- Could the step be done in a more efficient and less costly way?

- Is the step happening at the right time? (sequence)
- Could this step be done in parallel with another step to cut cycle time?
- Are the right people doing the right thing?
- Could this step be automated?
- Does the process contain proper feedback loops?
- Are roles and responsibilities for the process clear and well documented?
- Are there obvious bottlenecks, delays, waste, or capacity issues that can be identified and removed?
- What is the impact of the process on stakeholders?

Things to Remember When Creating and Analyzing Flowcharts

- Work with people who really know and live the process such as front-line employees. Managers may think they know how it works, or how it is supposed to work, but those on the front lines can tell you how it really works.
- Make people understand you are only there to help. The last thing you want people to think is that their jobs may be in jeopardy if the process gets improved so much that it is no longer necessary. You are there to help them do their jobs better, not to eliminate them. Explain they have employment security, not job security. Job security leads to redundancy and unnecessary work; for example, if a unionized electrician knocks a water pipe loose, she can't fix it due to union rules. A plumber must be called in to fix it and much damage could result in the factory in the interim. However, if employees are guaranteed employment and wage security, they are more open to cross-training and the preceding scenario would not happen.
- Start high level to identify major components of the process; then drill down.
- Keep asking questions, and question everything.
- Involve enough people so you get a complete understanding of the process.
- Validate and verify with key stakeholders to make sure the process is understood.
- Keep the flowchart as simple and understandable as possible so anyone can follow it.
- Walk the process from front to back (process owner's point of view) and from back to front (customer's point of view).
- Focus on the needs of the customer.
- Only improve processes with data and facts.

Takeaways from This Chapter

- Four steps are used to define a process:
    1. Identify the process.
    2. Identify the process owner.
    3. Identify where a process starts and where it ends.
    4. Identify the objectives and metrics to measure the success of a process.
- Flowcharts are used to document a process.
- The two main flowcharts used in process improvement activities are process flowcharts and deployment flowcharts.

References

Deming, W. E. (1982), *Quality, Productivity, and Competitive Position* (Cambridge, MA: Massachusetts Institute of Technology, Center for Advanced Engineering Study).

Deming, W. E. (1986), *Out of the Crisis* (Cambridge, MA: Massachusetts Institute of Technology, Center for Advanced Engineering Study).

Fitzgerald, J. M. and A. F. Fitzgerald (1973), *Fundamentals of Systems Analysis* (New York: John Wiley and Sons).

Gitlow, H., A. Oppenheim, R. Oppenheim, and D. Levine (2015), *Quality Management: Tools and Methods for Improvement*, 4th ed. (Naperville, IL: Hercher Publishing Company). This book is free online at hercherpublishing.com.

Gitlow, H. and D. Levine (2004), *Six Sigma for Green Belts and Champions: Foundations, DMAIC, Tools and Methods, Cases and Certification* (Upper Saddle River, NJ: Prentice-Hall).

Silver, G. A. and J. B. Silver (1976), *Introduction to Systems Analysis* (Englewood Cliffs, NJ: Prentice-Hall).

4
Understanding Data: Tools and Methods

What Is the Objective of This Chapter?
The objective of this chapter is to introduce you to some statistical tools that help you understand the data that you will use in the Six Sigma DMAIC model. The chapter is split into two sections. The first section is a high level overview of the tools and methods used in understanding data by looking at what each tool and method is, why each tool and method is used, and examples of each tool and method. The second section shows you how to use Minitab to utilize the different tools and methods you learn about in the first section.

What Is Data?
Data is information collected about a product, service, process, individual, item, or thing. Because no two things are exactly alike, data inherently varies. Each characteristic of interest is referred to as a *variable*. Data can also be words, sounds, pictures, to name a few types of data. We focus mainly on numerical data in this chapter and "word" data later in this book.

Types of Numeric Data
There are two basic types of numeric data; they are *attribute data* and *measurement data*. Each type of data is discussed in the following sections.

Attribute Data
Attribute data (also referred to as *classification* or *count data*) occurs when a variable is either classified into categories (defective or conforming) or used to count occurrences of a phenomenon (number of patient falls on a particular hospital floor in a particular month).

Attribute Classification Data
Attribute classification data places an item into one of two or more categories; for example, not defective (fit for use) or defective (not fit for use). Some examples of attribute classification data are the following:

- Percent of accounts receivable older than 90 days per month. Either the account is over 90 days or it isn't over 90 days; there are only two categories.
- Percent of employees off sick by supervisor by day. Either the employee is off sick or not; there are only two categories.
- Percent of occurrences of surgery delays in an operating room by month. Either the surgery is delayed or not; again, there are only two categories.

Table 4.1 shows the classification of defective items from a daily sample of 100 units for 11 days.

Table 4.1 Defective Items from a Daily Sample of 100 Units for 11 Days

Day	Number Defective	Sample Size	Proportion Defective
1	11	100	0.11
2	21	100	0.21
3	13	100	0.13
4	20	100	0.20
5	14	100	0.14
6	21	100	0.21
7	19	100	0.19
8	18	100	0.18
9	30	100	0.30
10	21	100	0.21
11	23	100	0.23

Attribute Count Data

Attribute count data consists of the number of defects in an item or area of opportunity (for example, a microscope, a room, a stretch of highway, a hospital, and so on). An item can have multiple defects and still not be defective. However, it is possible that one or more of the defects in an item make the item defective. For example, if a water bottle leaks it is defective. However, if the water bottle has a dent and a scratch on the label, it has two defects, neither of which makes the item defective (it is still fit for use).

Some examples of attribute count data are

- The number of data entry errors on a patient chart by chart
- The number of cars entering a hospital parking garage by day
- The number of surgeries performed on the wrong patient per year

The number of patient falls per week in a hospital is attribute count data and can be seen in Table 4.2.

Table 4.2 Patient Falls Per Week in a Hospital

Week	Falls
1	10
2	6
3	9
4	11
5	14
6	7
7	10
8	12
9	9
10	11

Measurement Data

Measurement data (also referred to as continuous or variables data) results from a measurement taken on an item of interest, or the computation of a numerical value from two or more measurements of variables data. Any value can theoretically occur, limited only by the precision of the measuring process. This type of data can have decimal points—for example, height, weight, temperature, waiting time, service time, diameter, revenues, costs, and cycle time.

Some examples of measurement data are

- Height by person
- Waiting time by patient
- Revenue by month
- Cost by line item by store by month

Other examples of measurement data include miles since refueling by ambulance, gallons consumed per ambulance, and a computation of miles per gallon per ambulance as shown in Table 4.3.

Table 4.3 Miles Since Refueling, Gallons Consumed, and MPG

Truck	Measurement of a Characteristic Miles Since Refueling	Measurement of a Characteristic Gallons Consumed	Computation of a Characteristic MPG
1	308	17.3	17.8
2	256	15.3	16.7
3	274	16.5	16.6
4	310	16.9	18.3
5	302	17.1	17.7
6	296	17.3	17.1

Graphing Attribute Data

When dealing with attribute data, responses are tallied into two or more categories, and the frequency or percentage in each category is obtained. Three widely used graphs—the bar chart, the Pareto diagram, and the line chart—are presented in this section.

Bar Chart

What: A bar chart presents each category of an attribute variable as a bar whose length is the frequency or percentage of observations falling into a particular category. The width of the bar is meaningless for a bar chart, but all bars should be the same width.

Why: A bar chart is used to graphically display the frequency or percentages of items that fall into two or more categories.

Example: To illustrate a bar chart we examine data from a hospital pharmacy regarding reasons on delays to orders for the hospital's chemotherapy treatment unit. Table 4.4 shows the data collected on the reasons for delays.

Table 4.4 Pharmacy Delay Reasons (January 1 Through June 30, 2014)

Delay Reason	Number	Percentage
Missing D.O.S.	74	41%
Missing height and weight	66	37%
Dose change	15	8%
Order clarification	9	5%
No consent form	7	4%
Labs pending	6	3%
Labs high	2	1%

The bar chart in Figure 4.1 shows that missing D.O.S. (date of service) and missing height and weight are the two most problematic reasons for delays in the pharmacy from the data in Table 4.4.

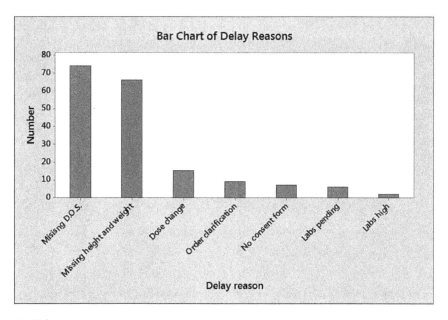

Figure 4.1 Delay reasons

Pareto Diagrams

What: Pareto diagrams are used to identify and prioritize issues that contribute to a problem we want to solve. Created by Italian economist Vilfredo Pareto, Pareto analysis focuses on distinguishing the vital few causes of problems from the trivial many causes of problems. The vital few are the few causes that account for the largest percentage of the problem, while the trivial many are the myriad of causes that account for a small percentage of the problem.

Why: A Pareto diagram is used to graphically display attribute data; specifically it is used to distinguish the significant few categories from the trivial many categories. Consequently, you are able to prioritize efforts on the most important causes (categories) of the problem. Pareto diagrams rank problematic categories from the largest to the smallest, except for the last category that may be "other." This is the source of the famous 80-20 rule. The general idea is that 80% of your problems come from 20% of your causes.

Example: The Director of a hospital pharmacy is interested in learning about the reasons on delays to orders for the hospital's chemotherapy treatment unit. Using the data from Table 4.4 he creates the Pareto diagram in Figure 4.2.

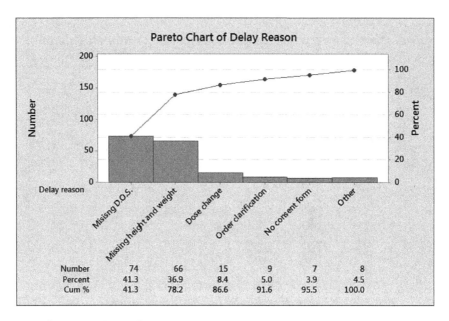

Figure 4.2 Delay reasons Pareto diagram

The Pareto diagram in Figure 4.2 shows that the major reasons for delays are missing (D.O.S.) date of service and missing height and weight on the orders. These two categories account for 78.2% of delays. There are six categories, so all things being equal you would expect one-sixth (16.6%) of the data to be in each category, but 78.2% are in the first two categories instead of 33.2%. Graphically displaying data by using a Pareto diagram promotes prioritization of effort that discourages micromanagement.

Line Graphs

What: A line graph is a graph of any type of variable plotted on the vertical axis and usually time plotted on the horizontal axis.

Why: A line graph is generally used to graphically display data over time.

Example: A line graph is illustrated by using data concerning a medical transcription service that enters medical data on patient files for hospitals. The service studied ways to improve the turnaround time (defined as the time between receiving data and time the client receives completed files). After studying the process, the service determined that transcription errors increased turnaround time. Table 4.5 presents the number and proportion of transcription with errors by day.

Table 4.5 Transmission Errors

Month	Date	Number of Errors	Proportion of Errors
August	1	6	0.048
August	2	3	0.024
August	5	4	0.032
August	6	4	0.032
August	7	9	0.072
August	8	0	0.000
August	9	0	0.000
August	12	8	0.064
August	13	4	0.032
August	14	3	0.024
August	15	4	0.032
August	16	1	0.008
August	19	10	0.080
August	20	9	0.072
August	21	3	0.024
August	22	1	0.008
August	23	4	0.032
August	26	6	0.048
August	27	3	0.024
August	28	5	0.040
August	29	1	0.008
August	30	3	0.024
September	3	14	0.112
September	4	6	0.048
September	5	7	0.056
September	6	3	0.024
September	9	10	0.080
September	10	7	0.056
September	11	5	0.040
September	12	0	0.000
September	13	3	0.024

Figure 4.3 presents the line chart for these data. The line graph clearly shows a great deal of fluctuation in the proportion of transcription errors from day to day. The highest number of errors occurred on day 23 (September 3), but a large number of errors also occurred on days 5, 8, 13, 14, and 27. The medical transcription service needs to study the process to determine the reasons for this variation. Are the variations due to special causes? Or are the variations due to common causes? Methods for studying these issues using control charts are covered in Chapter 5, "Understanding Variation: Tools and Methods," so stay tuned!

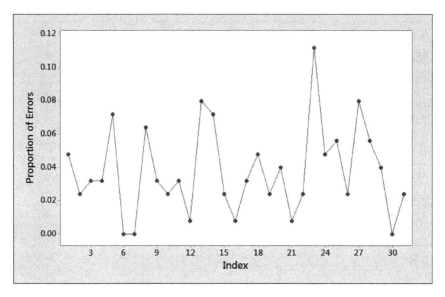

Figure 4.3 Line graph for transcription errors

Graphing Measurement Data

Many people do not have the ability to look at data and make much sense of it. Consequently, process improvers and other scientists create graphical or pictorial representations of data to help them understand it. In this section, we discuss several of these graphical or pictorial representations used for measurement data; they are histograms, dot plots, and run charts. You will find these tools extremely valuable in understanding your data; they let the data talk to you about what is going on in the process that generated them.

Histogram

What: A histogram is a special bar chart for measurement data. In the histogram, the data is grouped into adjacent numerical categories of equal size, for example, 100 to less than 200, 200 to less than 300, 300 to less than 400, and so on. The difference between a bar chart and a histogram is that the X axis on a bar chart is a listing of categories, whereas the X axis on a histogram is a measurement scale. In addition there are no gaps between adjacent bars.

Why: A histogram is used to graphically display measurement data to understand the distribution of the data.

Example: To illustrate a histogram we examine 100 patient wait times in minutes at an outpatient clinic in Figure 4.4. For these data, notice tick marks at 30, 40, 50, 60, 70, and 80. The distribution appears to be approximately bell-shaped with a heavy concentration of wait times between 40 and 70.

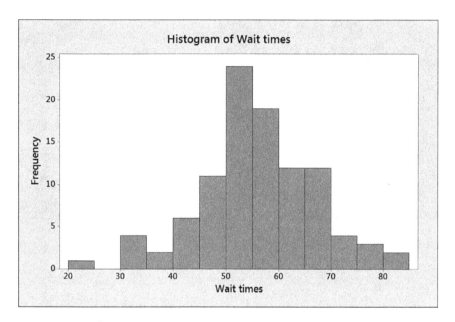

Figure 4.4 Histogram of patient wait times

Dot Plot

What: Similar to a histogram, a dot plot is a graph of measurement data in which dots that represent data values are stacked vertically on the horizontal axis for each value of the variable of interest.

Why: A dot plot is used to graphically display measurement data to understand the distribution of the data. Dots plots are generally used when you have smaller sets of data as once data sets become larger the dot plot becomes cluttered.

Example: To illustrate a dot plot we examine the same 100 patient wait times in minutes at an outpatient clinic in Figure 4.5 (refer to the histogram in Figure 4.4). Note that the dot plot looks different from the histogram. This occurs because the histogram groups the data into class intervals, whereas the dot plot presents each data value, with the height representing the frequency at each horizontal value. Nevertheless, the dot plot shows a concentration of values in the center of the distribution between 40 and 70 minutes.

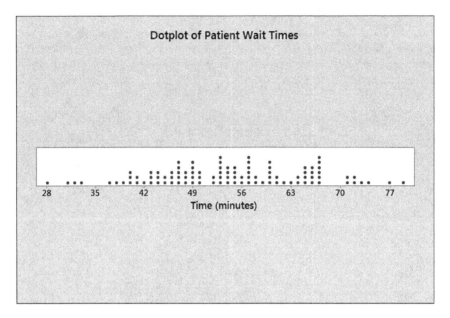

Figure 4.5 Dot plot of patient wait times

Run Chart

What: A run chart is a type of line chart that may have multiple measurements for each time period, called subgroups. Time is the x-axis and the variable of interest is plotted on the y-axis; remember there may be multiple data points on the y-axis for each time period on the x-axis.

Why: A run chart is used to graphically display measurement data where there may be multiple data points per subgroup. When data are collected over time, the variable of interest should be plotted in time order before any other graphs are plotted, descriptive statistics are calculated, or statistical analyses are performed.

Example: To illustrate a run chart, we examine a sample of three waiting times (in minutes) for patients per day in a medical clinic; we determine the waiting time for the first patient to enter at 9:00 a.m., then at noon, and then at 4:00 p.m. The data appears in Table 4.6.

Table 4.6 Waiting Times for a Sample of Three Patients per Day in a Medical Clinic (in minutes)

Day	Waiting Time (9:00 a.m.)	Waiting Time (noon)	Waiting Time (4:00 p.m.)
1	109.909	94.580	108.207
2	104.939	106.269	08.483
3	96.738	103.557	99.587
4	104.494	94.569	84.540
5	112.930	104.091	103.763
6	88.272	105.702	85.598
7	106.013	112.941	121.007
8	98.202	102.188	101.633
9	102.216	112.204	86.541
10	79.289	100.382	117.723
11	93.124	95.098	106.073
12	100.466	88.852	95.471
13	112.393	102.399	105.336
14	100.049	100.118	100.838
15	90.865	91.397	115.002
16	91.230	117.406	99.431
17	88.502	100.035	100.881
18	116.279	113.906	97.700
19	79.182	84.080	93.334
20	107.262	102.475	96.598
21	115.299	108.362	106.973
22	88.139	109.658	95.129
23	112.439	99.519	88.262
24	103.661	105.989	106.239
25	97.802	100.906	99.214
26	99.055	101.062	105.244
27	106.934	108.751	98.336
28	92.009	91.027	119.083
29	102.465	121.023	106.972
30	95.953	98.419	102.357
31	109.776	95.586	101.465
32	114.890	107.868	97.132
33	110.809	94.834	96.335
34	99.719	101.790	99.900
35	83.911	89.563	100.017

	Waiting Time	Waiting Time	Waiting Time
36	119.598	91.605	113.346
37	108.419	93.357	86.829
38	93.971	88.679	108.423
39	108.036	106.149	95.813
40	96.530	89.928	100.436
41	88.415	95.474	95.795
42	83.524	103.116	107.381
43	88.941	99.971	95.070
44	94.799	110.642	90.903
45	97.267	109.286	92.706
46	91.730	112.460	103.507
47	105.964	116.872	101.706
48	114.394	72.322	101.364
49	108.729	88.161	92.020
50	88.860	121.195	85.736

As you can see from Figure 4.6, there are the three data points per day, one for the 9:00 a.m. patient, one for the noon patient, and one for the 4:00 p.m. patient. We learn how to analyze this data further in Chapter 5. However, for now we can see that the data is not trending up or trending down, and it is not getting more variable (like a cone going from left [small] to right [large] or less variable (like a cone going from left [large] to right [small]).

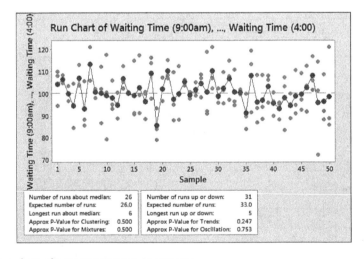

Figure 4.6 Run chart of patient wait times in minutes

Measures of Central Tendency for Measurement Data

Although the graphs studied thus far are useful for getting a visual picture of what the distribution of a variable looks like, it is important to compute measures of central tendency or location. Three measures of central tendency are discussed—the mean, the median, and the mode.

Mean

What: The most common numerical representation of central tendency is the arithmetic average or mean. It is the sum of the numerical values of the items being measured divided by the number of items. Because its computation is based on every observation, the arithmetic mean is greatly affected by any extreme value or values, so be careful because you may get a distorted representation of what the data are conveying!

Why: The mean is useful to convey the average value of measurement data especially if the data is symmetric around the average; for example, a bell shaped histogram or dot plot of data.

Example: To illustrate the computation of the mean, consider the following example related to your personal life: the time it takes to get ready for work in the morning. Many people wonder why it seems to take longer than they anticipate getting ready to leave for work, but virtually no one has measured the time it takes to get ready in the morning. Suppose the time to get ready in the morning is operationally defined as the time in minutes (rounded to the nearest minute) from when you get out of bed to when you leave your house. Suppose you collect these data for a period of two weeks, ten working days, with the results in Table 4.7.

Table 4.7 Time to Get Ready for Ten Days

Day	1	2	3	4	5	6	7	8	9	10
Time (minutes)	39	29	43	52	39	44	40	31	44	35

To calculate the arithmetic mean, you simply sum the observations and divide by the number of observations. In this case you would add up all the minutes and divide by the number of days, which is 396/10 = 39.6.

The mean (or average) time to get ready is 39.6 minutes, even though not one individual day in the sample actually had that value. Note that the calculation of the mean is based on all the observations in the set of data. No other commonly used measure of central tendency uses this characteristic.

Median

What: The median is the middle value in a set of data that has been ordered from the lowest to the highest value. Half the observations will be smaller than the median, and half will be larger. The median is not affected by any extreme values in a set of data.

Why: To convey the underlying character of measurement data by representing the middle value such that 50% of the observations are smaller and 50% of the observations are larger. Whenever an extreme value is present, the median is useful in describing the central tendency of the data.

Example: To calculate the median from a set of data, you must first organize the data into an ordered array that lists the values from smallest to largest. If there is an odd number of observations in the data set, the median is the middle most number in the data set. If there is an even number of observations in the data set, the median is the average of the two middle most numbers in the data set.

Going back to our example of the time it takes to get ready for work, we first arrange the values from smallest to largest as seen in Table 4.8.

Table 4.8 Ordered Time to Get Ready for Ten Days

Day	1	2	3	4	5	6	7	8	9	10
Time (minutes)	29	31	35	39	39	40	43	44	44	52

Since there is an even number of observations in the data set we then find the two middle most numbers, which happen to be 39 and 40; we then average them to get a median of 39.5.

Suppose our data set contained only the first nine observations as seen in Table 4.9.

Table 4.9 Time to Get Ready for Ten Days from Smallest to Largest

Day	1	2	3	4	5	6	7	8	9
Time (minutes)	29	31	35	39	39	40	43	44	44

In this case we would have an odd number of observations in the data set, so to calculate the median we would simply find the middle most observation, which would give us a median of 39.

Mode

What: The mode is the value in a set of data that appears most frequently in a data set. Unlike the arithmetic mean, the mode is not affected by the occurrence of any extreme values.

Why: The mode is used only to find the most commonly occurring value in a data set.

Example: Referring back to the times for the days, there are two modes: 39 minutes and 44 minutes because each of those values occurs twice. This is called a *bimodal distribution*.

Measures of Central Tendency for Attribute Data

Recall that there are two types of attribute data: classification data and count data. Classification and count data are subject to all the measures of central tendency used for measurement data, but classification data is a bit special because it consists of only 0s and 1s; that is, yes or no, and so on. This makes the computations a little simpler. Instead of the mean for classification data, we use the proportion.

Proportion

What: Often data are classified into two non-numerical conditions, such as broken versus not broken, defective versus conforming, operating versus not operating. The proportion or fraction of the data possessing one of two such conditions is a meaningful measure of central tendency.

Why: The proportion is used to understand the degree to which the output of a population or process is in either one or two possible states, for example, defective or conforming, and so on.

Example: A Chief Medical Officer at a large hospital was concerned with readmissions of a certain patient population. He took a random sample of 30 patients to look at the proportion that was readmitted to the hospital within 30 days; see Table 4.10.

Table 4.10 Hospital Readmissions Within 30 Days

Patient #	Condition
1	Readmitted
2	Not Readmitted
3	Not Readmitted
4	Not Readmitted
5	Not Readmitted
6	Not Readmitted
7	Readmitted
8	Not Readmitted
9	Not Readmitted
10	Not Readmitted
11	Not Readmitted
12	Readmitted

Patient #	Condition
13	Not Readmitted
14	Not Readmitted
15	Readmitted
16	Not Readmitted
17	Not Readmitted
18	Readmitted
19	Not Readmitted
20	Readmitted
21	Not Readmitted
22	Not Readmitted
23	Not Readmitted
24	Readmitted
25	Readmitted
26	Not Readmitted
27	Not Readmitted
28	Not Readmitted
29	Not Readmitted
30	Not Readmitted

To calculate the proportion (p), he would divide the number readmitted by the total number of patients in the sample. In this case 8/30 = .27 = 27%.

Measures of Variation

A second important property that describes a set of numerical data is variation. Variation is the amount of dispersion, or spread, in the data. Three measures of variation include the range, the variance, and the standard deviation.

Range

What: The range is simply the difference between the largest and the smallest data points in the data set. It is calculated by subtracting the smallest number from the largest number.

Why: The range is used to analyze the spread of the data. The range assumes that there are no extreme values in the data because that would distort the range and make it a meaningless measure of variation.

Example: Using the data on time to get ready in the morning (refer to Table 4.8), we would calculate the range as follows:

Range = largest value − smallest value

Range = 52 − 29 = 23 minutes

This means that the largest difference between any two days in the time to get ready in the morning is 23 minutes.

Sample Variance and Standard Deviation

What: To understand how the values in the data are spread around the mean we have to look at two other commonly used measures of variation called the *variance* and the *standard deviation*.

The variance and standard deviation measure how far the values in a data set are spread out around the average. A variance or standard deviation of zero indicates all the values in the data set are the same.

To calculate the sample variance you simply obtain the difference between each value and the mean, you square each difference, you add the squared differences, and then you divide this total by the number of observations minus 1. To calculate the sample standard deviation you simply take the square root of the variance.

Why: The variance and standard deviation tell us how the values are distributed around the mean. The variance and standard deviation are less affected by extreme value in the data set that is the range.

Example: Let's go back to our time to get ready in the morning example to help us understand variance and standard deviation. The times to get ready are listed in Table 4.11.

Table 4.11 Time to Get Ready for Ten Days Used to Calculate Variance and Standard Deviation

Time (x)	Difference Between x and Mean	Squared Differences around the Mean
39	-0.6	0.36
29	-10.6	112.36
43	3.4	11.56
52	12.4	153.76
39	-0.6	0.36
44	4.4	19.36
40	0.4	0.16
31	-8.6	73.96
44	4.4	19.36
35	-4.6	21.16
Mean = 39.6	Sum of differences = 0	Sum of squared differences = 412.4

To calculate the variance we take the following steps:

1. Subtract each value from the mean; see the middle column in Table 4.11.
2. Square those numbers and add them up to get the sum of squared differences, which is 412.4 as you see at the bottom of the right column in Table 4.11.
3. Divide the sum of squared differences by the number of observations (10) minus one which gives us 412.4/9 = 45.82.

To calculate the standard deviation simply take the square root of the variance, which is 6.77.

Understanding the Range, Variance, and Standard Deviation

So what does this all tell us? The following statements summarize what you need to know about the range, variance, and standard deviation:

- The more spread out, or dispersed, the data are, the larger will be the range, variance, and standard deviation.
- The more concentrated or homogeneous the data is, the smaller will be the range, variance, and standard deviation.
- If the values are all the same (so that there is no variation in the data), the range, variance, and standard deviation will all be zero.
- The range, variance, or standard deviation will always be greater than or equal to zero.
- The range can be a deceptive measure of dispersion if there are extreme values in the data set.

Here are some examples to help you understand the standard deviation.

Example #1: Distribution with Mean = 100 and Standard Deviation =10

Mean = 100; Standard Deviation = 10

In Figure 4.7 you see that the data is more spread out than in Figure 4.8 because the standard deviation = 10.

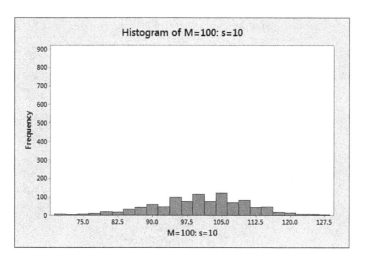

Figure 4.7 Histogram for Example #1

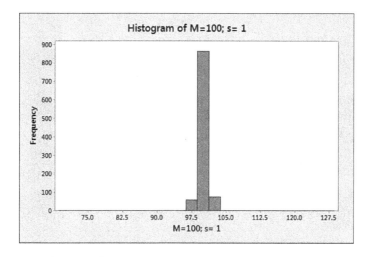

Figure 4.8 Histogram for Example #2

Example #2: Distribution with Mean = 100 and Standard Deviation = 1

In Figure 4.8 you see that the data is less spread than in Figure 4.7 because the standard deviation = 1.

Example #3: Distribution with Mean = 100 and Standard Deviation = 0

Mean = 100, Standard Deviation = 0

In Figure 4.9 you see that the standard deviation = 0, so all the data are the same.

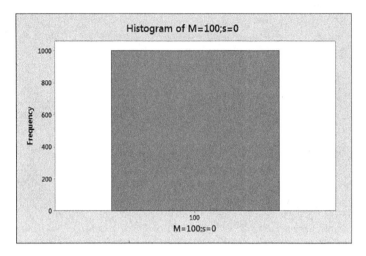

Figure 4.9 Histogram for Example #3

Measures of Shape

The third important property of data that we look at it is shape. The shape is the manner in which the data are distributed. Either a histogram or a dot plot can be used to study the shape of a distribution of data.

Skewness

Skewness is a measure of the size of the right or left tail of a unimodal (one hump) distribution. We examine three types of skewness: symmetrical, positive or right skewness, and negative or left skewness.

Symmetrical

A symmetrical distribution arises when the mean, median, and mode are all equal, see Figure 4.10.

Positive or Right Skewness

Positive or right skewness occurs when the data has some unusually high values, which causes the mean to be greater than the median as is seen in Figure 4.11.

When data is skewed to the right the mean is larger than the median, and the median is larger than the mode.

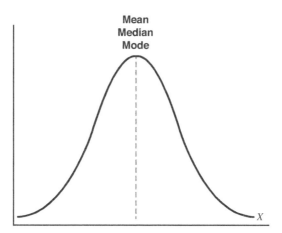

Figure 4.10 Symmetrical distribution

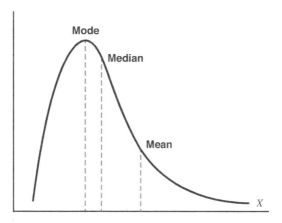

Figure 4.11 Positive or right skewed distribution

Negative or Left Skewness

Negative or left skewness occurs when the data has some unusually low values, which causes the mean to be less than the median as is seen in Figure 4.12.

When data is skewed to the left the mode is larger than the median, and the median is larger than the mean.

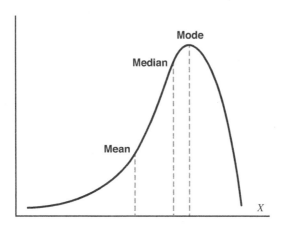

Figure 4.12 Negative or left skewed distribution

More on Interpreting the Standard Deviation

If the distribution of output from a process is unimodal, symmetric, and bell shaped, we call this the *normal distribution*. The normal distribution has some properties that are worth mentioning here to improve your understanding of the standard deviation.

First, if you create a region under the normal distribution that is plus or minus one standard deviation from the mean, then 68.26% of the output from the process that generated the normal distribution will lie in that area. Figure 4.13 shows a normal distribution with a mean of 100 and a standard deviation of 10. In this distribution 68.26% of the data will lie between 90 (Mean − 1 standard deviation = [100 − 10] = 90) and 110 (Mean + 1 standard deviation = [100 + 10] = 110).

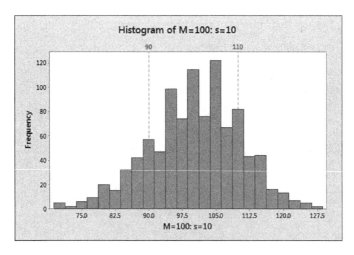

Figure 4.13 Mean +/- 1 standard deviation from the mean

Second, if you create a region under the normal distribution that is plus or minus two standard deviations from the mean, then 95.44% of the output from the process that generated the normal distribution will lie in that area. Figure 4.14 shows a normal distribution with a mean of 100 and a standard deviation of 10. In this distribution 95.44% of the data will lie between 80 (Mean − 2 standard deviations = [100 − 20] = 80) and 120 (Mean + 2 standard deviations = [100 + 20] = 120).

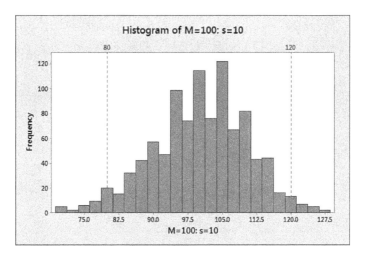

Figure 4.14 Mean +/- 2 standard deviations from the mean

Third, if you create a region under the normal distribution that is plus or minus three standard deviations from the mean, then 99.73% of the output from the process that generated the normal distribution will lie in that area. Figure 4.15 shows a normal distribution with a mean of 100 and a standard deviation of 10. In this distribution 99.73% of the data will lie between 70 (Mean − 3 standard deviations = [100 − 30] = 70) and 130 (Mean + 3 standard deviations = [100 + 30] = 130).

Statisticians calculated these probabilities a long time ago. It doesn't matter what the mean is or what the standard deviation is, the preceding probabilities apply.

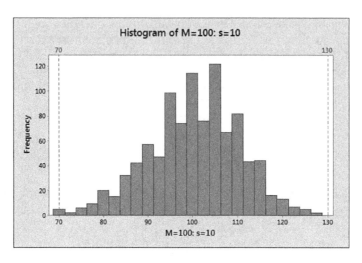

Figure 4.15 Mean +/- 3 standard deviations from the mean

How-To Guide for Understanding Data: Minitab 17 User Guide

Minitab is a statistical package designed to perform statistical analysis (Minitab 17, 2013). Today, Minitab is used both in academia and in industry. In Minitab, you create and open projects to store all your data and results. A session or log of activities, a Project Manager that summarizes the project contents, and any worksheets or graphs used are the components that form a project. Project components are displayed in separate windows inside the Minitab application window. By default, you see only the session and one worksheet window when you begin a new project in Minitab. (You can bring any window to the front by selecting the desired window in the Minitab Windows menu.) You can open and save an entire project or, as is done in this text, open and save worksheets.

Minitab's statistical rigor, availability for many different types of computer systems, and commercial acceptance makes this program a great tool for using statistics for quality improvement.

Go to www.ftpress.com/sixsigma to download the data files referenced in this chapter so you can practice with Minitab.

Using Minitab Worksheets

You use a Minitab worksheet (see Figure 4.16) to enter data for variables by column. Minitab worksheets are organized as numbered rows and columns numbered in the form Cn in which C1 is the first column. You enter variable labels in a special unnumbered row that precedes row 1. Unlike worksheets in programs such as Microsoft Excel, currently Minitab worksheets do not accept formulas and do not automatically recalculate themselves when you change the values of the supporting data.

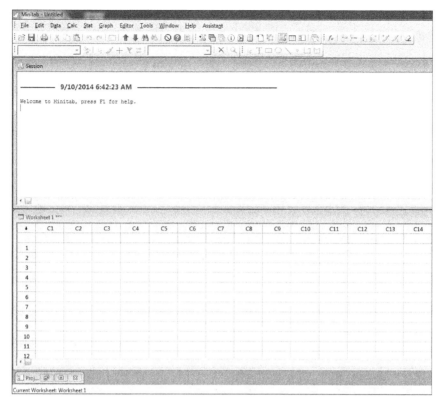

Figure 4.16 Minitab worksheet

By default, Minitab names opened worksheets serially in the form of Worksheet1, Worksheet2, and so on. Better names are ones that reflect the content of the worksheets, such as CHEMICAL for a worksheet that contains data for the viscosity of a chemical. To give a sheet a descriptive name, open the Project Manager window, right-click the icon for the worksheet, select Rename from the shortcut menu, and type in the new name.

Opening and Saving Worksheets and Other Components

You open worksheets to use data that have been created by you or others at an earlier time. To open a Minitab worksheet, first select **File | Open Worksheet**. In the Open Worksheet dialog box that appears (see Figure 4.17) perform the following steps:

1. Select the appropriate folder (also known as a directory) from the Look In drop-down list box.

2. Check, and select if necessary, the proper Files of Type value from the drop-down list at the bottom of the dialog box. Typically, you do not need to make this selection as the default choice Minitab lists all Minitab worksheets. However, to list all project

files, select Minitab Project; to list all Microsoft Excel files, select Excel (*.xls); and to list every file in the folder, select All.

3. If necessary, change the display of files in the central files list box by clicking the rightmost (View Menu) button on the top row of buttons and selecting the appropriate view from the drop-down list.

4. Select the file to be opened from the files list box. If the file does not appear, verify that steps 1, 2, and 3 were done correctly.

5. Click **OK**.

To open a Minitab Project that can include the session, worksheets, and graphs, select Minitab Project in the previous step 2 or select the similar **File | Open Project**. Individual graphs can be opened as well by selecting **File | Open Graph**.

Figure 4.17 Minitab Open Worksheet dialog box

You can save a worksheet individually to ensure its future availability, to protect yourself against a system failure, or to later import it into another project. To save a worksheet, select the worksheet's window and then select **File | Save Current Worksheet As**. In the Save Worksheet As dialog box that appears (see Figure 4.18), perform these steps:

1. Select the appropriate folder from the Save In drop-down list box.
2. Check, and select if necessary, the proper Save As Type value from the drop-down list at the bottom of the dialog box. Typically, you want to accept the default choice, Minitab, but select Minitab Portable to use the data on a different type of computer system or select an earlier version such as Minitab 13 to use the data in that earlier version.
3. Enter (or edit) the name of the file in the File Name box.
4. Optionally, click the Description button and in the Worksheet Description dialog box (not shown), enter documentary information and click **OK**.
5. Click **OK** (in the Save Worksheet As dialog box).

Figure 4.18 Save Worksheet As dialog box

To save a Minitab Project, select the similar **File | Save Project As**. The Save Project As dialog box (not shown) contains an Options button that displays the Save Project – Options dialog box in which you can indicate which project components other than worksheets (session, dialog settings, graphs, and Project Manager content) will be saved.

Individual graphs and the session can also be saved separately by first selecting their windows and then selecting the similar **File | Save Graph As** or **File | Save Session As**, as appropriate.

Minitab graphs can be saved in either a Minitab graph format or any one of several common graphics formats, and Session files can be saved as simple or formatted text files.

You can repeat a save procedure and save a worksheet, project, or other component using a second name as an easy way to create a backup copy that can be used should some problem make your original file unusable.

Obtaining a Bar Chart

To obtain a bar chart, open the **KEYBOARD** worksheet. Select **Graph | Bar Chart** and then follow these steps:

1. In the Bar Charts dialog box (see Figure 4.19), select **Simple**. Click **OK**.

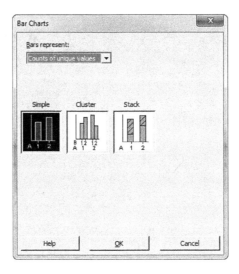

Figure 4.19 Bar Charts dialog box

2. In the Bar Chart dialog box (see Figure 4.20), enter **C2** or **Cause** in the Categorical Variables edit box.

3. Select the **Data Options** button. In the Bar Chart: Data Options dialog box (see Figure 4.21), select the **Frequency** tab since the data is in the form of frequencies in pre-specified classes. Enter **C2** or **Frequency** in the Frequency Variable(s) edit box. Click **OK** to return to the Bar Chart dialog box. Click **OK**.

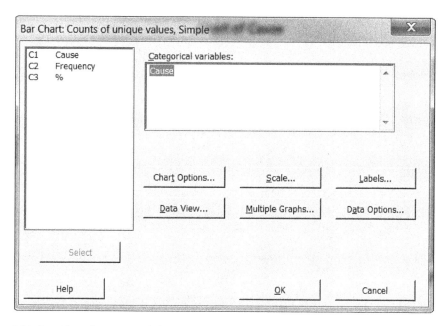

Figure 4.20 Bar Chart data source dialog box

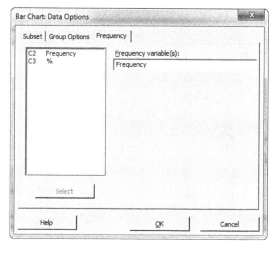

Figure 4.21 Bar Chart: Data Options dialog box

To select colors for the bars and borders in the bar chart, follow these steps:

1. Right-click on any of the bars of the bar chart.
2. Select **Edit Bars**.

3. In the Attributes tab of the Edit Bars dialog box, enter selections for Fill pattern and Border and Fill Lines.

Figure 4.22 shows the output for the bar chart.

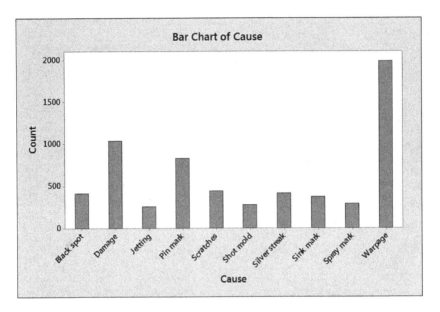

Figure 4.22 Minitab output for the bar chart

Obtaining a Pareto Diagram

To obtain a Pareto diagram, open the **KEYBOARD.MTW** worksheet. Note that this data set contains the causes of the defects in column C1 and the frequency of defects in column C2. Select **Stat | Quality Tools | Pareto Chart**; see Figure 4.23 for the dialog box. Follow these steps:

1. In the Defects or Attribute Data In edit box, enter **C1** or **Cause**.
2. In the Frequencies In edit box, enter **C2** or **Frequency.**
3. In the Combine Remaining Defects into One Category After This Percent edit box, enter **99.9.**
4. Click **OK**.

If the variable of interest was located in a single column and is in raw form with each row indicating a type of error, the Charts Defects Data In option button would be selected and the appropriate column number or variable name would be entered in the Defects or Attribute Data In edit box.

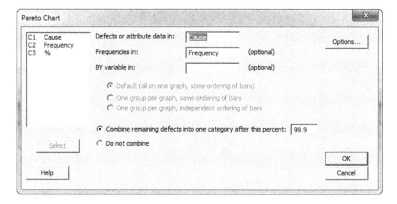

Figure 4.23 Pareto Chart dialog box

To select colors for the bars and borders in the Pareto diagram, perform these steps:

1. Right-click on any of the bars of the Pareto diagram.
2. Select **Edit Bars**.
3. In the Attributes tab of the Edit Bars dialog box, enter selections for Fill pattern and Border and Fill Lines.

Figure 4.24 shows the output for the Pareto diagram.

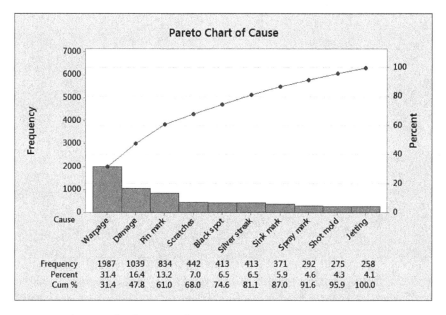

Figure 4.24 Minitab output for the Pareto diagram

Obtaining a Line Graph (Time Series Plot)

To obtain a line graph, open the **TRANSMIT** worksheet. To obtain a run chart of the percentage of errors, follow these steps:

1. Enter the label **Error%** in column C3.

2. Select **Calc | Calculator**.

3. In the Calculator dialog box (see Figure 4.25), enter **C3** or **Error%** in the Store Result in Variable edit box. To obtain the percentage of errors, enter **'Errors' / 125** in the Expression edit box. Click **OK**.

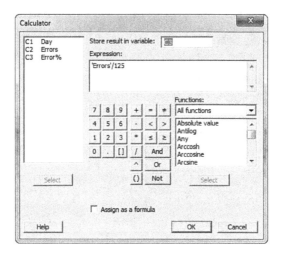

Figure 4.25 Calculator dialog box

4. Select **Graph | Time Series Plot**.

5. Select **Simple**, click **OK**.

6. In the Series edit box, enter **C3** or **'Error%'** (see Figure 4.26).

7. Click **OK**.

Figure 4.27 shows the output for the Time Series Plot.

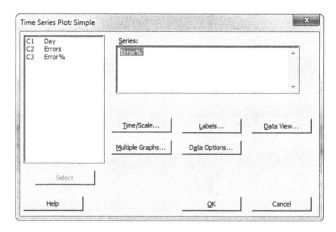

Figure 4.26 Time Series Plot: Simple data options dialog box

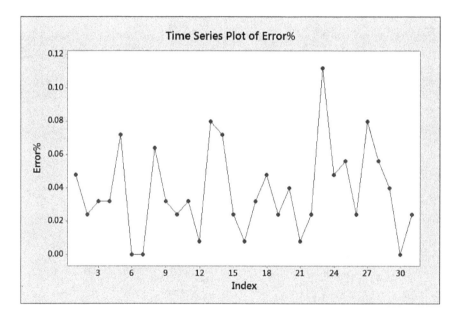

Figure 4.27 Minitab output for the Time Series Plot

Obtaining a Histogram

To obtain a histogram, open the **CHEMICAL.MTW** worksheet. Select **Graph | Histogram** and follow these steps:

Chapter 4 Understanding Data: Tools and Methods   79

1. In the Histograms dialog box (see Figure 4.28), select **Simple**. Click **OK**.

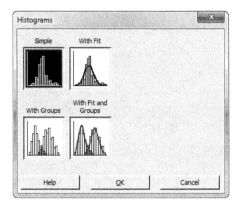

Figure 4.28 Histograms dialog box

2. In the Histogram data source dialog box (see Figure 4.29), enter **C2** or **Viscosity** in the Graph Variables edit box. To obtain reference lines, select the **Scale** button.

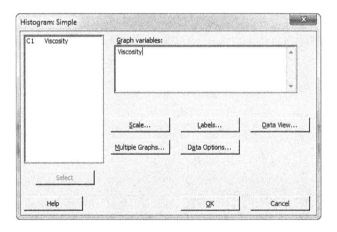

Figure 4.29 Histogram data source dialog box

3. In the Histogram: Scale dialog box (see Figure 4.30), select the **Reference Lines** tab. Enter **13** and **18** in the Show Reference Lines at Y Values box. Click **OK** to return to the Histogram data source dialog box. Click **OK** to obtain the histogram.

Figure 4.30 Histogram: Scale dialog box

To select colors for the bars and borders in the histogram, do the following:

1. Right-click on any of the bars of the histogram.
2. Select **Edit Bars**.
3. In the Attributes tab of the Edit Bars dialog box, enter selections for Fill pattern and Border and Fill Lines.

Figure 4.31 shows the output for the histogram.

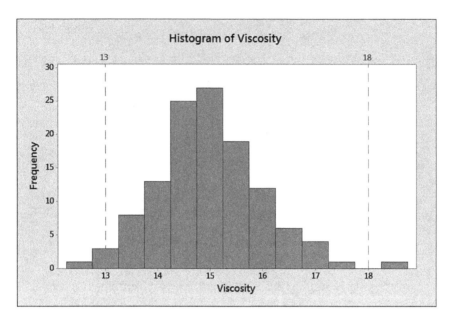

Figure 4.31 Minitab output for the histogram

Obtaining a Dot Plot

To obtain a dot plot using Minitab, open the **CHEMICAL.MTW** worksheet. Select **Graph | Dotplot**, and then do the following:

1. In the Dotplots dialog box (see Figure 4.32), select the **One Y Simple** choice. If dot plots of more than one group are desired, select the **One Y With Groups** Choice.

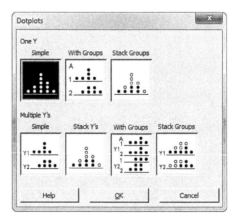

Figure 4.32 Dotplots dialog box

2. In the Dotplot data source dialog box (see Figure 4.33) in the Graph Variables edit box enter **C2** or **Viscosity**. Click **OK**.

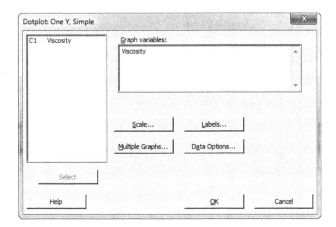

Figure 4.33 Dotplot data source dialog box

Figure 4.34 shows the output for the dot plot.

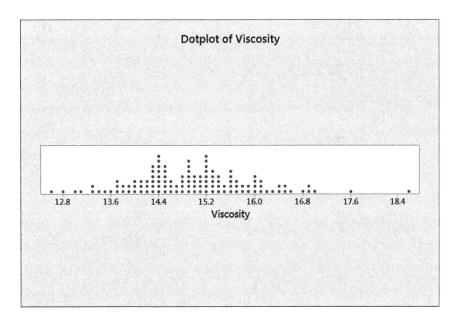

Figure 4.34 Minitab output for the dot plot

Chapter 4 Understanding Data: Tools and Methods

Obtaining a Run Chart

To obtain a run chart, open the **TRANSMIT** worksheet. To obtain a run chart of the percentage of errors, follow these steps:

1. Enter the label **Error%** in column C3.

2. Select **Calc | Calculator**.

3. In the Calculator dialog box (see Figure 4.35), enter **C3** or **Error%** in the Store Result in Variable edit box. To obtain the percentage of errors, enter **'Errors' / 125** in the Expression edit box. Click **OK**.

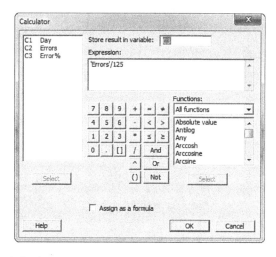

Figure 4.35 Calculator dialog box

4. Select **Stat | Quality Tools | Run Chart**.

5. In the Run Chart dialog box (see Figure 4.36), select the **Single Column** option button. Enter **C3** or **'Error%'** in the edit box. Enter **C1** or **Day** in the Subgroup Size edit box. Click **OK**.

Figure 4.37 shows the output for the run chart.

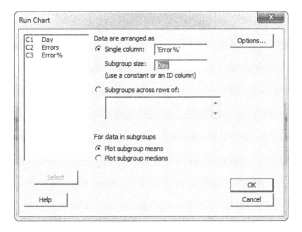

Figure 4.36 Run Chart dialog box

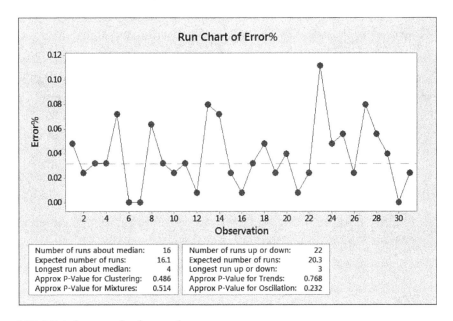

Figure 4.37 Minitab output for the run chart

Obtaining Descriptive Statistics

To obtain descriptive statistics for the viscosity of the chemical data set, open the **CHEMICAL.MTW** worksheet. Select **Stat | Basic Statistics | Display Descriptive Statistics** and follow these steps:

1. In the Display Descriptive Statistics dialog box (see Figure 4.38), enter **C2** or **Viscosity** in the Variables edit box.

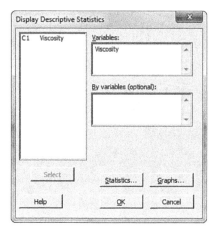

Figure 4.38 Display Descriptive Statistics dialog box

2. Select the **Statistics** button. In the Display Descriptive Statistics: Statistics dialog box (see Figure 4.39), select the **Mean, Standard Deviation, First Quartile, Median, Third Quartile, Minimum, Maximum, Range, Skewness,** and **N Total** (the sample size) check boxes (see Figure 4.39). Click **OK** to return to the Display Descriptive Statistics dialog box. Click **OK** again to obtain the descriptive statistics.

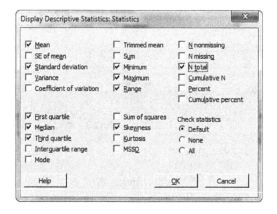

Figure 4.39 Display Descriptive Statistics: Statistics dialog box

Figure 4.40 shows the output for Descriptive Statistics.

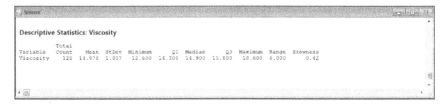

Figure 4.40 Minitab output for descriptive statistics

Takeaways from This Chapter

- When understanding data for process improvement, you need to be familiar with two types: attribute and measurement.
- Different graphs are used for displaying each type of data.
  - For attribute data:
    - Bar charts graphically display the frequency or percentages of items that fall into two or more categories.
    - Pareto diagrams help you distinguish between the critical few and the trivial many categories data exhibits.
    - Line graphs are used to graphically display attribute data over time.
  - For measurement data:
    - Histograms help you understand the distribution of the data.
    - Dot plots are similar to histograms, but you use them when you have smaller data sets.
    - Run charts are used to graphically display measurement data over time to spot trends.
- Various measures of central tendency help you find a single value that best represents a distribution of data.
- Various measures of variation help you find a single number that best represents the spread of the data.

References

Minitab Version 17 (State College, PA: Minitab 2013).

Additional Readings

Berenson, M. L. and D. M. Levine (2014), *Basic Business Statistics*, 13th ed. (Upper Saddle River, NJ: Prentice Hall).

Gitlow, H., A. Oppenheim, R. Oppenheim, and D. Levine (2015), *Quality Management: Tools and Methods for Improvement*, 4th ed. (Naperville, IL: Hercher Publishing Company). This book is free online at hercherpublishing.com.

Gitlow, H. and D. Levine (2004), *Six Sigma for Green Belts and Champions: Foundations, DMAIC, Tools and Methods, Cases and Certification* (Upper Saddle River, NJ: Prentice Hall).

5
Understanding Variation: Tools and Methods

What Are the Objectives of This Chapter?

This chapter has two objectives. The first objective is to introduce you to statistical control charts that help you to understand when and where special causes of variation are occurring in a process, and to understand when a process is exhibiting only common variation. It also explains how to reduce common causes of variation. The second objective is to explain measurement systems analysis; that is, how do you know if your data has too much variation (noise) in it to be of practical value in improving a process?

The chapter is split into two sections. The first section is a high level overview of the tools and methods used in understanding variation by looking at what each tool and method is, why each tool and method is used, and examples of each tool and method. The second section shows you how to use Minitab to utilize the different tools and methods that you learned about in the first section.

What Is Variation?

Recall from Chapter 2, "Process and Quality Fundamentals," that all processes have outputs and these outputs may be measured. The distribution of these measurements varies, and the differences between these measurements are called *process variation*.

Recall also that there are two types of variation: common cause variation and special cause variation (Gitlow et al., 2015; Gitlow and Levine, 2004).

Common Cause Variation

Common causes of variation create fluctuations or patterns in data that are due to the system itself—for example, the fluctuations (variation) caused by hiring, training, and supervisory policies and practices. Common causes of variation are the responsibility of management. Only management can change the policies and procedures that define the common causes of variation inherent in a system.

Special Cause Variation

Special causes of variation create fluctuations or patterns in data that are not inherent to a process; they come from outside the process. Special causes of variation are the responsibility of workers and engineers. Workers and engineers identify, and if possible, resolve special causes of variation. If they cannot resolve a special cause of variation, they enlist the aid of management.

Using Control Charts to Understand Variation

The control chart is a tool for distinguishing between the common and special causes of variation for a variable (Gitlow et al., 2015; Gitlow and Levine, 2004). They are used to assess and monitor the stability of a variable (presence of only common causes of variation). The data for a control chart is obtained from a subgroup or sample of items selected at each observation session, for example, by month, by patient, or by form.

There are two uses for control charts. The first one we discussed, that is, to distinguish special from common causes of variation. This happens on the factory floor, in a call center, and so on. The second one occurs once data is aggregated, say over several areas in an organization (for example, several sales teams), the control chart loses the ability to detect special causes of variation. Consequently, they are primarily used to stop management from overreacting to common causes of variation and treating common causes of variation as special causes of variation, and making the process more complex than is necessary. Think of it like this: Overreaction to common causes of variation can make a straw look like a bowl of spaghetti. Instead of it being a straight line from point A to point B in a process, it becomes a maze creating problems like increased waiting times, increased cycle times, and increased defects, to name a few problems.

The most common types of control charts can be divided into two categories determined by the type of data used to monitor a process. These two broad categories are *attribute control charts* and *variables control charts*.

Attribute Control Charts

Attribute control charts are used when you need to evaluate variables defined by attribute data. Either classification data (for example, defective or conforming) or count data (for example, number of defects per area of opportunity).

The attribute control charts covered in this chapter are

- **P charts**—For classification data with either equal or unequal subgroup size
- **C charts**—For count data with consistent areas of opportunity
- **U charts**—For count data with variable areas of opportunity

Variables Control Charts

Variables control charts are used when you need to evaluate variables measured on a continuous scale such as height, weight, temperature, cycle time, waiting time, revenue, costs, and so on. Variables control charts contain two sections; one studies process variability while the other studies process central tendency. When dealing with variables control charts we look at the subgroup size to determine which type of control chart to use. A subgroup is the group of measurements that make up each point that is plotted on the control chart.

The variables control charts covered in this chapter are

- **Individuals and moving range (I-MR) charts**—For subgroups of 1
- **X Bar and R charts**—For subgroups whose size is between 2 and 9
- **X Bar and S charts**—For subgroups whose size is greater than or equal to 10

An Anecdote of Using Control Charts

I was consulting in a factory that had just started to require statistical evidence of quality from their suppliers. Top management had just completed a course in statistical process control, but certainly were not experts in the field.

I was invited to a meeting with a supplier that presented the control charts of their products. The top management was impressed by the display of control charts. However, I pointed out that all the control charts were out of control (they showed special causes of variation) and were not producing within specification limits.

The moral to the story is that control charts do not equal quality. No amount of training can make people understand control charts; they must study them themselves, with the help of a master, to really understand them. The proper long-term use of control charts produces quality products.

Understanding Control Charts

Again, the principal focus of the control chart on the shop floor is the attempt to separate special or assignable causes of variation from random noise or common causes of variation. The distinction between the two causes of variation is crucial because special causes of variation are those that are not part of a process and are correctable or exploitable by workers, or sometimes they need the help of managers. Common causes of variation can be reduced only by changing the system. Only management can change the system or enlightened management can empower the associates to work on the system as long as the aim of management and needs of customers are clear.

Control charts allow you to monitor the process and determine the presence of special causes of variation. Control charts help prevent two types of errors: type one errors and type two errors (Gitlow et al., 2015):

- **Type one errors**—Involve the belief that an observed value represents special cause variation when in fact it is due to the common cause variation of the system. Treating common causes of variation as special cause variation can result in tampering with or overadjustment of a process with an accompanying increase in variation. This is the action discussed previously, which causes the process flow from point A to point B to look like a bowl of spaghetti instead of a straw, with all the ensuing problems.
- **Type two errors**—Involve treating special cause variation as if it is common cause variation and not taking corrective action when is necessary.

All control charts have a common structure:

- Measurement of a process characteristic, which is plotted in order of time, or some or the variable, for example, employee or location.
- A central line that represents the process average.
- Upper (UCL) ,and lower (LCL) control limits that provide information on the process variation. Control limits are calculated by adding and subtracting three standard deviations of the statistic of interest from the process mean.

Figure 5.1 shows an example of a control chart that exhibits only common causes of variation.

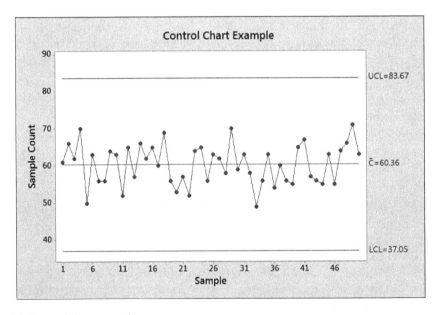

Figure 5.1 Control chart example

Once these control limits are computed from the data, the control chart is evaluated by discerning any nonrandom pattern that might exist in the data. Figure 5.2 illustrates three different patterns.

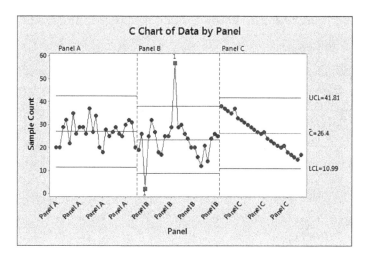

Figure 5.2 Three control chart patterns

In panel A of Figure 5.2, there does not appear to be any pattern in the ordering of values over time, and no points fall outside the three standard deviation control limits. It appears that the process is stable and contains only common cause variation.

Panel B, on the contrary, contains two points that fall outside the three standard deviation control limits. Each of these points should be investigated to determine the special causes that led to their occurrence. This process is not in control.

Although Panel C does not have any points outside the control limits, it has a series of consecutive points above the average value (the center line), as well as a series of consecutive points below the average value. In addition, a long-term overall downward trend in the value of the variable is clearly visible. This process is also not in control. We will explain why later in this book.

Control limits are often called *three-sigma limits*. In practice, virtually all of a process's output is located within a three-sigma interval of the process mean, provided that the process is stable; that is, no special causes of variation are present in the process.

Rules for Determining Out of Control Points

As you saw earlier, the simplest rule for detecting the presence of a special cause is one or more points falling beyond the mean plus or minus 3 standard deviation limits on the chart. The control chart can be made more sensitive and effective in detecting out of control points by considering other signals and patterns unlikely to occur by chance alone. For example, if only common causes are operating, you would expect the points plotted to approximately follow a bell-shaped normal distribution.

Figure 5.3 presents a control chart in which the area between the upper and lower control limits is subdivided into bands, each of which is one standard deviation wide. These additional limits or zone boundaries can be useful in detecting other unlikely patterns of data points.

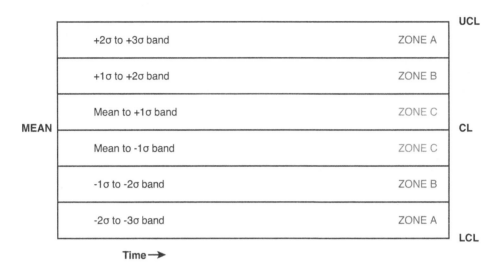

Figure 5.3 A control chart showing bands, each of which is one standard deviation wide

In a stable process, one would expect 68.26% of all the statistics to be between the upper and lower C bands, 95.44% of all the statistics to be between the upper and lower B bands, and 99.73% of all the statistics to be between the upper and lower A zones. This means that 0.27% of the statistics would be beyond the upper and lower A zones; 0.135% above upper zone A and 0.135% below lower zone A.

You can conclude that the process is out of control if any of the following events occur (Gitlow et al., 2015; Gitlow and Levine, 2004):

- **Rule 1**—A point falls above the UCL or below the LCL. If a point falls above the upper zone A limit then one of two things occurred: either a special cause occurred or a common cause occurred with the likelihood of 0.135% or 1,350 times per million opportunities. This is considered so unlikely by statisticians that they are willing to gamble the data point is a special cause of variation and act accordingly. See Figure 5.4 for an illustration of Rule 1.

- **Rule 2**—Nine or more consecutive points lie above the center line or nine or more consecutive points lie below the center line. See Figure 5.5 for an illustration of Rule 2.

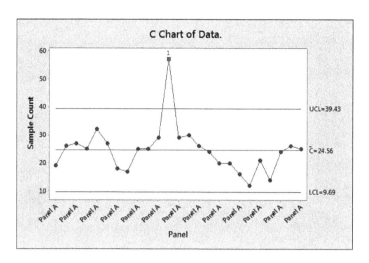

Figure 5.4 Illustration of Rule 1

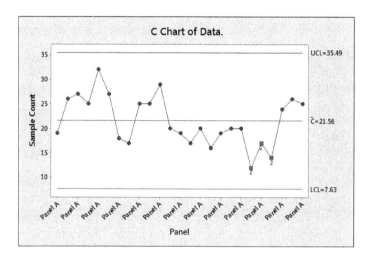

Figure 5.5 Illustration of Rule 2

- **Rule 3**—Six or more consecutive points move upward in value or six or more consecutive points move downward in value. See Figure 5.6 for an illustration of Rule 3.
- **Rule 4**—An unusually small number of consecutive points above and below the center line are present (a saw tooth pattern). Rule 4 is used to determine whether a process is unusually noisy (high variability) or unusually quiet (low variability). Figure 5.7 illustrates Rule 4.

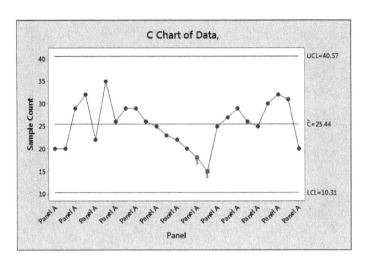

Figure 5.6 Illustration of Rule 3

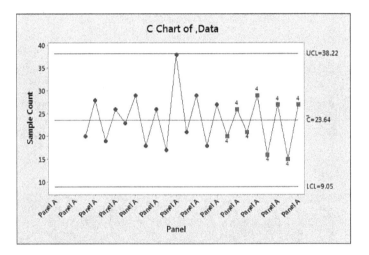

Figure 5.7 Illustration of Rule 4

- **Rule 5**—Two out of three consecutive points fall in the high Zone A or above, or in the low Zone A or below. See Figure 5.8 for an illustration of Rule 5.
- **Rule 6**—Four out of five consecutive points fall in the high Zone B or above, or in the low Zone B or below. See Figure 5.9 for an illustration of Rule 6.

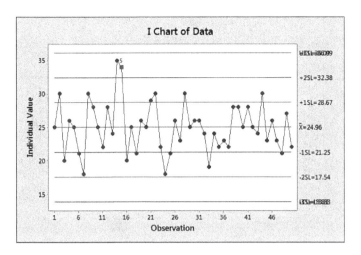

Figure 5.8 Illustration of Rule 5

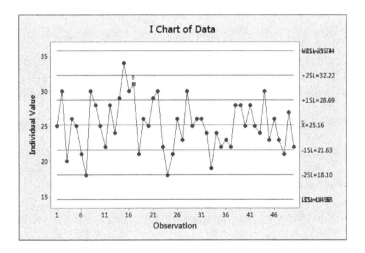

Figure 5.9 Illustration of Rule 6

- **Rule 7**—Fifteen consecutive points fall within Zone C on either side of the center line. Rule 7 is used to determine whether a process is unusually noisy (high variability) or unusually quiet (low variability). See Figure 5.10 for an illustration of Rule 7.

If only common causes are operating, each of the preceding seven patterns is statistically unlikely to occur. The presence of one or more of these low probability events indicates that one or more special causes may be operating, thereby resulting in a process that is out of a state of statistical control. Other rules for special causes have been developed and are incorporated within the control chart features of Minitab. Different rules may be considered appropriate for specific charts.

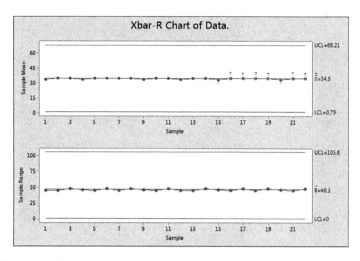

Figure 5.10 Illustration of Rule 7 (Source: minitab.com)

Control Charts for Attribute Data

Some control charts are specifically designed to operate on attribute data; they are p charts, c charts, and u charts. Each chart has a specific purpose that will be discussed in this section, and later in more detail.

P Charts

What: The p chart is a control chart used to study classification type attribute data, for example, the proportion of defective (nonconforming) items by month. An item is defective if it is not fit for use, for example, a water bottle with a hole in it. Subgroup sizes in a p chart may remain constant or may vary. A p chart may be used to help process improvers control defective versus conforming, go versus no-go, or acceptable versus not acceptable outputs from a process. Using the seven rules discussed previously, the control chart can be used to distinguish special from common causes of variation.

The p chart is used when

- There are only two possible outcomes for an event. An item must be found to be either conforming or nonconforming (defective).

- The probability, p, of a nonconforming item is constant over time.

- Successive items are independent over time.

- Subgroups are of sufficient size to detect an out of control event. A general rule for subgroup size for a p chart is that the subgroup size should be large enough to detect a special cause of variation if it exists. Frequently, subgroup sizes are between 50 and 500, per subgroup, depending on the metric being studied.

- Subgroup frequency, how often you draw a subgroup from the process under study, should be often enough to detect changes in the process under study. This requires expertise in the process under study. If the process can change very quickly, more frequent sampling is needed to detect special causes of variation. If the process changes slowly, less frequent sampling is needed to detect a special cause of variation.

Why: P charts are used to monitor stability of a process that generates classification attribute data; that is, for example, the percentage of process output that is defective by time period.

Example 1: P Chart with Equal Subgroup Sizes

As an illustration of the p chart, consider the case of a large health system that has had complaints from several pathologists concerning a supposed problem with a manufacturer sending cracked slides. Slides are pieces of glass that pathologists place biopsies on so they can be examined under a microscope. This allows the pathologist to assist the clinician in confirming a diagnosis of the disease from the biopsy. According to the pathologists, some slides are cracked or broken before or during transit, rendering them useless scrap. The fraction of cracked or broken slides is naturally of concern to the hospital administration. Each day a sample of 100 slides is drawn from the total of all slides received from each slide vendor. Table 5.1 presents the sample results for 30 days of incoming shipments for a particular vendor.

Table 5.1 30 Days of Incoming Slide Shipments for a Particular Vendor

Day	Sample Size	Number Cracked	Day	Sample Size	Number Cracked
1	100	14	16	100	3
2	100	2	17	100	8
3	100	11	18	100	4
4	100	4	19	100	2
5	100	9	20	100	5
6	100	7	21	100	5
7	100	4	22	100	7
8	100	6	23	100	9
9	100	3	24	100	1
10	100	2	25	100	3
11	100	3	26	100	12
12	100	8	27	100	9
13	100	4	28	100	3
14	100	15	29	100	6
15	100	5	30	100	9

These data are appropriate for a p chart because each slide is classified as cracked (defective) or not cracked (conforming), the probability of a cracked slide is assumed to be constant from slide to slide, and each slide is considered independent of the other slides.

Figure 5.11 illustrates the p chart obtained from Minitab for the cracked slides data.

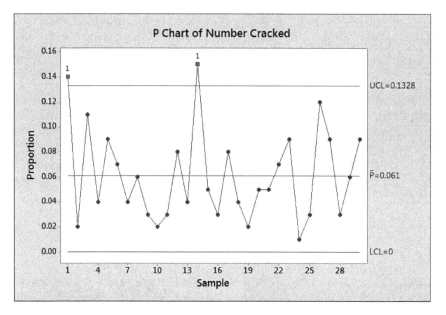

Figure 5.11 P chart for cracked slides

From Figure 5.11, we find a process that lacks control. On day 1, the proportion of cracked slides (14/100 = 0.14) is above the upper control limit and on day 14 the proportion of cracked slides (15/100 = 0.15) is above the upper control limit. None of the other rules for determining whether a process is out of control seems to be violated. That is, there are no instances when two out of three consecutive points lie in zone A on one side of the center line; there are no instances when four out of five consecutive points lie in zone B or beyond on one side of the center line; there are no instances when six consecutive points move upward or downward; nor are there nine consecutive points on one side of the center line.

Upon examination of the out of control points it was found that for those two shipments the boxes containing the slides were overpacked. Once this was realized, the vendor was contacted and the pathology department was assured that it would never happen again due to a changed protocol. Consequently, the two out of control data points were dropped, and the control chart was recomputed. The revised control chart is shown in Figure 5.12.

Now that the process is stable, a Pareto analysis is done on the causes of cracked tiles. Figure 5.13 shows that the major cause of cracked tiles is improper wrapping by the vendor. The vendor was notified and changed the wrapping protocol.

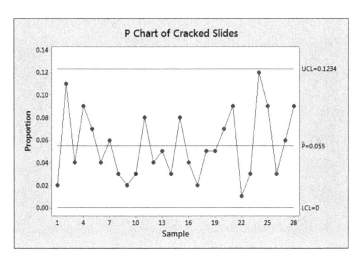

Figure 5.12 Revised control chart without special cause subgroups

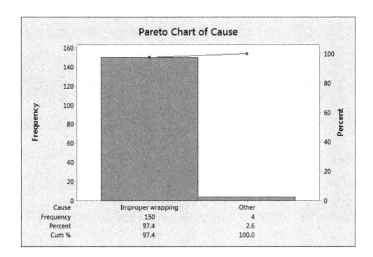

Figure 5.13 Pareto analysis of causes of cracked tiles

Figure 5.14 shows a before and after control chart of the cracked tile problem.

As, you can see from Figure 5.14, the common causes of variation were dramatically reduced due to good communication between the supplier and the pathology department.

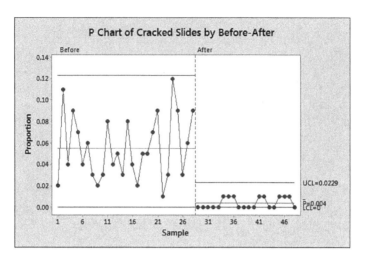

Figure 5.14 Before and after control chart

> **NOTE**
>
> Many statisticians believe that the Minitab zone rules are overly conservative. For example, some statisticians would argue that eight points in a row above or below the average is a special case of variation, while Minitab says nine points in a row indicates a special cause of variation; see the Western Electric handbook for more information on this point.

Example 2: P Chart with Unequal Subgroup Sizes

In many instances, unlike the example concerning the slides, the subgroup size varies from subgroup to subgroup. Consider the case of a hospital that has seen an increase in chemotherapy turnaround times. The pharmacy director suspects the increased turnaround times are due to various delays in the process. Based on carefully laid out operational definitions on what constitutes a delay, each day for 60 consecutive days data is collected on the proportion of orders with delays and is presented in Table 5.2.

Table 5.2 Delayed Orders for 60 Consecutive Days

Day	Delayed Orders	Total Orders	Day	Delayed Orders	Total Orders
1	21	94	31	22	91
2	22	82	32	21	105
3	25	95	33	24	107
4	25	86	34	20	107
5	26	106	35	22	100
6	11	109	36	21	96
7	21	113	37	19	123
8	20	112	38	15	119
9	23	86	39	21	126
10	12	112	40	18	95
11	11	99	41	17	147
12	20	97	42	24	105
13	19	104	43	18	96
14	16	129	44	21	103
15	22	91	45	15	87
16	19	124	46	21	99
17	26	106	47	23	83
18	16	105	48	17	91
19	24	89	49	7	74
20	16	79	50	19	113
21	20	104	51	18	120
22	25	97	52	21	74
23	20	96	53	16	107
24	24	79	54	23	90
25	20	115	55	22	116
26	17	95	56	20	123
27	20	104	57	13	90
28	21	112	58	20	95
29	14	100	59	21	100
30	20	103	60	21	93

Figure 5.15 shows the p chart based on the delays data. Due to the different subgroup sizes, the values obtained from Minitab for the upper and lower control limits and the zones are different for each subgroup. None of the points are outside the control limits and none of the rules concerning consecutive points have been violated. Thus, any improvement in the number of order delays must come from management changing the system.

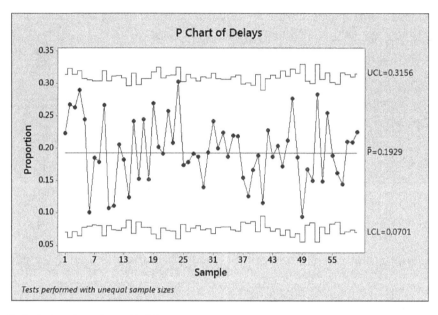

Figure 5.15 P chart for orders with delays

C Charts

What: When there are multiple opportunities for defects or imperfections in a given continuous unit (such as a large sheet of fabric, a hospital ward, a stretch of highway, a refrigerator, an automobile, a factory with a constant number of employees, to name a few examples), each unit is called an area of opportunity; each area of opportunity is a subgroup. The c chart is used when the areas of opportunity are of constant size.

A defective item is a nonconforming unit. It must be discarded, reworked, returned, sold, scrapped, or downgraded. It is unusable for its intended purpose in its present form. A defect, however, is an imperfection of some type that does not necessarily render the entire item unusable, yet is undesirable. One or more defects may not make an entire item defective. An assembled piece of machinery such as a car, dishwasher, or air conditioner may have one or more defects that may not render the entire item defective but may cause it to be downgraded or may necessitate its being reworked, for example a scratch, or a dent.

The c chart is used when

- Areas of opportunity are of constant size.
- Subgroups should be of sufficient size to detect an out of control event. A general rule for subgroup size for a c chart is that the subgroup size should be large enough to detect a special cause of variation if it exists. This is operationally defined as the average number of defects per area of opportunity being at least 20.0.
- The frequency of collected subgroups should be often enough to detect changes in the process under study. This requires expertise in the process under study. If the process can change quickly, more frequent sampling is needed to detect special causes of variation. If the process changes slowly, less frequent sampling is needed to detect a special cause of variation.

Why: C charts are used to monitor stability of count attribute data where the area of opportunity is constant from observation to observation.

Example: To illustrate the c chart, consider the number of add-ons (unscheduled patients) in an outpatient clinic in a hospital. Many times patients are added on to the regular schedule at the last minute. This is problematic as capacity of the clinic is limited and add-ons create problematic wait times, usually too long for patient satisfaction to be at an acceptable level. A process improvement team collected data on the number of add-ons per day at one of its outpatient clinics. Results of these data collections produce the results in Table 5.3.

Table 5.3 Number of Add-Ons Per Day for 50 Consecutive Days in an Outpatient Clinic

Day	Add-ons	Day	Add-ons	Day	Add-ons	Day	Add-ons
1	13	15	16	29	18	43	10
2	19	16	12	30	19	44	5
3	9	17	14	31	10	45	17
4	21	18	14	32	14	46	14
5	18	19	14	33	14	47	11
6	13	20	13	34	12	48	23
7	18	21	14	35	16	49	13
8	19	22	11	36	12	50	8
9	19	23	17	37	8		
10	15	24	11	38	12		
11	11	25	15	39	20		
12	9	26	21	40	24		
13	16	27	16	41	11		
14	12	28	15	42	10		

The assumptions necessary for using the c chart are well met here, as the clinic days are considered to be the areas of opportunity; add-ons are discrete events and seem to be independent of one another. Even if these conditions are not precisely met, the c chart is fairly robust, or insensitive to small departures from the assumptions.

From Figure 5.16, the number of add-ons appears to be stable around a center line or mean of 14.32. None of the add-ons are outside the control limits, and none of the rules concerning zones have been violated.

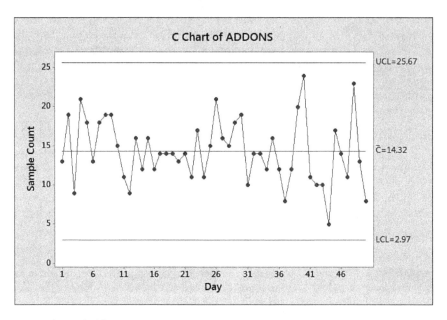

Figure 5.16 C chart of add-ons

U Charts

What: The u chart is similar to the c chart in that it is a control chart for the count of the number of defects in a given area of opportunity. The fundamental difference between a c chart and a u chart lies in the fact that during construction of a c chart, the area of opportunity remains constant from observation to observation, while this is not a requirement for the u chart. Instead, the u chart considers the number of events (such as number of defects) as a fraction of the total size of the area of opportunity in which these events were possible; for example, the number of accidents in a factory where the number of employees changes dramatically from time period to time period. The characteristic used for the control chart, U_i, is the ratio of the number of events in the ith subgroup to the area of opportunity in the ith subgroup.

The u-chart is used when

- You are considering the number of events (such as defects) as a fraction of the total size of the area of opportunity in which these events were possible, thus circumventing the problem of having different areas of opportunity for different observations.

- Subgroups are of sufficient size to detect an out of control event. A general rule for subgroup size for a u chart is that the subgroup size should be large enough to detect a special cause of variation if it exists.

- The frequency with which subgroups are selected for study should be often enough to detect changes to the process under study. This requires expertise in the process under study. If the process can change quickly, more frequent sampling is needed to detect special causes of variation. If the process changes slowly, less frequent sampling is needed to detect a special cause of variation.

Why: U charts are used to monitor stability of count attribute data where the area of opportunity is not constant from subgroup to subgroup.

Example: To illustrate a u chart, consider the case of patient falls in a hospital. A large metropolitan hospital wants to examine the number of patient falls to see whether they are a problem. Since the census (the number of patients in the hospital) is constantly changing from day to day the u chart should be used. Table 5.4 shows the data on the number of patient falls for the past 30 days.

Table 5.4 Patient Falls in a Hospital (Note: One patient may fall multiple times in a day.)

Day	Census	Falls	Day	Census	Falls
1	978	6	16	1009	7
2	1040	5	17	976	2
3	1101	7	18	998	1
4	990	3	19	1016	8
5	956	8	20	987	5
6	1004	3	21	1092	3
7	1025	10	22	1056	6
8	999	5	23	983	2
9	1013	7	24	1025	5
10	1045	4	25	987	7
11	994	2	26	1012	3
12	1105	4	27	1091	7
13	1043	6	28	1034	4
14	987	2	29	1024	8
15	992	4	30	1011	3

Figure 5.17 illustrates the u chart obtained from Minitab for the patient falls data.

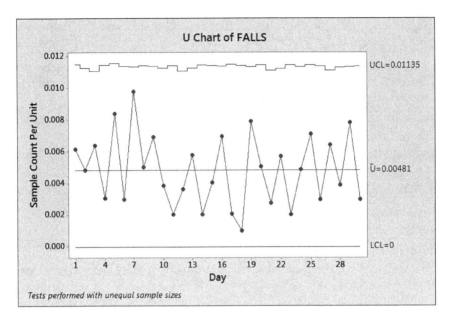

Figure 5.17 U chart of patient falls in a hospital

From Figure 5.17, the number of defects per unit (falls per day) appears to be stable around a center line or mean of .00481. No points indicate a lack of control, so there is no reason to believe that any special variation is present. If sources of special variation were detected, we would identify the source or sources of the special variation, eliminate them from the system if detrimental, or incorporate them into the system if beneficial; drop the data points from the data set; and reconstruct and reanalyze the control chart.

Control Charts for Measurement Data

Variables control charts are used to study a metric from a process when the metric is measurement data, for example, height, area, temperature, cost, revenue, cycle time, service time, or waiting time (Gitlow et al., 2015; Gitlow and Levine, 2004). Variables charts are typically used in pairs. One chart studies the variation in a process, while the other chart studies the process average. The chart that studies process variability must be examined before the chart that studies the process average. This is because the chart that studies the process average assumes that the process variability is stable.

It is important to note that once the percentage of defective items or the number of defects per item gets low, attribute charts become increasingly less useful in improving a process. This is because as the percentage of defectives or the number of defects gets small, the sample size required to find a defective item or items with almost no defects gets so large that attribute charts become not useful. For example, if a p chart is used on a metric that has one defective per 10,000 items, the sample size necessary to find one defect on average is 10,000. As you can see, this creates a situation that is not practical. In this case process improvers switch from attribute charts to variables control charts that measure how far off nominal an item is, even if it is conforming. This means moving from the goal post view of quality to the Taguchi Loss Function view of quality.

Individuals and Moving Range (I-MR) Charts

What: Often a situation occurs where only a single value is observed per subgroup. Perhaps measurements are destructive and/or expensive; or perhaps they represent a single batch where only one measurement is appropriate, such as the total yield of a homogeneous chemical batch process; or the measurements are monthly or quarterly revenue or cost data. Whatever the case, there are circumstances when data must be taken as individual units that cannot be conveniently divided into subgroups. In this case I-MR charts are used.

Individuals and moving range charts have two parts, one that charts the process variability and the other that charts the process average. The two parts are used in tandem. Stability must first be established in the portion charting the variability because the estimate of the process variability provides the basis for the control limits of the portion charting the process average.

Single measurements of variables are considered a subgroup of size one (n=1 per subgroup). Hence, there is no variability within the subgroups themselves. An estimate of the process variability must be made in some other way. The estimate of variability used for individual value charts is based on the point-to-point variation in the sequence of single values, measured by the moving range (the absolute value of the difference between each data point and the one that immediately preceded it). For example, the first moving range is the absolute value of the difference between the first subgroup and the second subgroup, the second moving range is the absolute value of the difference between the second subgroup and the third subgroup, and so on. Consequently, there will always be one less moving range than individual values.

As before, subgroup frequency should be often enough to detect changes in the process under study. This requires expertise in the process under study. If the process can change quickly, more frequent sampling is needed to detect special causes of variation. If the process changes slowly, less frequent sampling is needed to detect a special cause of variation.

Why: I-MR charts are used to monitor process control and stability of measurement data where subgroup size is equal to 1.

Example: To illustrate the individual value chart, consider a pathology department of a hospital that needs to provide diagnosis of GI biopsies in a timely manner. There are many steps in the process including extracting the biopsy from the patient, lab processing, reading and sign-out by the faculty. The hospital administration is concerned with the turnaround times for GI biopsies taking too long. Table 5.5 shows turnaround times for 100 consecutive GI biopsy cases.

Table 5.5 Turnaround Times for 100 Consecutive GI Biopsy Cases

Case	GI Biopsy Turnaround Times in Days	Case	GI Biopsy Turnaround Times in Days
1	4.3	32	5.4
2	5.1	33	2.6
3	2.9	34	4.5
4	3.3	35	3.1
5	4.4	36	4.8
6	3.8	37	4.1
7	5.8	38	3.1
8	3.7	39	3.3
9	4.3	40	4.3
10	5.2	41	2.9
11	4.2	42	2.4
12	3.6	43	2.4
13	4	44	4.5
14	2.5	45	3.4
15	5.3	46	5
16	4.9	47	2.9
17	4.8	48	4.7
18	6.4	49	3.9
19	4.2	50	4.2
20	3.4	51	2.2
21	4	52	4.6
22	4.2	53	2.4
23	3.9	54	5.5
24	2.2	55	4.1
25	5.5	56	4.6
26	4.6	57	4.7
27	2.8	58	3.7
28	2.6	59	4.6
29	3.6	60	3.6
30	4	61	3.8
31	6	62	4.2

Case	GI Biopsy Turnaround Times in Days
63	5.4
64	4.8
65	5.9
66	3.9
67	3.9
68	4
69	4.9
70	3.6
71	4
72	2.6
73	3.2
74	5.6
75	4.4
76	5.1
77	5.8
78	4.4
79	4.1
80	2.5
81	1.9
82	3.6
83	4.9
84	3.8
85	4.5
86	5.2
87	5
88	1.8
89	4.8
90	3.8
91	5.1
92	3.6
93	5.3
94	3.6

Case	GI Biopsy Turnaround Times in Days
95	4.3
96	4.3
97	3.7
98	2.9
99	2
100	4.5

Figure 5.18 illustrates the moving range and individual value charts obtained from Minitab for the GI biopsy turnaround time data.

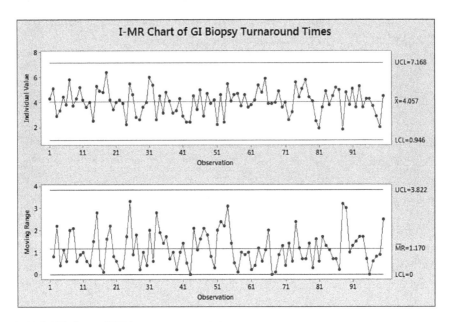

Figure 5.18 I-MR chart of GI biopsy turnaround times

In Figure 5.18, the bottom portion is the moving range chart, and the top portion is the individual value chart. First, the moving range chart is examined for signs of special variation. None of the points on the moving range chart are outside the control limits, and there are no other signals indicating a lack of control. Thus, there are no indications of special sources of variation on the moving range chart. Now the individual value chart can be examined. There are no indications of a lack of control, so the process can be considered to be stable and the output predictable with respect to time as long as conditions remain the same.

After the team brainstormed what could be changed in the process flowchart to decrease the number of days to turn around biopsies (TAT or turnaround time), several changes become obvious; for example, changing the frequency with which slides are delivered to pathology to do the biopsies. The changes are implemented in the flowchart, and the results of the revised flowchart are shown in the right panel of Figure 5.19.

X Bar and R Charts

What: X Bar and R charts are used when the subgroup size is between 2 and 9. The subgroup range, R, is plotted on the R chart, which monitors variability of the process, while the subgroup average, X Bar, is plotted on the X Bar chart, which monitors the central tendency of the process.

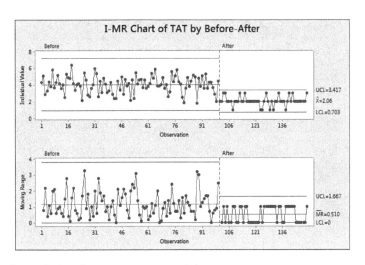

Figure 5.19 Before and after change concept control chart

Why: X Bar and R charts are used if the subgroup size is small (2 to 9) because it is unlikely that an extreme value will be selected into a subgroup. If an extreme value is accepted into a subgroup, the range gives a distorted view of the variability in the process because the range is computed solely from extreme values.

Example: To illustrate the X Bar and R charts, consider a large pharmaceutical firm that provides vials filled to a specification of 52.0 grams. The firm's management has embarked on a program of statistical process control and has decided to use variables control charts for this filling process to detect special causes of variation. Samples of six vials are selected every 5 minutes during a 105-minute period. Each set of six measurements makes up a subgroup. Table 5.6 lists the vial weights for 22 subgroups.

Table 5.6 Vial Weights for Six Vials Selected During 22 Time Periods

Observation	Time	1	2	3	4	5	6
1	9:30	52.22	52.85	52.41	52.55	53.10	52.47
2	9:35	52.25	52.14	51.79	52.18	52.26	51.94
3	9:40	52.37	52.69	52.26	52.53	52.34	52.81
4	9:45	52.46	52.32	52.34	52.08	52.07	52.07
5	9:50	52.06	52.35	51.85	52.02	52.30	52.20
6	9:55	52.59	51.79	52.20	51.90	51.88	52.83
7	10:00	51.82	52.12	52.47	51.82	52.49	52.60
8	10:05	52.51	52.80	52.00	52.47	51.91	51.74
9	10:10	52.13	52.26	52.00	51.89	52.11	52.27
10	10:15	51.18	52.31	51.24	51.59	51.46	51.47

Observation	Time	1	2	3	4	5	6
11	10:20	51.74	52.23	52.23	51.70	52.12	52.12
12	10:25	52.38	52.20	52.06	52.08	52.10	52.01
13	10:30	51.68	52.06	51.90	51.78	51.85	51.40
14	10:35	51.84	52.15	52.18	52.07	52.22	51.78
15	10:40	51.98	52.31	51.71	51.97	52.11	52.10
16	10:45	52.32	52.43	53.00	52.26	52.15	52.36
17	10:50	51.92	52.67	52.80	52.89	52.56	52.23
18	10:55	51.94	51.96	52.73	52.72	51.94	52.99
19	11:00	51.39	51.59	52.44	51.94	51.39	51.67
20	11:05	51.55	51.77	52.41	52.32	51.22	52.04
21	11:10	51.97	51.52	51.48	52.35	51.45	52.19
22	11:15	52.15	51.67	51.67	52.16	52.07	51.81

X Bar and R charts obtained from Minitab for the vials filled data are shown in Figure 5.20.

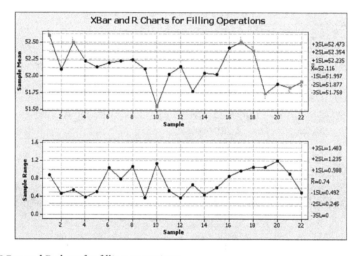

Figure 5.20 X Bar and R chart for filling operations

In Figure 5.20, the bottom portion is the R chart, and the top portion is the X Bar chart. First, the R chart is examined for signs of special variation. None of the points on the R chart is outside the control limits, and there are no other signals indicating a lack of control. Thus, there are no indications of special sources of variation on the R chart.

Now the X Bar chart can be examined. Notice that a total of five points on the X Bar chart are outside the control limits (1, 3, 10, 17, and 19), and points 16, 17, and 18 are above +2 standard deviations (2 out of 3 points 2 sigma or beyond rule). Also notice that 18 through 22 show the same type of pattern as points 16 through 18. This indicates a lack of control. An interesting point comes up here. If the high points are resolved, the low points may be in control and vice versa. This type of analysis requires a process expert.

Further investigation is warranted to determine the source(s) of these special variations. The next step is to eliminate the bad special causes of variation and instill the good special causes of variation. Once this is done, the next step is to reduce the common causes of variation in the process. This is covered in Chapters 10 through 14, which are all about the Six Sigma DMAIC model that is used to improve a process.

X Bar and S Charts

What: As the sample size (subgroup size) n increases, the range becomes increasingly less efficient as a measure of variability. Since the range ignores all information between the two most extreme values, as the sample size increases, the range will use a smaller proportion of the information available in a sample. In addition, the probability of observing an extreme value in a sample increases as n gets larger. A single extreme value will result in an unduly large value for the sample range and will inflate the estimate of process variability. Thus, as the subgroup size increases (n is equal to or greater than 10), the individual subgroup standard deviations provide a better estimate of the process standard deviation than the range.

Subgroups should be of sufficient size to detect an out of control event, as with X Bar and R charts. The common subgroup sizes for X Bar and S charts are 10 or more items. Additionally, subgroup frequency should be often enough to detect changes in the process under study. This requires expertise in the process under study. If the process can change quickly, more frequent sampling is needed to detect special causes of variation. If the process changes slowly, less frequent sampling is needed to detect a special cause of variation.

Why: X Bar and S charts are used to monitor process control and stability of measurement data where subgroup size is greater than or equal to 10.

Example: To illustrate X Bar and S charts, consider a hospital studying the length of time patients spend in the admitting process. Samples of 12 patients are selected each day for a 20 day period. The first patient to arrive on the hour is sampled, and the hospital clinic is open 12 hours per day. Admitting time has been operationally defined to all stakeholders' satisfaction. It is measured in seconds and is seen in Table 5.7.

Table 5.7 Admitting Process Time in Seconds

Day	Patient Time
1	362
1	468
1	553
1	390
1	460
1	910
1	707
1	829
1	955
1	705
1	884
1	904
2	611
2	873
2	768
2	807
2	476
2	816
2	567
2	833
2	521
2	959
2	315
2	414
3	320
3	944
3	593
3	857
3	710
3	724
3	545
3	526
3	348

Day	Patient Time
3	456
3	576
3	855
4	621
4	927
4	948
4	817
4	641
4	764
4	986
4	430
4	743
4	451
4	645
4	996
5	680
5	794
5	650
5	780
5	442
5	372
5	627
5	882
5	756
5	548
5	767
5	745
6	759
6	665
6	730
6	930
6	369
6	635

Day	Patient Time
6	313
6	843
6	264
6	663
6	991
6	431
7	372
7	835
7	884
7	930
7	667
7	747
7	390
7	644
7	339
7	664
7	245
7	893
8	370
8	294
8	480
8	558
8	502
8	595
8	847
8	544
8	853
8	876
8	744
8	816
9	530
9	881
9	943

Day	Patient Time	Day	Patient Time	Day	Patient Time
9	383	12	754	14	535
9	316	12	428	15	997
9	611	12	811	15	205
9	778	12	916	15	893
9	531	12	332	15	734
9	896	12	765	15	474
9	772	12	961	15	631
9	719	12	437	15	746
9	670	12	692	15	642
10	494	12	380	15	484
10	914	12	566	15	525
10	870	13	797	15	685
10	272	13	253	15	358
10	662	13	829	16	242
10	348	13	857	16	474
10	447	13	898	16	966
10	306	13	387	16	823
10	751	13	918	16	515
10	445	13	900	16	617
10	717	13	691	16	894
10	387	13	600	16	519
11	659	13	450	16	636
11	919	13	775	16	547
11	603	14	678	16	993
11	897	14	679	16	858
11	319	14	351	17	594
11	499	14	663	17	817
11	799	14	638	17	381
11	482	14	928	17	462
11	615	14	258	17	429
11	497	14	338	17	786
11	430	14	446	17	901
11	765	14	936	17	278
12	274	14	584	17	472

Day	Patient Time
17	885
17	991
17	557
18	368
18	850
18	510
18	688
18	201
18	795
18	977
18	715
18	253
18	310

Day	Patient Time
18	412
18	813
19	806
19	575
19	348
19	298
19	487
19	697
19	249
19	668
19	533
19	985
19	284

Day	Patient Time
19	707
20	497
20	785
20	806
20	263
20	435
20	337
20	659
20	537
20	786
20	607
20	466
20	564

In Figure 5.21, the bottom portion is the S chart, and the top portion is the X Bar chart. First, the S chart is examined for signs of special variation. None of the points on the S chart is outside the control limits, and there are no other signals indicating a lack of control. Thus, there are no indications of special sources of variation on the S chart. Now the X Bar chart can be examined. There are no indications of a lack of control, so the process can be considered to be stable and the output predictable with respect to time as long as conditions remain the same.

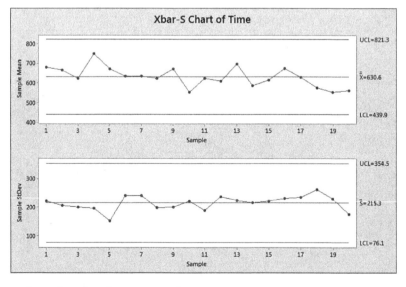

Figure 5.21 X bar S chart for admitting time data

The next step is to reduce the common causes of variation in the process. This is covered in Chapters 10 through 14, which are all about the Six Sigma DMAIC model, which is used to improve a process.

Which Control Chart Should I Use?

Often the most difficult part of using control charts is knowing which control chart to use in which situation. Figure 5.22 is a flow diagram that can help in deciding which type of chart is most appropriate for a given situation.

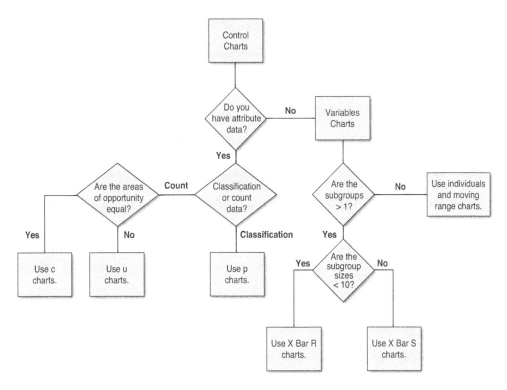

Figure 5.22 A flow diagram to help decide which type of control chart is most appropriate for a given situation

Control Chart Case Study

A manufacturer of surgical screws wants to improve the capability of his process. Each day, for 120 days, he counts the number of defective surgical screws from the defective bin. The company manufactures 100,000 surgical screws of a given type per day. In this case study we are examining one of many types of surgical screws. The different classifications of defective surgical screws for the type of surgical screws under study are improper threads, non-round heads, improper Philips head indentation, burr(s), too short, too long, and bent. Table 5.8 shows the number of defective surgical screws in each classification by day.

Table 5.8 Defective Surgical Screws in Each Classification by Day

Row	Improper Threads	Non-Round Head	Improper Philips Head Indentation	Burr(s)	Too Long	Too Short	Bent	Total Defectives
1	8	14	9	6	22	10	7	76
2	18	18	9	7	32	16	14	114
3	12	13	10	13	35	7	9	99
4	12	7	12	5	27	12	9	84
5	12	8	11	5	27	13	19	95
6	8	11	16	11	26	13	18	103
7	5	12	14	14	22	16	10	93
8	16	10	11	9	27	15	15	103
9	14	9	10	11	25	3	7	79
10	12	7	13	13	17	15	9	86
11	11	6	15	10	27	9	12	90
12	13	8	14	8	27	11	13	94
13	10	12	8	10	25	9	4	78
14	14	8	15	8	23	9	9	86
15	8	6	10	10	24	7	8	73
16	11	13	10	11	21	12	9	87
17	16	10	14	11	26	11	7	95
18	12	13	17	15	36	31*	13	137**operator sick on the machine that cuts length
19	12	11	10	12	20	10	8	83
20	7	6	14	3	16	14	10	70
21	10	14	12	8	28	18	13	103
22	7	12	9	5	21	12	8	74
23	7	13	16	11	24	7	16	94
24	9	4	8	9	32	8	7	77
25	8	12	12	13	17	7	10	79
26	8	17	10	10	23	16	10	94
27	7	12	19	13	28	14	6	99

Row	Improper Threads	Non-Round Head	Improper Philips Head Indentation	Burr(s)	Too Long	Too Short	Bent	Total Defectives
28	16	11	16	12	23	15	7	100
29	12	9	11	9	27	14	10	92
30	10	11	9	10	33	14	6	93
31	10	12	7	16	33	18	15	111
32	15	10	12	4	22	15	8	86
33	12	18	14	4	24	6	14	92
34	8	11	15	8	23	11	16	92
35	15	14	11	13	30	16	7	106
36	10	10	11	7	25	9	11	83
37	11	15	10	10	32	11	13	102
38	11	12	9	5	30	13	4	84
39	7	14	18	9	28	7	10	93
40	2	7	12	4	26	11	14	76
41	11	13	18	13	26	10	7	98
42	5	12	19	3	27	13	6	85
43	10	11	13	13	24	9	12	92
44	9	11	11	4	22	13	17	87
45	7	6	7	10	31	12	17	90
46	7	11	14	11	32	13	13	101
47	15	10	10	9	22	13	8	87
48	5	15	7	9	33	13	8	90
49	11	10	10	11	26	8	10	86
50	13	8	17	6	20	10	10	84
51	12	10	6	13	23	5	18	87
52	4	18	14	9	20	8	6	79
53	10	10	16	9	24	11	10	90
54	9	8	13	5	28	16	9	88
55	14	9	21	10	26	11	14	105
56	11	9	12	10	17	7	17	83
57	5	20	10	12	26	8	17	98
58	12	15	12	11	34	13	7	104
59	15	16	14	7	19	15	5	91
60	14	13	14	9	21	14	14	99
61	11	11	6	6	20	13	11	78

Row	Improper Threads	Non-Round Head	Improper Philips Head Indentation	Burr(s)	Too Long	Too Short	Bent	Total Defectives
62	2	13	14	11	22	15	6	83
63	6	6	16	4	20	13	9	74
64	12	16	9	8	17	9	17	88
65	6	12	14	7	20	14	10	83
66	9	12	14	5	18	12	5	75
67	10	17	14	11	30	20	7	109
68	13	14	7	8	27	13	11	93
69	11	8	14	3	23	10	17	86
70	6	11	16	9	26	14	16	98
71	10	12	12	6	30	16	7	93
72	8	13	16	6	24	8	11	86
73	8	9	17	9	27	9	12	91
74	6	20	11	13	23	15	10	98
75	10	19	5	8	22	10	7	81
76	8	9	13	8	26	15	9	88
77	6	6	12	7	23	11	15	80
78	11	10	16	12	27	5	10	91
79	10	11	8	8	26	11	13	87
80	8	15	15	7	35	18	5	103
81	4	13	16	4	29	5	11	82
82	13	9	12	8	33	14	8	97
83	16	15	5	13	20	11	16	96
84	8	9	13	10	30	13	12	95
85	11	13	14	13	27	9	7	94
86	10	14	10	10	13	14	7	78
87	10	14	16	8	27	14	12	101
88	10	8	10	10	21	6	6	71
89	9	15	9	9	26	15	14	97
90	9	9	8	11	23	16	18	94
91	9	11	13	17	27	9	19	105
92	8	14	15	9	27	11	7	91
93	10	12	13	12	12	12	8	79
94	13	9	13	12	25	18	14	104
95	8	17	10	15	27	6	16	99

Row	Improper Threads	Non-Round Head	Improper Philips Head Indentation	Burr(s)	Too Long	Too Short	Bent	Total Defectives
96	9	13	11	8	21	10	20	92
97	10	13	14	12	26	14	8	97
98	11	13	19	10	28	16	14	111
99	14	9	10	12	25	5	12	87
100	15	8	13	8	27	7	7	85
101	4	11	13	9	14	11	17	79
102	8	7	19	14	21	8	8	85
103	8	3	6	7	25	17	10	76
104	11	11	17	11	25	12	11	98
105	6	11	19	9	21	8	15	89
106	13	12	12	8	26	16	14	101
107	13	10	11	13	23	6	13	89
108	9	11	13	6	24	19	15	97
109	12	13	12	3	22	8	9	79
110	10	12	13	12	27	5	6	85
111	6	9	8	7	22	6	17	75
112	11	19	12	19	28	12	9	110
113	18	12	12	12	21	8	9	92
114	11	10	6	6	24	11	8	76
115	8	9	11	6	21	18	10	83
116	7	11	13	12	31	14	13	101
117	6	15	7	12	35	8	11	94
118	13	12	13	7	21	15	11	92
119	7	11	13	7	29	6	9	82
120	4	11	14	9	19	10	8	75

As you can see from the p chart in Figure 5.23 there is a special cause of variation on the 18th subgroup. Team members went back to the log sheet for that day and saw that the operator who cuts the length of the surgical screws was out sick. The team members assumed that this was the cause of the special variation. Consequently, they added into the flowchart of the job, to have a trained backup operator for each machine in case this special cause happened again. Next time they would be prepared. Consequently, they dropped the out of control point and recalculated the control chart. Figure 5.24 shows the result.

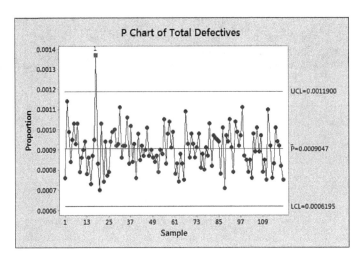

Figure 5.23 P chart of total defectives

As you can see in Figure 5.24, the process is now in statistical control. Since the process is stable the team members took all the data points and put them into a Pareto diagram, shown in Figure 5.25.

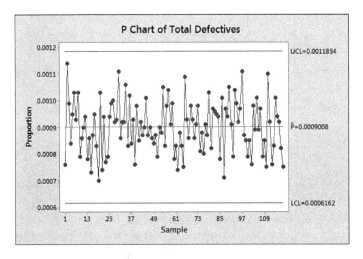

Figure 5.24 P chart after special cause removed

Team members realized that the same problem caused too short and too long screws. Consequently, they combined the categories; see the Pareto diagram in Figure 5.26.

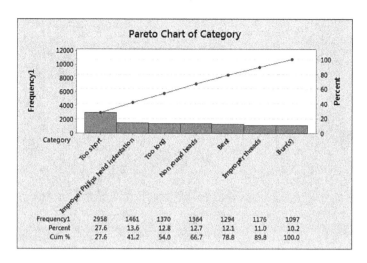

Figure 5.25 Pareto diagram of defect reasons

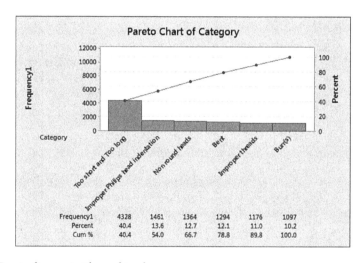

Figure 5.26 Pareto diagram with combined reasons

Team members studied the problem and discovered that a setting on the surgical screw machine was off. They corrected it, and the problem of too short and too long screws completely disappeared; see the control chart in Figure 5.27.

The preceding case study is an example of how to use control charts and change concepts to resolve special causes in a process and how to remove common causes from a process.

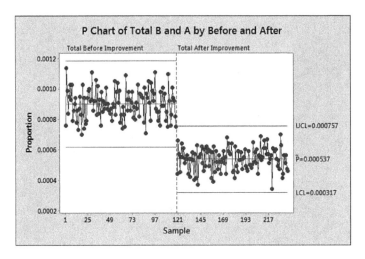

Figure 5.27 P chart after correcting problem with too short and too long screws

Measurement Systems Analysis

Measurement systems analysis studies are used to calculate the capability of a measurement system for a variable to determine whether it can deliver accurate data to team members. There are three parts to a measurement systems study:

1. Measurement systems analysis (MSA) checklist
2. Test-retest study
3. Gage Repeatability and Reproducibility (R&R) study

We want to make sure that when we are measuring and collecting data that our measurement methods, instruments, and the process of collecting data are correct to ensure the integrity of the data we are using in our process improvement efforts.

We only cover the basics of measurement systems analysis in this book. For a more advanced understanding see Gitlow and Levine (2004). We discuss measurement systems analysis checklists and Gage R&R studies. *R&R* stands for repeatability and reproducibility.

Measurement System Analysis Checklist

A measurement system analysis checklist involves determining whether the following tasks have been completed (Gitlow and Levine, 2004):

1. Description of the ideal measurement system (flowchart the process)
2. Description of the actual measurement system (flowchart the process)
3. Identification of the causes of the differences between the ideal and actual measurement systems

4. Identification of the accuracy (bias) and precision (repeatability) of the measurement system

5. Estimation of the proportion of observed variation due to unit-to-unit variation and R&R variation using a Gage R&R study

Gage R&R Study

A Gage R&R study is used to estimate the proportion of observed total variation due to unit-to-unit variation and R&R variation. R&R variation includes repeatability, reproducibility, and operator-part interaction (different people measure different units in different ways). If R&R variation is large relative to unit-to-unit variation, the measurement system must be improved before collecting data.

The data required by a Gage R&R study should be collected so that it represents the full range of conditions experienced by the measurement system. For example, the most senior inspector and the most junior inspector should repeatedly measure each item selected in the study. The data should be collected in random order to prevent inspectors from influencing each other.

An example of a Gage R&R study follows. Two inspectors independently use a light gauge to measure each of five units four separate times; this results in 40 measurements. The data to be collected by the two inspectors is shown in Table 5.9. Table 5.9 presents the standard order, or the logical pattern, for the data to be collected by the team members.

Table 5.9 Standard Order for Collecting Data for the Gage R&R Study

Row	Unit	Inspector	Measurement	Row	Unit	Inspector	Measurement
1	1	Enya	To be collected	14	2	Lucy	To be collected
2	1	Enya	To be collected	15	2	Lucy	To be collected
3	1	Enya	To be collected	16	2	Lucy	To be collected
4	1	Enya	To be collected	17	3	Enya	To be collected
5	1	Lucy	To be collected	18	3	Enya	To be collected
6	1	Lucy	To be collected	19	3	Enya	To be collected
7	1	Lucy	To be collected	20	3	Enya	To be collected
8	1	Lucy	To be collected	21	3	Lucy	To be collected
9	2	Enya	To be collected	22	3	Lucy	To be collected
10	2	Enya	To be collected	23	3	Lucy	To be collected
11	2	Enya	To be collected	24	3	Lucy	To be collected
12	2	Enya	To be collected	25	4	Enya	To be collected
13	2	Lucy	To be collected	26	4	Enya	To be collected

Row	Unit	Inspector	Measurement
27	4	Enya	To be collected
28	4	Enya	To be collected
29	4	Lucy	To be collected
30	4	Lucy	To be collected
31	4	Lucy	To be collected
32	4	Lucy	To be collected
33	5	Enya	To be collected

Row	Unit	Inspector	Measurement
34	5	Enya	To be collected
35	5	Enya	To be collected
36	5	Enya	To be collected
37	5	Lucy	To be collected
38	5	Lucy	To be collected
39	5	Lucy	To be collected
40	5	Lucy	To be collected

Table 5.10 shows the random order used by team members to collect the measurement data required for the measurement study. Random order is important because it removes any problems induced by the structure of the standard order. Table 5.10 is an instruction sheet to the team members actually collecting the data.

Table 5.10 Random Order for Collecting Data for the Gage R&R Study

Random Order	Standard Order	Unit	Inspector
1	36	5	Enya
2	5	1	Lucy
3	30	4	Lucy
4	29	4	Lucy
5	26	4	Enya
6	28	4	Enya
7	6	1	Lucy
8	8	1	Lucy
9	4	1	Enya
10	3	1	Enya
11	18	3	Enya
12	20	3	Enya
13	40	5	Lucy
14	9	2	Enya
15	31	4	Lucy
16	24	3	Lucy
17	38	5	Lucy
18	17	3	Enya
19	32	4	Lucy
20	11	2	Enya
21	27	4	Enya
22	19	3	Enya
23	10	2	Enya
24	33	5	Enya
25	37	5	Lucy
26	2	1	Enya
27	35	5	Enya
28	23	3	Lucy
29	13	2	Lucy
30	7	1	Lucy
31	15	2	Lucy
32	22	3	Lucy
33	14	2	Lucy
34	34	5	Enya
35	1	1	Enya
36	12	2	Enya
37	16	2	Lucy
38	39	5	Lucy
39	21	3	Lucy
40	25	4	Enya

Table 5.11 shows the data collected in the Gage R&R study in random order.

Table 5.11 Data for Gage R&R Study

Random Order	Standard Order	Unit	Inspector	Measure
1	36	5	Enya	21.85
2	5	1	Lucy	21.19
3	30	4	Lucy	23.14
4	29	4	Lucy	23.09
5	26	4	Enya	23.28
6	28	4	Enya	23.23
7	6	1	Lucy	21.29
8	8	1	Lucy	21.24
9	4	1	Enya	21.24
10	3	1	Enya	21.33
11	18	3	Enya	22.28
12	20	3	Enya	22.34
13	40	5	Lucy	21.78
14	9	2	Enya	21.65
15	31	4	Lucy	23.02
16	24	3	Lucy	22.17
17	38	5	Lucy	21.84
18	17	3	Enya	22.31
19	32	4	Lucy	23.19
20	11	2	Enya	21.67
21	27	4	Enya	23.24
22	19	3	Enya	22.31
23	10	2	Enya	21.60
24	33	5	Enya	21.84
25	37	5	Lucy	21.76
26	2	1	Enya	21.29
27	35	5	Enya	21.93
28	23	3	Lucy	22.14
29	13	2	Lucy	21.50
30	7	1	Lucy	21.21
31	15	2	Lucy	21.51
32	22	3	Lucy	22.23
33	14	2	Lucy	21.55

Random Order	Standard Order	Unit	Inspector	Measure
34	34	5	Enya	21.89
35	1	1	Enya	21.34
36	12	2	Enya	21.56
37	16	2	Lucy	21.55
38	39	5	Lucy	21.81
39	21	3	Lucy	22.18
40	25	4	Enya	23.27

A visual analysis of the data in Table 5.11 using a Gage R&R run chart from Minitab reveals the results shown in Figure 5.28. In Figure 5.28, each dot represents Enya's measurements and each square represents Lucy's measurements. Multiple measurements by each inspector are connected with lines. Good repeatability is demonstrated by the low variation in the squares connected by lines and the dots connected by lines, for each unit. Figure 5.28 indicates that repeatability (within group variation) is good. Good reproducibility is demonstrated by the similarity of the squares connected by lines and the dots connected by lines, for each unit. Reproducibility (one form of between group variation) is good. The gage run chart shows that most of the observed total variation in light gage readings is due to differences between units. This data looks good.

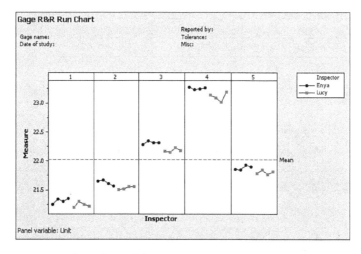

Figure 5.28 Gage R&R run chart obtained from Minitab

An Anecdote of Measurement Systems

Several years ago I visited a chocolate factory. We began the tour, and when we got to the chocolate bean roasting operation, I asked the owner: "How do you know when the beans are properly roasted?" He said: "I have two roaster tasters, and each has at least ten years of experience. No problem!" I asked how he knew that each roaster was consistent with himself over time and consistent with the other roaster over time. He got angry with me because he thought my question was stupid. I asked if he would consider a test where ten beans were sliced into four sections and randomly assigned to the roasters. My idea was for each roaster to taste each of the ten beans twice in random order, and then to use a gage run chart to determine whether they were consistent with themselves for each bean, and consistent with each other for each bean. He got furious and kicked me out of the factory. I don't know if his measurement system for roasting beans was effective in providing quality information, but his chocolate did taste pretty good to me. But then again, I am a chocoholic.

How-To Guide for Understanding Variation: Minitab User Guide (Minitab Version 17, 2013)

The following section lists the steps necessary to create the statistical tools and methods described in this chapter using Minitab 17. It is important that you get comfortable with Minitab so it becomes part of your regular routine as you do your job and improve your job using process improvement theory and practices.

Go to www.ftpress.com/sixsigma to download the data files referenced in this chapter so you can practice with Minitab.

Using Minitab to Obtain Zone Limits

To plot zone limits on any of the control charts discussed in this chapter, open to the Data Source dialog box for the control chart being developed and do the following:

1. Click the **Scale** button. Click the **Gridlines** tab. Select the **Y major ticks, Y minor ticks**, and **X major ticks** check boxes. Click **OK** to return to the Data Source dialog box.

2. Select the **Options** button. Select the **S limits** tab. In the Standard Deviation Limit Positions edit box, select **Constants** in the drop-down list box and enter **1 2 3** in the edit box. Click **OK** to return to the Data Source dialog box.

Using Minitab for the P Chart

To illustrate how to obtain a p chart, refer to the data in Table 5.1 concerning the number of broken slides. Open the **SLIDES.MPJ** worksheet and do the following:

1. Select **Stat | Control Charts | Attribute Charts | P**. In the P Chart dialog box (see Figure 5.29) enter **C3** or **'Number Cracked'** in the Variables edit box. Since the subgroup sizes are equal, select **Size** in the Subgroup Sizes drop-down list box and enter **100** in the edit box. Click the **P Chart Options** button.

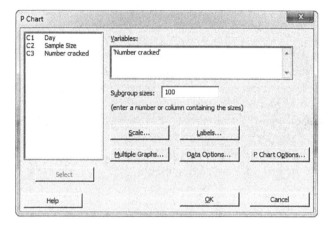

Figure 5.29 Minitab P Chart dialog box

2. In the P Chart: Options dialog box, click the **Tests** tab (see Figure 5.30). Select **Perform All Tests for Special Causes** from the drop-down list. Click **OK** to return to the P Chart dialog box. (These values stay intact until Minitab is restarted.)

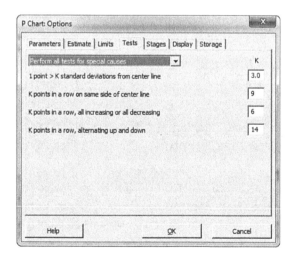

Figure 5.30 Minitab P Chart: Options dialog box, Tests tab

3. If there are points that should be omitted when estimating the center line and control limits, click the **Estimate** tab in the P Chart: Options dialog box (see Figure 5.31). Enter the points to be omitted in the edit box shown. Click **OK** to return to the P Chart dialog box. In the P Chart dialog box, click **OK** to obtain the p chart.

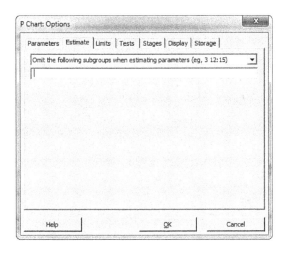

Figure 5.31 Minitab P Chart: Options dialog box, Estimate tab

Figure 5.32 shows the output for the p chart.

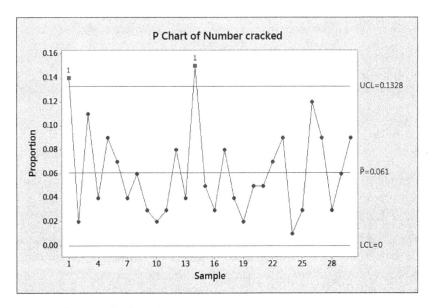

Figure 5.32 Minitab output for the p chart

Chapter 5 Understanding Variation: Tools and Methods   133

Using Minitab for the C Chart

To illustrate how to obtain a c chart, refer to the data in Table 5.3 concerning the number of add-ons in an outpatient clinic. Open the **ADDONS.MPJ** worksheet and follow these steps:

1. Select **Stat | Control Charts | Attribute Charts | C**. In the C Chart dialog box (see Figure 5.33), enter **C2** or **ADDONS** in the Variables edit box.

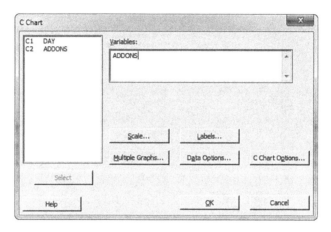

Figure 5.33 Minitab C Chart dialog box

2. Click the **C Chart Options** button. In the C Chart: Options dialog box, click the **Tests** tab. Select **Perform All Tests for Special Causes** from the drop-down list. Click **OK** to return to the C Chart dialog box. (These values stay intact until Minitab is restarted.) Click **OK** to obtain the c chart.

3. If there are points that should be omitted when estimating the center line and control limits, click the **Estimate** tab in the C Chart: Options dialog box. Enter the points to be omitted in the edit box shown. Click **OK** to return to the C Chart dialog box.

Figure 5.34 shows the output for the c chart.

Using Minitab for the U Chart

To illustrate how to obtain a u chart, refer to the data in Table 5.4 concerning the number of patient falls in a hospital. Open the **FALLS.MPJ** worksheet and perform these steps:

1. Select **Stat | Control Charts | Attribute Charts | U**. In the U Chart dialog box (see Figure 5.35), enter **C3** or **FALLS** in the Variables edit box. In the Subgroup Sizes drop-down list box, select **Indicator Column** and enter **C2** or **CENSUS** in the edit box.

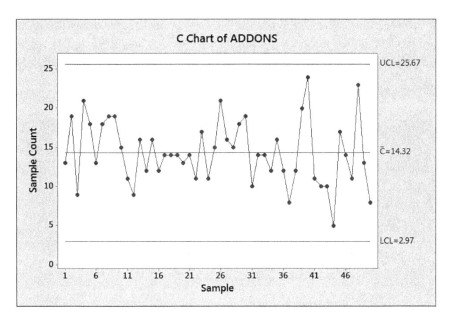

Figure 5.34 Minitab output for the c chart

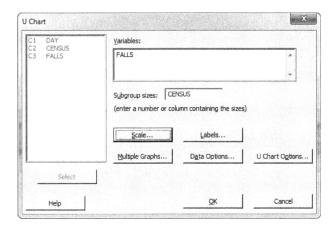

Figure 5.35 Minitab U Chart dialog box

2. In the U Chart: Options dialog box, click the **Tests** tab. Select the **Perform All Tests for Special Causes** from the drop-down list. Click **OK** to return to the U Chart dialog box. (These values stay intact until Minitab is restarted.) Click **OK** to obtain the U chart.

3. If there are points that should be omitted when estimating the center line and control limits, click the **Estimate** tab in the U Chart: Options dialog box. Enter the points to be omitted in the edit box shown. Click **OK** to return to the U Chart dialog box.

Figure 5.36 shows the output for the u chart.

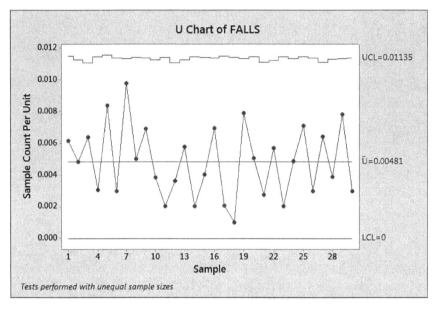

Figure 5.36 Minitab output for the u chart

Using Minitab for the Individual Value and Moving Range Charts

Individual Value and Moving Range charts can be obtained from Minitab by selecting **Stat | Control Charts | Variable Charts for Individuals | I-MR** from the menu bar. To illustrate how to obtain Individual Value and Moving Range charts, refer to the data in Table 5.5 concerning the turnaround times of GI biopsies. Open the **TURNAROUND.MPJ** worksheet and follow these steps:

1. Select **Stat | Control Charts | Variable Charts for Individuals | I-MR**. In the Individuals-Moving Range Chart dialog box (see Figure 5.37), enter **'GI BIOPSY TURNAROUND TIMES'** in the Variables edit box. Click the I-MR Options button.

2. In the I-MR Chart: Options dialog box, click the **Tests** tab. Select **Perform All Tests for Special Causes** from the drop-down list. Click **OK** to return to the I-MR Chart dialog box. (These values stay intact until Minitab is restarted.)

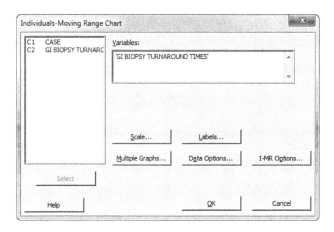

Figure 5.37 Minitab I-MR Chart dialog box

3. If there are points that should be omitted when estimating the center line and control limits, click the **Estimate** tab in the I-MR Chart: Options dialog box. Enter the points to be omitted in the edit box shown. Click **OK** to return to the I-MR Chart dialog box. (Note: When obtaining more than one set of Individual Value and Moving Range charts in the same session, be sure to reset the values of the points to be omitted before obtaining new charts.)

4. In the I-MR Chart dialog box, click **OK** to obtain the individual value and moving range charts.

Figure 5.38 shows the output for the I-MR chart.

Using Minitab for the X Bar and R Charts

X Bar and R charts can be obtained from Minitab by selecting Stat | Control Charts | Variable Charts for Subgroups | Xbar-R from the menu bar. The format for entering the variable name is different, depending on whether the data are stacked down a single column or unstacked across a set of columns with the data for each time period located in a single row. If the data for the variable of interest are stacked down a single column, choose All Observations for a Chart Are in One Column from the drop-down list and enter the variable name in the edit box below. If the subgroups are unstacked with each row representing the data for a single time period, choose Observations for a Subgroup Are in One Row of Columns from the drop-down list and enter the variable names for the data in the edit box below.

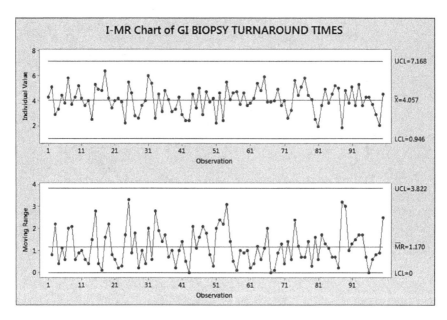

Figure 5.38 Minitab output for the I-MR chart

To illustrate how to obtain X Bar and R charts, refer to the data in Table 5.6 concerning the weight of vials. Open the **VIALS.MPJ** worksheet and follow these steps:

1. Select **Stat | Control Charts | Variable Charts for Subgroups | Xbar-R**. In the Xbar-R Chart dialog box (see Figure 5.39) enter **C3** or '**1**', **C4** or '**2**', **C5** or '**3**', **C6** or '**4**', **C7** or '**5**', and **C8** or '**6**', in the edit box. Click the **Xbar-R Options** button.

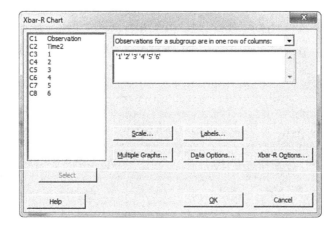

Figure 5.39 Minitab Xbar-R Chart dialog box

2. In the Xbar-R Chart: Options dialog box (see Figure 5.40), click the **Tests** tab. Select the **Perform All Tests for Special Causes** from the drop-down list. Click **OK** to return to the Xbar-R Chart dialog box. (These values stay intact until Minitab is restarted.)

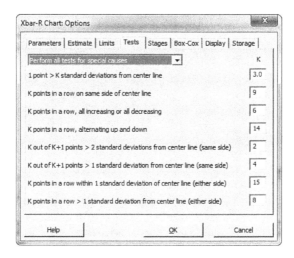

Figure 5.40 Minitab Xbar-R Chart: Options dialog box, Tests tab

3. If there are points that should be omitted when estimating the center line and control limits, click the **Estimate** tab in the Xbar-R Chart: Options dialog box (see Figure 5.41). Enter the points to be omitted in the edit box shown. Click **OK** to return to the Xbar-R Chart dialog box. (Note: When obtaining more than one set of X Bar and R charts in the same session, be sure to reset the values of the points to be omitted before obtaining new charts.)

4. In the Xbar-R Chart dialog box, click **OK** to obtain the X Bar and R charts.

Figure 5.42 shows the output for the X Bar and R chart.

Using Minitab for the X Bar and S Charts

X Bar and S charts can be obtained from Minitab by selecting **Stat | Control Charts | Variable Charts for Subgroups | Xbar-S** from the menu bar. The format for entering the variable name is different, depending on whether the data are stacked down a single column or unstacked across a set of columns with the data for each time period located in a single row. If the data for the variable of interest are stacked down a single column, choose **All Observations for a Chart Are in One Column** from the drop-down list and enter the variable name in the edit box below. If the subgroups are unstacked with each row representing the data for a single time period, choose **Observations for a Subgroup Are in One Row of Columns** from the drop-down list and enter the variable names for the data in the edit box below.

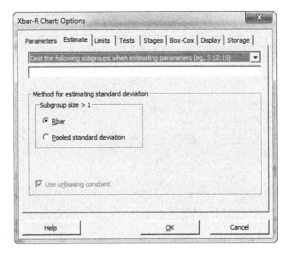

Figure 5.41 Minitab Xbar-R Chart: Options dialog box, Estimate tab

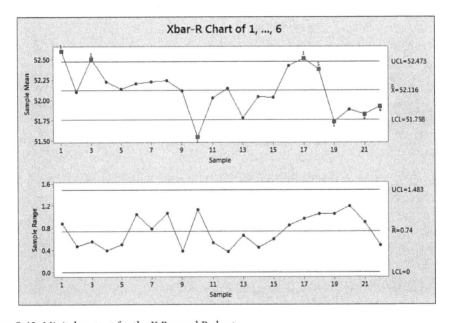

Figure 5.42 Minitab output for the X Bar and R chart

To illustrate how to obtain X Bar and S charts, refer to the data in Table 5.7 concerning admitting processing time. Open the **ADMITTING.MPJ** worksheet and follow these steps:

1. Select **Stat | Control Charts | Variable Charts for Subgroups | Xbar-S**. In the Xbar-S Chart dialog box (see Figure 5.43) enter **TIME** in the edit box. In the Subgroup Sizes drop-down list box, select **All Observations for a Chart Are in One Column**. Enter **10** in the edit box. Click the **Xbar-S Options** button.

2. In the Xbar-S Chart: Options dialog box, click the **Tests** tab. Select **Perform All Tests for Special Causes** from the drop-down list. Click **OK** to return to the Xbar-S Chart dialog box. (These values stay intact until Minitab is restarted.)

3. If there are points that should be omitted when estimating the center line and control limits, click the **Estimate** tab in the Xbar-S Chart: Options dialog box. Enter the points to be omitted in the edit box shown. Click **OK** to return to the Xbar-S Chart dialog box. (Note: When obtaining more than one set of X Bar and S charts in the same session, be sure to reset the values of the points to be omitted before obtaining new charts.)

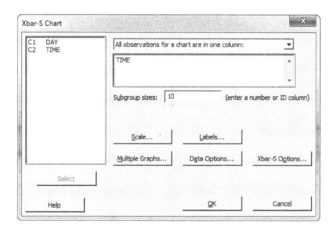

Figure 5.43 Minitab Xbar-S Chart dialog box

4. In the Xbar-S Chart dialog box, click **OK** to obtain the X Bar and S charts.

Figure 5.44 shows the output for the X Bar and S chart.

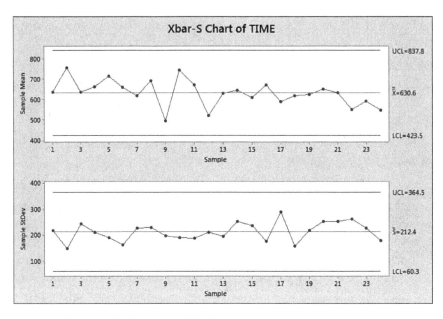

Figure 5.44 Minitab output for the X Bar and S chart

Takeaways from This Chapter

- Control charts are used to distinguish between common causes of variation and special causes of variation, which lets you know whether your process is stable and under control.
- The type of control chart you use depends on the type of data you have.
- Different rules help identify special causes of variation.
- If you have attribute data you use attribute control charts as follows:
  - **P charts**—If you have attribute classification data
  - **C charts**—If you have attribute count data with constant areas of opportunity
  - **U charts**—If you have attribute count data with nonconstant areas of opportunity
- If you have measurement data you use variables control charts as follows:
  - **X bar R charts**—If your subgroup size is between 2 and 9
  - **X bar S charts**—If your subgroup size is 10 or greater
  - **Individuals and moving range (I-MR) charts**—If your subgroup size is 1

References

Gitlow, H. G., A. Oppenheim, R. Oppenheim, and D. M. Levine (2015), *Quality Management: Tools and Methods for Improvement*, 4th ed. (Naperville, IL: Hercher Publishing Company). This book is free online at hercherpublishing.com.

Gitlow, H. and D. Levine (2004), *Six Sigma for Green Belts and Champions: Foundations, DMAIC, Tools and Methods, Cases and Certification* (Upper Saddle River, NJ: Prentice-Hall).

Minitab Version 17 (State College, PA: Minitab, Inc., 2013).

Additional Readings

Montgomery, D. C. (2000), *Introduction to Statistical Quality Control*, 4th ed. (New York: John Wiley).

Shewhart, W. A. (1931), *Economic Control of Quality of Manufactured Product* (New York: Van Nostrand-Reinhard), reprinted by the American Society for Quality Control, Milwaukee, 1980.

Western Electric (1956), *Statistical Quality Control Handbook* (Indianapolis, IN: Western Electric).

6

Non-Quantitative Techniques: Tools and Methods

What Is the Objective of This Chapter?

The objective of this chapter is to introduce you to the non-quantitative tools and methods used in process improvement. The chapter is split into two sections. The first section is a high level overview of the non-quantitative tools and methods used in process improvement with examples, and the second section is a more in-depth step-by-step how-to guide on using the different non-quantitative tools and methods.

The idea is for you to read the first part now so you understand the tools and methods utilized in the Six Sigma DMAIC model discussed in Chapters 10 through 14. You can either read the detailed how-to section now, or you can come back and read it later while working on your project.

Before we proceed any further, we need to define some notation. CTQ is an acronym for Critical to Quality characteristic. It is the problematic metric that your process improvement project is aimed at improving; it is called Y in Statistics, or the dependent variable. X refers to the steps in the process that can be manipulated to get better output for the CTQ. For example, suppose the total cycle time to process a student in the Accounts Receivable department at a university (CTQ) is a function of the time the student waits for service (X_1), the time it takes the clerk to pull the student's records (X_2), the time it takes to process the student's material (X_3), and the time it takes to refile the student's records and to call on the next student waiting in line (X_4). In this case we can say that the CTQ is an additive function of X_1, X_2, X_3, and X_4; in mathematical terms: CTQ = $f(X_1, X_2, X_3, X_4)$, or more specifically CTQ = $X_1 + X_2 + X_3 + X_4$.

High Level Overview and Examples of Non-Quantitative Tools and Methods

This section introduces you to non-quantitative tools and methods used in process improvement with examples to ensure understanding of each tool. Have fun!

Flowcharting

What: A flowchart is a tool used to map out (draw a picture of) a process (Gitlow et al., 2015). The two types of flowcharts introduced in Chapter 3, "Defining and Documenting a Process," are the following:

- **Process flowcharts**—Process flowcharts describe the steps and decision points of a process in a downward direction from the start (top of the page) to the stop (bottom of the page) of the process.

- **Deployment flowcharts**—Deployment flowcharts are organized into "lanes" that show processes that involve various departments, people, stages, or other categories.

Why: Process flowcharts are used when you want to depict a process at a high level or when you want to drill down into a detailed portion of a process. Deployment flowcharts are used when you want to show who is responsible for different parts of a process, as well as track the number and location of handoffs within the process.

Example: Figure 6.1 shows an example of a process flowchart for a typical inpatient cardiology consult process.

Figure 6.2 is an example of a deployment flowchart for a surgical biopsy process.

The starbursts in Figure 6.2 show the locations of problematic areas in the process. These areas may be examined at a later point in time as possible areas for improving the process. The starbursts indicate steps in the process that may become potential Xs for improving the problematic CTQ.

Voice of the Customer (VoC)

What: Voice of the Customer analysis involves surveying stakeholders of a process to understand their requirements and needs (Gitlow and Levine, 2004). Stakeholders can include customers, employees, investors, regulatory agencies, building and grounds, the legal system, and the environment, to name a few possible stakeholders of a process.

There are typically four stages when conducting the VoC analysis:

1. Define/segment the market. (Identify the stakeholder groups.)
2. Plan the VoC. (Identify who will be interviewed and who will interview them, establish a time schedule, and prepare the questions.)
3. Collect the data by stakeholder group.
4. Organize and interpret the data.

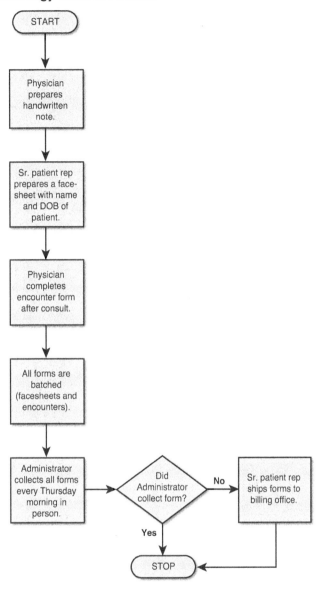

Figure 6.1 Process flowchart

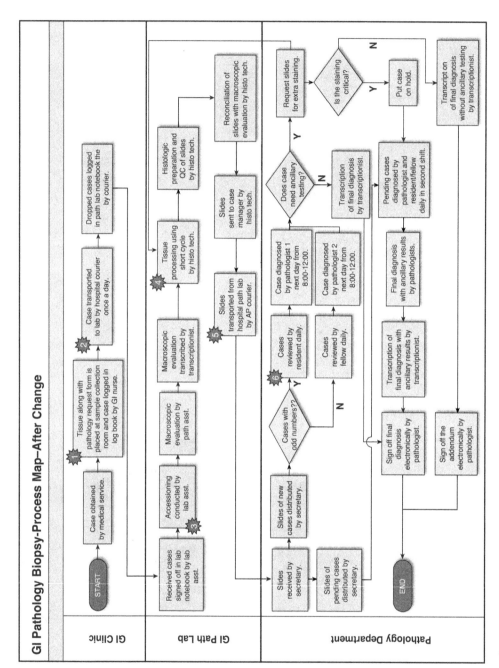

Figure 6.2 Deployment flowchart

Why: The data we collect from the VoC analysis helps us take the needs of our stakeholders into consideration when defining the metrics (variables) of concern to them.

Example: Due to the complex nature and length of VoC analysis, an example is given at the same time we show you how to conduct VoC analysis later in the second part of this chapter.

> ### An Anecdote on Voice of the Customer Analysis
>
> A few years ago a friend of mine decided to perform a personal Voice of the Customer analysis. At the time his main customers were his wife, his son, his mother (his dad had passed), and his friends. He decided to begin the VoC data collection with his wife. He was surprised that she quickly listed out 20 items that she thought he needed to attend to, his CTQs if you will. Well, he realized that it was impossible for him to do all 20 items. So, he created a 1 to 10 dynamite scale that his wife used to rate the importance of each of the 20 items. One stick of dynamite meant that the item was not so important, while 10 sticks of dynamite meant the item was very important. Three of the 30 items involved spending more time with the children; this accounted for 70% of his wife's sticks of dynamite. Therefore, "spending time with the children" was his wife's critical area of concern (CTQ). He decided to play with the children for at least two hours every Saturday. His wife was happy with the results. His new best practice method saved many fights with his wife by allowing him to prioritize his wife's issues; think Pareto diagram.

Supplier-Input-Process-Output-Customer (SIPOC) Analysis

What: A SIPOC analysis is a simple tool for identifying the suppliers and their inputs into a process, the high level steps of a process, the outputs of the process, and the customer segments interested in the outputs of the process (Gitlow and Levine, 2004).

Why: The SIPOC analysis helps team members define and give scope to a project. It helps the team understand the process at a high level, and it helps identify customer (stakeholder) segments and their needs and wants (outputs: both intended and unintended), and suppliers and their inputs. It also helps to clarify who will be interviewed during the VoC analysis.

Example: Table 6.1 shows an example of a SIPOC analysis for the patient scheduling process at an outpatient clinic in a hospital.

Table 6.1 SIPOC Analysis

Process Name: Patient Scheduling Process in the Outpatient Clinic at XYZ Hospital Process Owner: Assistant Vice President of Outpatient Services				
SUPPLIERS	**INPUTS**	**PROCESS**	**OUTPUTS**	**CUSTOMERS**
(providers of required resources)	(resources required by the process)		(deliverables from the process; these can be CTQs or Xs)	(stakeholders who put requirements on the outputs)
Referring physicians Patients/families Appointment schedulers Insurance providers	Referral Phone call Patient information Patient and physician availability Insurance plan	Call comes in ↓ Patient entered into electronic medical record system ↓ Appointment scheduled ↓ Insurance verified ↓ Reminder call if time ↓	New patient record in electronic medical record system Scheduled appointment Verified insurance Reminder Arrived visit Patient no show	Patient registration reps in clinics Nurses Physicians Insurance providers Finance office Patients/families

An Anecdote about a SIPOC Analysis

A SIPOC analysis is the first real opportunity that team members have to look at their process from a 360-degree point of view. They may learn some surprising things about dysfunctional suppliers or customers they didn't know existed. For example, I was consulting at a factory that had a problem with down time on machines. So, the process of concern was the request for maintenance process. One of the customers of the request for maintenance process was the employees who operated machines that needed maintenance. One of the suppliers to the request for maintenance process was the maintenance personnel that fixed the machines.

The request for maintenance protocol had a 1 to 4 scale for prioritizing requests for maintenance:

- 1 = low priority, get to it as soon as is convenient.
- 2 = moderate priority, get to it within the next day or two, or else it might turn into a bigger problem.
- 3 = high priority, get to it sometime today because the machine is not working at the necessary speed.

> - 4 = emergency priority, drop whatever you are doing and get to the machine because it is not running and is essentially taking the factory down until it is fixed.
>
> The problem was that all requests for maintenance were coded as "4" by the machine operators. They did this because they had quotas to meet so they would not get in trouble. This meant that the maintenance personnel were running to fix machines that were low priority, while high priority machines were not attended too in a timely fashion. So, here you see how a SIPOC analysis helps identify a dysfunction in the request for maintenance process.
>
> As a side bar, resolving this problem is difficult in the traditional paradigm of management that relies on quotas to reward and punish employees. This remained a problem after I stopped consulting for the factory because they never really adopted a process orientation to their business.

Operational Definitions

What: An operational definition promotes understanding between people by putting communicable meaning into words (Gitlow et al., 2015; Gitlow and Levine, 2004). An operational definition contains three parts: a criterion to be applied to an object or group, a test of the object or group, and a decision as to whether the object or group meets the criterion.

- **Criteria**—Operational definitions establish VoC specifications for each CTQ that are compared with the Voice of the Process outputs in the test step of an operational definition.
- **Test**—A test involves comparing Voice of the Process data with VoC specifications for each CTQ for a given unit of output.
- **Decision**—A decision involves making a determination whether a given unit of output meets VoC specifications.

Why: Operational definitions are required to give communicable meaning to terms such as *late, clean, good, red, round, 15 minutes,* or *3:00 p.m.* Problems, such as endless bickering and ill-will, can arise from the lack of an operational definition. A definition is operational if all relevant users of the definition agree on the definition.

Example: A firm produces washers. One of the critical quality characteristics is roundness. The following procedure is one way to arrive at an operational definition of roundness, as long as the buyer and seller agree on it.

Step 1: Criterion for roundness.

> Buyer: "Use calipers (a measuring device) that are in reasonably good order." (You perceive at once the need to question every word.)
>
> Seller: "What is 'reasonably good order'?"

(We settle the question by letting you use your calipers.)

Seller: "But how should I use them?"

Buyer: "We'll be satisfied if you just use them in the usual way."

Seller: "At what temperature?"

Buyer: "The temperature of this room."

Buyer: "Take six measures of the diameter about 30 degrees apart. Record the results."

Seller: "But what is 'about 30 degrees apart'? Don't you mean exactly 30 degrees?"

Buyer: "No, there's no such thing as exactly 30 degrees in the physical world. So try for 30 degrees. We'll be satisfied."

Buyer: "If the range between the six diameters doesn't exceed .007 centimeters, we'll declare the washer to be round."

(They have determined the criterion for roundness.)

Step 2: Test of roundness.
   a. Select a particular washer.
   b. Take the six measurements and record the results in centimeters: 3.365, 3.363, 3.368, 3.366, 3.366, and 3.369.
   c. The range is 3.369 to 3.363, or a 0.006 difference. They test for conformance by comparing the range of 0.006 with the criterion range of 0.007 (*Step 1*).

Step 3: Decision on roundness.

Because the washer passed the prescribed test for roundness, it is declared to be round.

If a seller has employees who understand what round means and a buyer who agrees, many of the problems the company may have had satisfying the customer will disappear.

An Anecdote about Operational Definitions

If you think about the times in your life when you got the angriest at someone (or yourself), it may have involved a failure to operationally define a critical item in a plan. For example, 50 years ago (before cell phones) my mother was picking up my father at the airport. At that time it was a 2-hour drive to the airport each way. My parents thought they agreed on the time and place for the pickup, but the place wasn't specific enough, and they missed each other. My mother drove home and my father had to take a 2-hour taxi ride. They didn't speak for a week because they were so angry at each other for messing up what seemed like a simple situation. This certainly made the situation much worse. Beware of poorly defined operational definitions.

> **Anecdote on the Importance of Operational Definitions**
>
> A philosophy teacher gave his students a test. He took his chair and put it onto the desk and said simply, "your assignment is to prove that this chair does not exist." The students worked furiously writing long, detailed explanations, except for one student who completed his test in 1 minute before handing it in to the surprised looks of his classmates and teacher. The next week when the teacher gave the tests back, the student who finished the test in 1 minute got the highest grade. His answer? "What chair?"

Failure Modes and Effects Analysis (FMEA)

What: FMEA is a tool used to identify, estimate, prioritize, and reduce risk of damage being caused by potential CTQs and Xs or to identify areas of risk to the success of a project (Gitlow and Levine, 2004).

Why: To study systems to identify what may go wrong with a product, service, or process and to mitigate the risks of potential problems.

Example: Table 6.2 shows the structure of an FMEA.

Check Sheets

What: Check sheets are used for collecting or gathering data on CTQs or Xs in a logical format, which will be analyzed by the tools and methods discussed in this book (Gitlow et al., 2015; Gitlow and Levine, 2004).

Why: Check sheets have several purposes, the most important being to enable the user(s) to gather and organize data in a format that permits efficient and easy analysis. The check sheet's design should facilitate data gathering.

Example: Table 6.3 shows an attribute check sheet of pharmacy delay reasons in a chemotherapy treatment unit lab.

Table 6.2 Format for an FMEA

1	2	3	4	5	6	7	8	9	10	11	12	13	14	15	16
Critical Parameter	Potential Failure mode	Potential Failure Effect	Severity	Potential Causes	Occurrence	Current Controls	Detection	RPN	Recommended Action for an X (Alternative setting of an X)	Responsibility, Contingency Plan if Alternative Setting for X Fails, and Target Date	Date Action Taken	Severity	Occurrence	Detection	RPN
Before RPN =									After RPN =						

Table 6.3 Attribute Check Sheet

	Monday				Tuesday				
Type of Defect	8-10 a.m.	10-12 a.m.	12-2 p.m.	2-4 p.m.	8-10 a.m.	10-12 a.m.	12-2 p.m.	2-4 p.m.	Total
Missing dose	||		|||			|||||					12
Wrong date of service				|||	|||	|||	|||			|||||	19
Missing height and weight	|||	|||	||	|||||	||		||		17
Order expired	|||		||||| |||||	||	||	|||	||||| |||||	||	32
Total	9	3	18	11	12	6	14	7	80

As you can see from Table 6.3, order expired is the most common defect.

Brainstorming

What: Brainstorming is a process used to elicit a large number of ideas from a group of people in a brief amount of time (Gitlow et al., 2015; Gitlow and Levine, 2004). Team members use their collaborative thinking power to generate unlimited ideas and thoughts.

Why: Brainstorming is used to

- Identify relevant problems to address.
- Identify causes of a particular problem.
- Identify solutions to a particular problem.
- Identify strategies to implement solutions.

Example: Consider a group of eight people, one from each department of a hospital, who brainstorm about the problem of excessive employee absenteeism. They have already decided on the topic to be discussed, so they can proceed to making their lists of causes. After completing their lists, they take turns reading their ideas, sequentially, one at a time. The designated leader records the ideas on a flip chart.

The first person's list of possible causes of excessive employee absenteeism is

1. Low pay.
2. No repercussions for missing work.
3. Job is boring.
4. Family problems.

The second person's list is

1. Supervisor is a jerk.
2. Drug problems in the organization.
3. Don't like their job.
4. Unsafe environment.
5. Don't like coworkers.

The rest of the team have similar lists. Everyone reads their lists aloud and the causes are recorded. The leader then asks whether any new ideas have been sparked by piggybacking on one of the first person's causes. "Family problems" might elicit another cause, "personal problems." Asking for wild ideas might generate a response such as "addiction to the Internet" or "crappy food in the cafeteria."

Once all the ideas have been discussed, each member of the team receives a copy of the list of ideas to study. The team meets again and evaluates the ideas by adding or dropping ideas.

Affinity Diagrams

What: An affinity diagram is a tool used by teams to organize and consolidate a substantial and unorganized amount of verbal and/or pictorial data relating to a problem; frequently from a brainstorming session or from VoC interviews (Gitlow et al., 2015).

Why: Affinity diagrams help organize data that comes from brainstorming sessions, VoC interviews, or other sources into natural clusters that help expose the latent structure of the problem being studied.

Example: A team of employees in an organization addressed the question: "Why are call center employees leaving the job?" The team members recorded their ideas on cards. They then placed the cards on a table and everyone simultaneously moves the cards into clusters that are thematically similar: *in silence!* Silence is important so the team members don't exert influence over which card should go in which cluster. Figure 6.3 shows the resulting affinity diagram.

In Figure 6.3 the team's view of the problems in call center employees leaving the job are given by the header cards:

1. Chairs not comfortable.
2. Not satisfied with job.
3. Poor management.

A detailed study of these three categories will help the group understand why call center employees keep leaving the job.

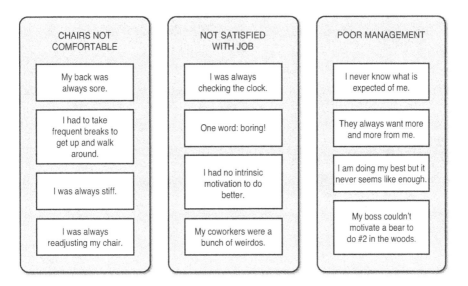

Figure 6.3 Affinity diagram

Cause and Effect (Fishbone) Diagrams

What: A cause and effect (C&E) diagram (Gitlow et al., 2015), also known as a *fishbone diagram*, is a tool that is used to

- Organize the potential causes of a problem (called an *effect*), usually because there are so many potential causes of a given problem.
- Select the most probable causes of the problem (effect).
- Verify the cause and effect relationship between the problem (effect) and the most probable cause(s).

Why: People are frequently overwhelmed by the number of causes related to a problem, a cause and effect diagram helps a team identify the various causes and narrow them down to the most probable cause(s).

Example: Figure 6.4 is an example of a cause and effect (fishbone) diagram to understand reasons that patients are not showing up to their appointments in a hospital clinic. The causes in the fishbone diagram could have come from a brainstorming session on the causes of no shows to a hospital clinic. The team studies the C&E diagram and circles the most likely causes (Xs) of the effect/problem (CTQ); see Figure 6.4. Figure 6.4 shows that the team believes that "transportation to the clinic," "physician," and "new vs. established patients" are the most likely causes (Xs) of "no shows at the clinic" (CTQ).

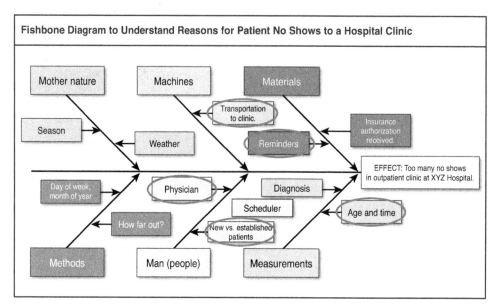

Figure 6.4 Cause and effect (fishbone) diagram

Another example: Figure 6.5 is an example of a cause and effect (fishbone) diagram to understand why our friend Joe is still single:

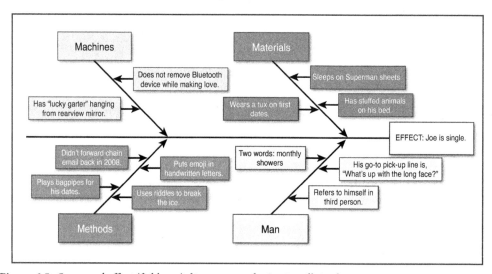

Figure 6.5 Cause and effect (fishbone) diagram on why Joe is still single

Please note that Figures 6.4 and 6.5 use materials, machines, mother nature, methods, man (people), and measurements as the names of the major fishbones. Another method for naming the major fishbones is to use the names of the clusters from an affinity diagram. If you refer back to Figure 6.3, you could construct a C&E diagram with the major fishbone categories of "chairs not comfortable," "not satisfied with job," and "poor management."

Pareto Diagrams

What: Pareto diagrams are used to identify and prioritize issues (Xs) that contribute to a problem (CTQs); it is mainly a prioritization tool. The Pareto principle is frequently called the 80-20 rule; 80% of your problems (CTQs) come from 20% of your potential causes (Xs). Pareto analysis focuses on distinguishing the vital few significant causes of problems from the trivial many causes of problems. The vital few are the few causes that account for the largest percentage of the problem (80%), while the trivial many are the myriad of causes that account for a small percentage of the problem (20%). The causes of the problem become Xs, and the problem is the CTQ(s) (Gitlow et al., 2015; Gitlow and Levine, 2004).

Why: The Pareto principle focuses attention on the significant few causes (Xs) instead of the trivial many causes. Consequently, we are able to prioritize efforts on the most important causes (Xs) of the problem to make the most efficient use of our time and resources.

Example: The director of a hospital pharmacy is interested in learning about the reasons on delays to orders for the hospitals chemotherapy treatment unit. He collects data for a month on the reasons for delays that are seen in Table 6.4.

Table 6.4 Pharmacy Delay Reasons

Delay Reason	Number
Missing D.O.S.	74
Missing height and weight	66
Dose change	15
Order clarification	9
No consent form	7
Labs pending	6
Labs high	2

After collecting data and creating a Pareto diagram in Figure 6.6, he sees that missing date of service (D.O.S.) and missing height and weight on the orders account for 78.2% of delays. He has the IT folks put a "hard stop" on the order entry system that prohibits orders being placed without a date of service or height and weight. He collects more data the next month and as expected sees a drastic reduction in order delays.

Gantt Charts

What: A Gantt chart is a simple scheduling tool. It is a bar chart that plots tasks and subtasks against time. It clarifies which tasks can be done in parallel and which tasks must be done serially (Gitlow et al., 2015).

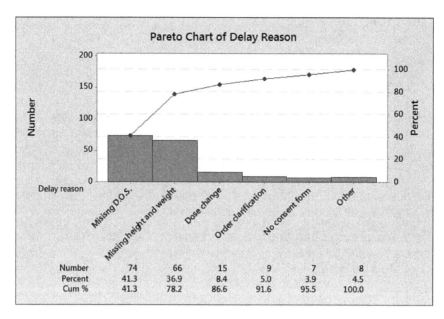

Figure 6.6 Delay reasons Pareto diagram

Why: Once a list of tasks and subtasks has been created for a project, responsibilities can be assigned for each. Next, beginning and finishing dates can be scheduled for each task and subtask. Finally, any comments relevant to a task or subtask are indicated on the Gantt chart; see Table 6.5.

Example: Table 6.5 shows task 3 begins in March and ends in April, while task 4 begins May and ends in November. This Gantt chart shows that three tasks begin in May.

Change Concepts

What: Change concepts are approaches to changing the Xs (steps in a process flowchart) that have been found to be useful in developing solutions that lead to improvements in processes. By creatively combining these change concepts with process knowledge, teams can develop changes that lead to substantial improvement. The change concepts were developed by Associates in Process Improvement (Langley et al., 1996; Gitlow et al., 2015).

Table 6.5 Gantt Chart

Tasks	Responsibility	J	F	M	A	M	J	J	A	S	O	N	D	J	F	M	A	M	J	Comments
1	HG	B	E																	
2	BJ			B				E												
3	RM			B	E															
4	HG					B							E							
5	RM					B								E						
6	BJ					B									E					
7	HG																B	E		
8	RM																	B	E	

The 70 change concepts listed here are organized into the following nine general groupings:

Grouping	Number of Concepts in Grouping
A. Eliminate Waste	11
B. Improve Work Flow	11
C. Optimize Inventory	4
D. Change the Work Environment	11
E. Enhance the Producer/Customer Relationships	8
F. Manage Time	5
G. Manage Variation	8
H. Design Systems to Avoid Mistakes	4
I. Focus on the Product/Service	8

All 70 change concepts are listed and described in the second part of this chapter.

Why: Change concepts allow process improvers to develop, test, and implement changes (new Xs or new settings for Xs). This ability is paramount to the success of anyone who wants to constantly improve a process. Typically a specific and unique change is needed to obtain improvement in a specific set of circumstances.

Example: College students need access to accurate and timely information on their accounts payable with the university they are attending.

The goals of Student Account Services are to first satisfy student account information needs electronically, then by phone, and finally, if necessary, in person. This approach enables the university to improve both customer service quality and employee productivity.

A quality improvement team analyzed the total number of calls and corresponding abandoned calls, which were at an all-time high, through the call center. The team studied the list of 70 change concepts and concluded that the 13th change concept, "schedule into multiple processes," could be helpful in reducing the percentage of abandoned calls.

The team developed this change concept into hiring inexpensive undergraduate student assistants who could answer basic phone questions and perform triage for complex inquiries. If an inquiry required a higher level of account expertise than that possessed by the student assistant, the assistant would forward the call to a full-time customer representative. The change concept was put into place in August, and the number of abandoned calls dropped substantially as can be seen in Table 6.6 and the line graph in Figure 6.7.

Table 6.6 Abandoned Calls Per Month

Month	Abandoned Calls
January	132
February	150
March	146
April	166
May	141
June	139

Month	Abandoned Calls
July	173
August	35
September	29
October	31
November	22
December	19

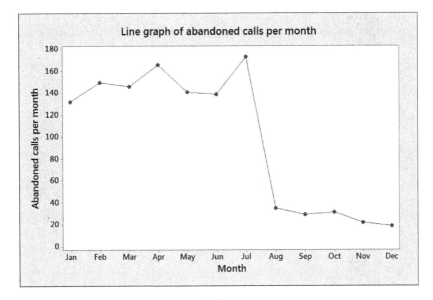

Figure 6.7 Line graph of abandoned calls per month

Communication Plans

What: A communication plan is created for a project to identify, inform, and appease concerns regarding a process improvement event because many people typically are affected by it.

Why: Implementing new processes typically involves the need for employees to change their behaviors. Having a communication plan allows the team to be proactive in informing people in the organization as to what is going on so there are no surprises and so no one is caught off guard when it is time to implement solutions.

Example: Table 6.7 shows a communication plan.

Table 6.7 No Shows Project Communication Plan

#	Event/Communication	Participants/Audience	Medium	Frequency	When	Lead	Scheduled?	Status	Notes
1	General announcement memo to staff	All staff in Department of Behavioral Health	Email and fax blast	Once	TBD	AVP, BH	n/a	TBD	Draft communication to inform staff of project, sent by AVP and CEO
2	Weekly email to stakeholders	All project stakeholders	Email	Weekly	Every Friday	Black Belt	n/a	Ongoing	Black Belt to delegate to team member
3	Meetings with opinion leaders	Opinion leaders	Face to face	Once	TBD	Black Belt	N	TBD	Identify opinion leaders and work with them to promote the project
4	Poster placed in clinic lunch room	All staff in Department of Behavioral Health	Printed poster	Once	TBD	Communications Director	N	TBD	Need to create and place poster in clinic lunch room
5	Presentation at physician staff meeting	All physicians in Department of Behavioral Health	Face to face	Once	TBD	Black Belt	N	TBD	Need to create presentation and schedule time to present
6	Update hospital intranet site	All staff in Department of Behavioral Health	Intranet website	Once	TBD	Black Belt	N	TBD	Create content and meet with webmaster to post to intranet

How-To Guide for Using Non-Quantitative Tools and Methods

This section explains in detail how to use in practice the tools and methods described briefly in the first section of this chapter. Section one of this chapter is the "what" and "why," with an example, for each tool and method, whereas section two is the "how to" for each of the tools and methods discussed in this chapter.

How to Do Flowcharting

As discussed in Chapter 3, the steps in creating a flowchart are described in Table 6.8. The American National Standards Institute, Inc. (ANSI) approved a standard set of flowchart symbols used for defining and documenting a process. The shape of the symbol and the information written within the symbol provide information about that particular step or decision in a process.

Table 6.8 Flowchart Symbols and Functions

Symbol	Function
Start/stop symbol	The general symbol used to indicate the beginning and end of a process is an oval.
Basic processing symbol	The general symbol used to depict a processing operation is a rectangle.
Decision symbol	A diamond is the symbol that denotes a decision point in the process. This includes attribute type decisions such as pass-fail, yes-no. It also includes variable types of decisions such as into which of several categories a process measurement falls.
Flowline symbol	A line with an arrowhead is the symbol that shows the direction of the stages in a process. The flowline connects the elements of the system.

- Successful flowcharting efforts have the commitment (not only support) of top management, the process owner, the employees who work in the process, and any other key process stakeholders. Commitment is required because if you are going to flowchart a process there must be a reason for the effort, such as documentation or improvement of the process. Both documentation and improvement require the commitment of all the stakeholders of the process.

- There must be a rationale for defining, documenting, or improving a process. The employees who work in the process and the process owner must understand that flowcharting the process is a positive exercise aimed at making things better and that no one will be blamed for problems in the process.

- Stakeholders of a process must agree on the starting/stopping points of the process, as well as the objectives and metrics used to measure success.

- The employees who are flowcharting the process must have enough time so that they are not rushed and can do a thorough job.

- The employees who are flowcharting the process start by having the whole team walk the process from front to back (process owner's point of view) and back to front (customer's point of view).

- The employees who are flowcharting the process need to determine how detailed the flowchart should be, as well as the type of flowchart to be used by the employees. A flowchart needs to be detailed enough to be able to uncover problems. This means that some sections of a flowchart will be at a high level and others will be at a detailed level.

- One possible method for flowcharting a process is to write the process steps on Post-it notes and then place them on large sheets of paper taped to the wall for all to see. Next, the steps are arranged in the order that reflects the flow of the process, and more detail and decision symbols are added until the "as is" process is reflected on the wall. Finally, connector lines showing the direction of flows in the process are added onto the large sheet of paper. Next, the flowchart on the wall is replicated in an electronic format such as Visio. Finally, all the stakeholders of the process verify and validate to obtain full agreement that the flowchart captures the essence of the process.

Refer to Figure 6.1, which shows an example of a process flowchart created using the symbols in Table 6.8.

Deployment flowcharts differ from process flowcharts because they add either rows or columns with headings listing departments, employees, or stages of the process. Refer back to Figure 6.2; it shows that the different steps in the process that take place in different areas of the organization are delineated by the rows labeled GI Clinic, GI Path Lab, and Pathology Department. The handoffs in the process from one department/area to another are visible by the lines that cross from one row to another. Again, the starbursts indicate potential problematic areas to be improved in the process.

How to Do a Voice of the Customer (VoC) Analysis

How to Define/Segment the Market

The first thing we need to do is figure out from whom are we going to get our VoC data; that is, who are the stakeholder segments to be interviewed that can provide input on the CTQs of interest.

The simplest method for segmenting a market is to study the SIPOC analysis and focus attention on the outputs and on the customers by asking

- What are the outputs of your process, both intended and unintended?
- Who are the customers of those outputs?
- Are there particular groups of customers whose needs focus on specific outputs? Today? Tomorrow?

How to Plan the VoC

Next we have to plan the VoC. There are two types of VoC data: reactive data and proactive data. *Reactive data* arrives regardless of whether the organization collects it, for example, customer complaints, product returns or credits, contract cancellations, market share changes, customer defections and acquisitions, customer referrals, closure rates of sales calls, web page hits, technical support calls, and sales, to name a few. *Proactive data* arrives only if it is collected by personnel in the organization: This is the type of VoC data we need to collect. It is data obtained through positive action, for example, data gathered through interviews, focus groups, surveys, comment cards, sales calls, market research, customer observations, and benchmarking. Some of the steps involved in planning the VoC are the following:

- Choosing stakeholders within each stakeholder segment to interview; for example, customer segments, employee segments, investor segments, regulatory segments, environmental segments, building and grounds, and so on.
- Creating the interview questions.
- Setting up the interviews.
- Assigning interviews to team members.

How to Collect the Data

This step involves conducting the interviews to collect VoC data from your stakeholder segments. A few things to keep in mind when conducting the interviews:

- Record their answers in bullet point form; it makes the data easier to organize later.
- Let the interviewee talk; this is time for you to understand her perspective on the process. However, don't let the interview become a therapy session; gently stay on task.
- Write down exactly what they say in their words; you don't want anything to get lost in translation. This is important! If profanity is used, record it. It may be an indicator of the degree of emotion felt by a person being interviewed.

How to Organize and Interpret the Data

The last step involves organizing and interpreting all the VoC data you have collected. Focus points are the underlying themes for one or more raw VoC statements. Affinity diagrams are used to create focus points.

Example: Voice of the Customer Analysis for Patient No Shows at an Outpatient Psychiatric Clinic

Define/Segment the Market

To define and segment the market for the outpatient psychiatric clinic discussed earlier, the team went back to the SIPOC analysis they had created and looked at the suppliers and customers. They decided to focus on the following segments during the VoC to help them better understand the process. The team members decided that each of the following segments form a stakeholder group; in other words, there are no subsegments of patients or physicians. This is frequently not the case.

- Patients/family
- Appointment schedulers
- Insurance providers
- Patient registration reps in clinics
- Nurses
- Physicians
- Finance office

Plan the VoC

Next up was planning the VoC. The team decided to focus on collecting proactive VoC data since there was no reactive VoC data available for this process. Next, they did the following:

Choosing Stakeholders within Each Market Segment to Interview

The team decided that they would take a judgment sample (expert opinion) of the following stakeholder groups to interview. The stakeholders selected in each group were deemed most appropriate to reflect the views of that group by the team doing the VoC, perhaps with the help of an expert. The team members decided that they could afford to interview the following number of stakeholders within each segment; it would give them a great understanding of the process:

- Five patients and family members
- Two appointment schedulers
- Two insurance providers

- Two patient registration reps in clinics
- Five nurses
- Five physicians
- One finance office

As stated previously, the sample size in each segment is a function of the budget available to perform the VoC. It is an arbitrary number that in the opinion of an expert yields reliable information about that segment.

Creating Interview Questions

The team then decided on questions to ask each interviewee related to the patient scheduling process in the Outpatient Psychiatric Clinic at XYZ Hospital. The questions are designed to understand the process as much as possible:

- How do you feel about the patient scheduling process as it relates to patient no shows?
- What issues/problems do you see with the patient scheduling process that may lead to patient no shows?
- What solutions/recommendations do you have to decrease the number of patient no shows?
- What feelings or images come to mind when you think about patient no shows? This last question is important because it captures the circumstance surrounding the process.

Setting Up Interviews

The process expert assigned one of the team members the task of setting up the interviews with the various stakeholders making sure to follow the timeline set out by the Black Belt in the project plan.

Assigning Interviews to Team Members

The process expert then assigned interviews to different team members based on his understanding of each team member, their personality, and their relationships with various stakeholders.

Collect the Data

The various team members then went out and conducted the interviews to collect VoC data from their respective stakeholders. The process expert gave the team specific instructions:

- Record their answers in bullet point form to make the data easier to organize later.
- Let them talk. This is time for you to understand their perspective on the process.
- Write down exactly what they say; you don't want anything to get lost in translation.

Table 6.9 Voice of the Customer Summary Table

Selected Market Segment	Raw VoC Data	Affinity Diagram Theme (Focus Point)	Driving Issue	CTQ
Patients	"I can't get an appointment for 3 months. I get here and the waiting room is empty. No wonder I am crazy!" (1) **Note: The number one (1) indicates the thematic group the VoC raw data point falls into; in this case variation in patients showing up for their appointments. As you will see, several raw VoC data points fall into this thematic group.**	Variation in patients showing up for their appointments. (1)	Patients not showing up for their appointments affects everyone.	Patient no show rate, by month.
	"They don't take no shows seriously, so I don't feel bad if something else comes up." (1)	Variation in scheduling patients for appointments. (2)	Time between date patient calls to schedule appointment and date of appointment is too long.	Turnaround time to schedule appointments, by patient.
	"Everyone else seems to miss appointments and they aren't held accountable, so I don't feel bad if I do." (1)			
	"Why does it take so long to get an appointment here?" (2)			
Appointment schedulers	"These no shows make our job so hard." (1)			
	"Tough to predict who will no show and who won't." (1)			
	"The variation in who shows and who doesn't makes it tough to schedule." (1)			
	"We hear complaints from everyone. I am glad they are finally doing something about it!" (1)			
Insurance providers	"No shows seem to be a problem that a lot of hospitals struggle with." (1)			
	"They aren't good for anybody." (1)			

Selected Market Segment	Raw VoC Data	Affinity Diagram Theme (Focus Point)	Driving Issue	CTQ
Patient registration representatives	"We see a lot of frustration when patients no show—from the docs to the nurses to other patients to staff. We need to decrease them asap!" (1)			
	"Sometimes we are busy, sometimes we aren't." (1)			
	"It makes it hard for our boss to schedule us. I feel bad for her sometimes." (1)			
	"Sometimes we get sent home if there are too many no shows. I get it but it still sucks." (1)			
Nurses	"I have physicians giving me crap for times they have nothing to do, like it's my fault!" (1)			
	"Sometimes I am super busy, sometimes it's slow." (1)			
	"I hear patients complaining that they had to wait a few months to get in. I have no idea why." (2)			
Physicians	"I am the most expensive resource and half the time I feel like I am sitting on my butt." (1)			
Finance office	"It definitely has a financial impact, and with the way things have been going we need every penny." (1)			
	"They are directly affecting our bottom line." (1)			
	"We need to figure out how to decrease them and fast!" (1)			

Organize and Interpret the Data

Team members analyzed the VoC data in Table 6.9 by the market segments (see column 1 in Table 6.9). Next, they used all the raw VoC data points (see column 2 of Table 6.9) to create thematic groups indicated by the numbers in parentheses in each cell that are summarized in column 3 of Table 6.9. Next, team members identified the issue underlying each focus point, called driving issues (see column 4 in Table 6.9). Then team members converted each cognitive issue into one or more quantitative variable, called critical-to-quality (CTQ) variables (see column 5 in Table 6.9).

The team identified two CTQs via the VoC interviews: patient no show rate by month (which they kind of knew all along would be a CTQ) and turnaround time to schedule appointments by patient. They decided to focus on the first CTQ, patient no show rate, because they agreed that turnaround time to schedule appointments could be a function of the no show rate.

How to Do a SIPOC Analysis

To create a SIPOC analysis the team fills in each of the five columns in the SIPOC table in Table 6.10 by answering the questions in the following sections.

Suppliers

Team members identify relevant suppliers by asking the following questions:

- Where does information and material come from?
- Who are the suppliers?
- Is the Human Resources department a source of inputs into the process (employees)?

Inputs

Team members identify relevant inputs by asking the following questions:

- What do your suppliers give to the process?
- What effect do the inputs (Xs) have on the process?
- What effect do the inputs (Xs) have on the CTQs (outputs)?
- What effect do employees (Xs) have on the CTQ(s)?

Process

Team members create a high level flowchart of the process taking particular care to identify the beginning and ending points of the process.

Outputs:

Team members identify outputs of the process by asking the following questions:

- What products or services does this process make, both intended and unintended?
- What are the outputs that are critical to the customer's perception of quality? Or lack of quality? These outputs are called critical-to-quality (CTQ) characteristics of the process.
- Are there any unintended outputs from the process that may cause problems?

Customers

Team members identify relevant customers (market segments) by asking the following questions:

- Who are the customers (stakeholders) or market segments of this process?
- Have we identified the outputs (CTQs) for each market segment?

See Table 6.10 for the layout of a SIPOC analysis.

Table 6.10 SIPOC Analysis

Process name:				
Process owner:				
SUPPLIERS	INPUTS	PROCESS	OUTPUTS	CUSTOMERS
(Providers of required resources)	(Resources required by the process)		(Deliverables from the process)	(Stakeholders who put requirements on the outputs)
Here we list all the suppliers of the inputs to the process.	Here we list all the inputs into the process.	This is a high level flowchart of the process.	Here we list all the outputs from the process, both intented and unintended.	Here we list all the customers of the outputs.

How to Create Operational Definitions

The creation of operational definitions requires the following steps. Remember, all users of the operational definition must agree with the operational definition or it is of no value in eliminating misunderstandings and arguments.

Step 1: Determine the criterion for the item being operationally defined. The team (customer and supplier or key stakeholders) must establish the criteria on the item being operationally defined. For example "the blanket must be at least 50% wool." In other words, a random sample of 15 one inch by one inch pieces of blanket are selected, and if all 15 have at least 50% wool, the blanket is considered a wool blanket.

Step 2: Test for the item being operationally defined. Next, the team selects the sample of 15 one inch by one inch pieces of the blanket and has them tested by a wool expert to analyze the proportion of wool in each piece.

Step 3: Decision for the item being operationally defined. Finally the team must decide whether the test to determine if the item being operationally defined meets the criteria established. For example, if all 15 one inch pieces of wool analyzed by the wool expert are 50% wool or greater, then the blanket is 50% wool.

> **An Anecdote about Operational Definitions**
>
> At one time I was consulting for a paper mill. The nearest lodgings to the paper mill was a pretty low class motel. I walked into my room and the maid was cleaning the toilet with a toilet bowl brush. I was glad to see that. However, then she used the same brush to clean the toilet seat and used a rag to dry the seat. Finally, she put the strip of paper around the toilet seat that says sanitized. Clearly, the maid and I had different operational definitions of sanitized.

How to Do a Failure Modes and Effects Analysis (FMEA)

There are nine steps to conducting an FMEA:

1. Identify the critical parameters (Xs) and their potential failure modes identified in the cause and effects matrix or diagram through brainstorming or other tools—that is, ways in which the process step (X) might fail (columns 1 and 2 of Table 6.11).

2. Identify the potential effect of each failure (consequences of that failure) and rate its severity (columns 3 and 4 of Table 6.11). The definition of the severity scale is shown in Table 6.12.

3. Identify causes of the effects and rate their likelihood of occurrence (columns 5 and 6 of Table 6.11). The definition of the likelihood of occurrence scale is shown in Table 6.13.

4. Identify the current controls for detecting each failure mode and rate the organization's ability to detect each failure mode (columns 7 and 8 of Table 6.11). The definition of the detection scale is shown in Table 6.14.

5. Calculate the *RPN (risk priority number)* for each failure mode by multiplying the values in columns 4, 6, and 8 (column 9 of Table 6.11).

6. Identify the action(s) and/or contingency plans for an alternative setting of an X that will improve the distribution of the CTQ, identify person(s) responsible to implement and maintain the alternative setting of the X, and target completion dates for reducing or eliminating the RPN for each failure mode (columns 10 and 11 of Table 6.11). Actions are the process changes needed to reduce the severity and likelihood of occurrence, and increase the likelihood of detection, of a potential failure mode; they are change concepts. Contingency plans are the alternative actions immediately available to a process owner when a failure mode occurs in spite of process improvement actions. A contingency plan might include a contact name and phone number in case of a failure mode.

7. Identify the date the action was taken to reduce or eliminate each failure mode (column 12 of Table 6.11).

8. Rank the severity (column 13 of Table 6.11), occurrence (column 14 of Table 6.11), and detection (column 15 of Table 6.11) of each failure mode after the recommended action (column 10 of Table 6.11) has been put into motion.

Table 6.11 Failure Modes and Effects Analysis

1	2	3	4	5	6	7	8	9	10	11	12	13	14	15	16
Critical Parameter	Potential Failure Mode	Potential Failure Effect	SEV	Potential Causes	OCC	Current Controls	DET	RPN	Recommended Action	Responsibility and Target Date	Date Action Taken	SEV	OCC	DET	RPN
Lack of buy in by top management.	Project doesn't get supported.	Project unexecutable.	10	Lack of commitment.	10	None	10	1000	Top management must have skin in the game.	CEO.	6/2/2019	2	4	10	80
Project exceeds budget.	Black Belt gets fired!	Need to hire new BB.	3	BB lacks financial expertise.	8	BB does his best.	5	120	Finance rep on team responsible for budget.	Finance rep.	6/2/2019	2	5	5	50
Takes too long to get data.	Project is delayed.	Heads roll.	8	BB lacks IT expertise.	8	BB does his best.	5	320	IT rep on team responsible for data.	IT rep.	6/2/2019	4	4	5	80
Lack of buy in from physicians.	Implementation issues.	Unsuccessful implementation.	7	Entitled physicians.	8	Lots of bitching back and forth.	8	448	Physician Champion part of team.	Physician Champion.	6/2/2019	4	6	8	192
Lack of buy in from nurses.	Implementation issues.	Unsuccessful implementation.	7	Entitled physicians.	8	Lots of bitching back and forth.	8	448	Nursing Champion part of team.	Nursing Champion.	6/2/2019	4	6	8	192
Lack of awareness.	Lack of cooperation.	Misaligned team.	5	Lack of communication.	5	Rumor mill in overdrive.	2	50	Communication plan created.	Black Belt.	6/2/2019	1	5	2	10
Scope too broad.	Project becomes unmanageable.	Project unexecutable.	7	Trying to boil the ocean.	7	Planning by Black Belt.	3	147	Strong planning effort by Black Belt.	Black Belt.	6/2/2019	3	3	3	27

Chapter 6 Non-Quantitative Techniques: Tools and Methods

9. Multiply the values in columns 13, 14, and 15 of Table 6.11 to recalculate the RPN for each failure mode after the recommended action (column 10 of Table 6.11) has been put into motion.

If the after RPN is significantly lower than the before RPN for the RPNs with large before values, you have likely hit upon a good change concept, or modification to an X (step in the flowchart). You can see this occurring in "lack of physician buy-in" and "lack of nurse buy-in."

Table 6.12 Definition of "Severity" Scale = Likely Impact of Failure

Impact	Rating	Criteria: A Failure Could...
Bad	10	Injure a customer or employee
	9	Be illegal
	8	Render the unit unfit for use
	7	Cause extreme customer dissatisfaction
	6	Result in partial malfunction
	5	Cause a loss of performance likely to result in a complaint
	4	Cause minor performance loss
	3	Cause a minor nuisance; can be overcome with no loss
	2	Be unnoticed; minor effect on performance
Good	1	Be unnoticed and not affect the performance

Table 6.13 Definition of "Occurrence" Scale = Frequency of Failure

Impact	Rating	Time Period	Probability of Occurrence
Bad	10	More than once per day	> 30%
	9	Once every 3-4 days	<= 30%
	8	Once per week	<= 5%
	7	Once per month	<= 1%
	6	Once every 3 months	<= .3 per 1,000
	5	Once every 6 months	<= 1 per 10,000
	4	Once per year	<= 6 per 100,00
	3	Once every 1-3 years	<= 6 per million (approx. Six Sigma)
	2	Once every 3-6 years	<= 3 per ten million
Good	1	Once every 6-100 years	<= 2 per billion

Table 6.14 Definition of "Detection" Scale = Ability to Detect Failure

Impact	Rating	Definition
Bad	10	Defect caused by failure is not detectable.
	9	Occasional units are checked for defects.
	8	Units are systematically sampled and inspected.
	7	All units are manually inspected.
	6	Manual inspection with mistake proofing modifications.
	5	Process is monitored with control charts and manually inspected.
	4	Control charts used with an immediate reaction to out of control condition.
	3	Control charts used as above with 100% inspection surrounding out of control condition.
	2	All units automatically inspected or control charts used to improve the process.
Good	1	Defect is obvious and can be kept from the customer or control charts are used for process improvement to yield a no inspection system with routine monitoring.

How to Do Check Sheets

Check sheets are used in Six Sigma projects to collect data on CTQs and Xs in a format that permits efficient and easy data collection and analysis by team members. Three types of check sheets are discussed: attribute check sheets, measurement check sheets, and defect locations check sheets.

Attribute Check Sheets

An *attribute check sheet* is used to gather data about defects in a process. The logical way to collect data about a defect is to determine the number and percentage of defects generated by each cause. To create an attribute check sheet simply list the different types of defects and then tally the number of occurrences for each over a relevant test time period.

Table 6.15 is an example of a defect check sheet for an outpatient scheduling call center for a health system. This check sheet was created by tallying each type of call defect during four 2-hour time periods for one week. Keeping track of these data provides management with information on which to base improvement actions, assuming a stable process.

Table 6.15 Outpatient Call Center Defects

Type of Defect	Frequency	Percentage
Improper use of English language	2	2.9
Grammatical errors in speech	6	8.6
Inappropriate use of words	3	4.3
Rude response	3	4.3
Didn't know answer to patient's question	25	35.7
Call took too much time	30	42.8
Not available	1	1.4
Total	70	100

Measurement Check Sheets

Gathering measurement type data about a product, service, or process involves collecting information such as revenue per month, cost per quarter, cycle time per unit produced, waiting time by customer, temperature, size, length, weight, and diameter by item. This data is best represented on a frequency distribution, also called a *measurement check sheet*.

Table 6.16 is a measurement check sheet showing the frequency distribution of cycle times to answer 508 patient calls to make appointments (how long the patient is kept on hold with horrible muzak playing on the phone) in an outpatient scheduling call center for a health system that came into the call center between 8:00 a.m. and 5:00 p.m. on January 16, 2015.

Table 6.16 Frequency Distribution of Cycle Times

Cycle Time (Minutes)	Frequency
05 < 10 minutes	36
10 < 15 minutes	178
15 < 20 minutes	233
20 < 25 minutes	53
25 < 30 minutes	8
Total	508

This type of check sheet is a simple way to examine the distribution of a CTQ or X and its relationship to specification limits (the boundaries of what is considered an acceptable cycle time). The number and percentage of items outside the specification limit are easy to identify so that appropriate action can be taken to reduce the number of defective calls, perhaps using a process improvement project. For example, suppose the call center has a policy that no call can last more than 20 minutes. If this policy is in place, 12% (61/508) of call cycle times are out of specification. This may help employees and management improve the process, assuming it is stable, while both keeping calls under 20 minutes and giving callers satisfactory answers.

Defect Location Check Sheet

Another way to gather information about defects in a product is to use a defect location check sheet. A *defect location check sheet* is a picture of a product (or a portion of it) on which a relevant employee indicates the location and nature of a defect.

Figure 6.8 is an example of a check sheet for collecting data regarding defects on a cube. It shows a defect in the top left-hand corner of the cube (in this case a dent); the location of the defect is marked with an "x." Suppose that an analysis of multiple check sheets reveals that many x's are in the upper left corner on the front face of the cube. If this is so, further analysis might shed light on the type of defect in the upper left corner. In turn, this might lead employees to identify the root cause of the defects. This, of course, leads to improvements in the cube production process.

Figure 6.8 Defect location check sheet example

Brainstorming

How: Brainstorming is a way to elicit a large number of ideas from a group of people in a short period of time about an idea, topic, or problem. Members of the group use their collective thinking power to generate ideas and unrestrained thoughts.

- Effective brainstorming should take place in a structured session.
- The group should be between 3 and 12 people. No animals, babies, or electronic devices are allowed; they are too distracting.
- Composition of the group should include a variety of people who have different points of view about the idea, topic, or problem.
- The group leader should keep the group focused, prevent distractions, keep ideas flowing, and record the outputs.
- The brainstorming session should be a closed-door meeting with no distractions.
- Seating should promote the free flow of ideas; a circle is the best seating arrangement.
- The leader should record the ideas so everyone can see them, preferably on a flip chart, blackboard, or illuminated transparency. Or, each participant can write his own ideas on 3x5 cards; one idea per 3x5 card.

Procedure

The following steps are recommended prior to a brainstorming session:

1. Select the topic or problem.
2. Send out the topic to all brainstorming participants to ensure all agree on the topic *before* arriving at the session.
3. Conduct research on the topic in a library or on the Internet.
4. Prepare a list of the identified ideas and provide a copy to each of the brainstorm participants before the brainstorming session. There is no reason to reinvent the wheel. The brainstorm participants should add ideas to what is already easily accessible through research.
5. Establish a time and place for the brainstorming session.
6. Invite all attendees to the session.
7. Remind all attendees to study the list of ideas provided to them on the topic.
8. Ask all participants to add additional ideas for discussion in the brainstorming session and bring them to the session, say one idea per 3x5 card.

The following steps are recommended *at* a brainstorming session:

1. Post the topic or problem.
2. Each group member makes an additional list of ideas about the problem.
3. Each person reads one idea at a time from her list of ideas, sequentially, starting at the top of the list. Group members continue in this circular reading fashion until all the ideas on everyone's lists are read.
4. If a member's next idea is a duplication of a previously stated idea, that member goes on to the subsequent idea on his list.
5. After each idea is read by a group member, the leader requests all other group members to think of "new ideas."
6. Members are free to pass on each go-round but should be encouraged to add something.
7. If the group reaches an impasse, the leader can ask for everyone's "wildest idea."

Rules

Certain rules should be observed by the participants to ensure a successful brainstorming session—otherwise, participation may be inhibited.

1. Don't criticize, by word or gesture, anyone's ideas.
2. Don't discuss any ideas during the session, except for clarification.
3. Don't hesitate to suggest an idea because it sounds silly. Many times a "silly" idea can lead to the problem's solution.

4. Don't allow any group member to present more than one idea at a time.
5. Don't allow the group to be dominated by one or two people.
6. Don't let brainstorming become a gripe session.

How to Do Affinity Diagrams

Affinity diagrams are used to analyze verbal or pictorial data. In this text, we analyze verbal data. For example, affinity diagrams take verbal data (on, say, 3x5 cards) and place them into thematic groups, such that the data (3x5 cards) in a group are similar thematically to the other data points (3x5 cards) in the group. They can be used to evaluate data from brainstorming or from VoC Customer interviews.

Constructing an affinity diagram begins with identifying a problem. Team composition usually consists of the same people who participated in the brainstorming session about the problem under study or conducted the VoC interviews.

A team should take the following steps to construct an affinity diagram:

1. Select a team member to serve as the group's facilitator.
2. The facilitator transfers all the ideas generated from a brainstorming session to 3x5 cards; one idea per 3x5 card.
3. The facilitator spreads all the 3x5 cards on a large surface (table) in no particular order, but all cards face the same direction.
4. *In silence*, all group members simultaneously move the 3x5 cards into groups (clusters) so the 3x5 cards in a group seem to be related; that is, they have an unspoken underlying theme or affinity for each other.
5. After the group agrees the clusters are complete (usually 3 to 15 clusters emerge), the group discusses all the 3x5 cards in each cluster and prepares a header card that sums up the information for each cluster.
6. The facilitator transfers the information from the cards onto a flip chart, or "butcher paper" and draws a circle around each cluster.
7. The underlying structure of the problem, usually typified by the names of the header cards, is used to understand the product or process problem.

Refer to Figure 6.3; that example was created using the preceding steps to understand why call center employees are leaving the job.

In Figure 6.3 the team's view of the problems in call center employees leaving the job are given by the header cards:

- Chairs not comfortable
- Not satisfied with job
- Poor management

A detailed study of these three categories may help team members understand why call center employees keep leaving the job.

How to Do Cause and Effect Diagrams (C&E Diagrams)

The following steps are recommended for constructing a cause and effect diagram.

1. State the problem, which is the effect.
2. Select the team; usually the brainstorming team or the VoC interviewer team(s).
3. Affinitize the data (one data point per 3x5 card) into major causes ("fishbones") or use the 6 Ms: materials, machines, mother nature, methods, man, and measurement.
4. Put the major causes on a C&E diagram as the major categories.
5. Add subcauses for each major cause and place them on the C&E diagram.
6. Allow time to ponder the subcauses before evaluating them. You may find some holes in your cause and effect diagram; fill in the holes by brainstorming or just being thoughtful.
7. Circle subcauses that are most likely contributing to the problematic CTQ (Y).
8. Verify each potential subcause with data or relevant literature and so on.

Refer to Figure 6.4; that figure was created to understand reasons that patients are not showing up to their appointments in a hospital clinic.

How to Do Pareto Diagrams

The following steps are recommended for constructing a Pareto diagram. We illustrate a Pareto diagram with an example concerning sources of defective data entries for a particular data entry operator.

1. Establish categories for the data being analyzed. Data should be classified according to defects, products, work groups, size, and other appropriate categories; a check sheet listing these categories should be developed. In the following example, data on the type of defects for a data entry operator will be organized and tabulated.
2. Specify the time period during which data will be collected. Three factors important in setting a time period to study are (1) selection of a convenient time period, such as one week, one month, one quarter, one day, or four hours, (2) selection of a time period that is constant for all related diagrams for purposes of comparison, and (3) selection of a time period that is relevant to the analysis, such as a specific season for a certain seasonal product. In the data entry example, the time period is four months—January through April 2014. In the example, the types of defects are recorded as they occur during the time period and are totaled, as shown in Table 6.17.
3. **Note:** Pareto diagrams are most useful in analyzing the data from a control chart only if the control chart is in a state of statistical control.

Table 6.17 Record of Defects for Data Entry Operator: Check Sheet to Determine the Sources of Defects

(Assumption: If the following data is from a control chart, the percentage of defective entries must be stable over time for proper use of the Pareto diagram.)					
Major Causes of Defective Entries	**Month 1/12**	**2/12**	**3/12**	**4/12**	**Total**
Transposed numbers	7	10	6	5	28
Out of field	1		2		3
Wrong character	6	8	5	9	28
Data printed too lightly		1	1		2
Torn document	1	1		2	4
Creased document			1	1	2
Illegible source document			1		1
Total	15	20	16	17	68

4. Construct a frequency table arranging the categories from the one with the largest number of observations to the one with the smallest number of observations. The frequency table should contain a category column; a frequency column indicating the number of observations per category with a total at the bottom of the column; a cumulative frequency column indicating the number of observations in a particular category plus all frequencies in categories above it; a relative frequency column indicating the percentage of observations within each category with a total at the bottom of the column; and a relative cumulative frequency column indicating the cumulative percentage of observations in a particular category plus all categories above it in the frequency table. An "other" category, if there is one, should be placed at the extreme right of the chart. If the "other" category accounts for as much as 50% of the total, the breakdown of categories must be reformulated. A rule of thumb is that the "other" bar should be smaller than the category with the largest number of observations. The frequency table for the data entry example, in Table 6.18, shows that two types of defects (transposed numbers and wrong characters) are causing 82.4% of the total number of defective entries.

Table 6.18 Frequency Table of Defects for Data Entry Operator: Pareto Analysis to Determine Major Sources of Defects

Major Cause of Defective Entries	Frequency	Relative %	Cumulative Frequency	Cumulative %
Transposed numbers	28	41.2	28	41.2
Wrong character	28	41.2	56	82.4
Torn document	4	5.9	60	88.3

Major Cause of Defective Entries	Frequency	Relative %	Cumulative Frequency	Cumulative %
Out of field	3	4.4	63	92.7
Data printed too lightly	2	2.9	65	95.6
Creased document	2	2.9	67	98.5
Illegible source document	1	1.5	68	100.0
Total	68	100.0		

5. Draw a Pareto diagram:

- Draw horizontal and vertical axes on graph paper and mark the vertical axis with the appropriate units, from zero up to the total number of observations in the frequency table.

- Under the horizontal axis, write the most frequently occurring category on the far left, and then the next most frequent to the right, continuing in decreasing order to the right. In the data entry example, "transposed numbers" and "wrong character" are the most frequently occurring defects and are positioned to the far left; "illegible source document" accounts for the fewest defective cards and appears at the far right of the chart. The rightmost category will be "other." It may very well be a significant size bar on the Pareto diagram; remember the rules for the "other" category we discussed previously.

- Draw in the bars for each category. For some applications, this may provide enough information on which to base a decision, but often the percentage of change between the columns must be determined. Figure 6.9 displays the bars for the data entry example.

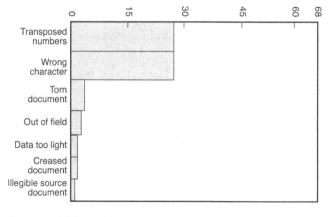

Figure 6.9 Pareto diagram of defective data entries for an operator

- Plot a cumulative percentage line on the Pareto diagram. Indicate an approximate cumulative percentage scale on the right side of the chart and plot a cumulative percentage line on the Pareto diagram. To plot the cumulative percentage line, or **cum line**, start at the lower left (zero) corner and move diagonally to the top-right corner of the first column. In our example, the top of the line is now at the 28 level, as in Figure 6.10(a). Repeat the process, adding the number of observations in the second column. In our example, the line rests on the 56 level, as in Figure 6.10(b). The process is repeated for each column, until the line reaches the total number of observations level that includes 100% of the observations, as in Figure 6.10(c).

- Title the chart and briefly describe its data sources. Without information on when and under what conditions the data were gathered, the Pareto diagram will not be useful.

Typically a Pareto diagram is drawn using Minitab. See Chapter 4, "Understanding Data: Tools and Methods," for instructions on how to create a Pareto diagram using Minitab.

How to Do Gantt Charts

Recall that a Gantt chart is a tool for scheduling projects. Each task or subtask for a project is listed on the vertical axis, as are the person(s) or area(s) responsible for its completion; see columns 1 and 2 in Table 6.5. The horizontal axis is time, see columns 3 and beyond in Table 6.5. It shows the anticipated and actual duration of each task by a bar of the appropriate length. The left end of the bar indicates the earliest start time, and the right end of the bar indicates the latest stop time for the task.

The Gantt chart is especially useful in determining which tasks can be done in parallel or which must be serially. This can save a lot of time in completing the project.

A great use of a Gantt chart is to provide an executive charged with a big project a tool to monitor the project so she doesn't lose sleep over it. For example, the COO of a hospital is in charge of building a new wing on time and within budget, but she is not a civil engineer and has no idea what is supposed to happen and when. All she knows is that her butt is on the line if the project is late and over budget. One way to calm this executive down is to have a project manager draw a Gantt chart for the project so all the executive has to do is to check whether every task is on time and under budget. If not, she can call the person responsible (see column 2 in Table 6.5) for that step and find out what is going on. This turns into better sleep for the executive!

How to Use Change Concepts

The change concepts presented here are not specific enough to be applied directly to making improvements. Rather, the concept must be considered within the context of a particular situation and then turned into an idea for a process improvement. The idea needs to be specific enough to describe how the change can be developed, tested, and implemented in the particular situation. When describing the change concepts, we tried to be consistent in the degree of specificity or generality of the concepts. Sometimes, a new idea seems at first to be a new change concept; but often, upon further reflection, it is seen to be an application of one of the existing change concepts.

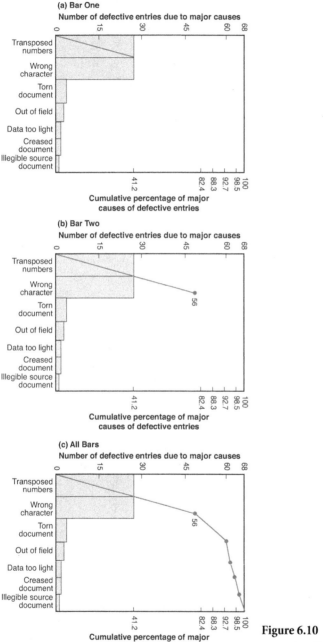

Figure 6.10 Pareto diagram: Cumulative percentage line plotting

The primary purpose of this discussion is to provide help to individuals and teams who are trying to answer the question: What change can we make (in the Xs) that will result in improvement of our problematic process (the CTQ)? Change concepts can serve to provoke a new idea for an individual or team. A team leader can choose one or more of the 70 change concepts and then the team can explore some ideas for possible application of this concept to a step in the problematic process (X). The list of ideas should be recorded. After the generation of ideas is complete, the ideas can be discussed and critiqued. Any of the ideas that show promise can be further explored by the team to obtain a specific idea for a change to a step in the process, or an X.

Some of the change concepts appear to offer conflicting advice for developing changes. For example, concept 25 "reduce choice of features" and concept 63 "mass customize" appear to be aimed in opposite directions. Change concept 4 "reduce controls on the system" and change concept 51 "standardization (create a formal process)" also suggest conflicting directions. The important consideration is the context in which the change concept is being applied. The change concepts are listed in Langley J., Nolan K., Nolan T., Norman C., and Provost L. (1996), *The Improvement Guide,* Jossey-Bass, Inc. (San Francisco, CA), or Gitlow, H., Oppenheim, A., Oppenheim, R., and Levine, D. (2015), *Quality Management*, 4th ed., Hercher Publishing Company (Naperville, IL).

Eliminate Waste

In a broad sense, any activity or resource in an organization that does not add value to an external customer can be considered waste. This section provides some concepts for eliminating waste.

1. **Eliminate things that are not used**. Constant change in organizations results in less demand for specific resources and activities that were once important to the business. Unnecessary activities and unused resources can be identified through surveys, audits, data collection, and analysis of records. The next step is to take the obvious actions to remove the unused elements from the system.

2. **Eliminate multiple entry**. In some situations, information is recorded in a log or entered into a database more than one time, creating no added value. This practice is also called *data redundancy*. Changing the process to require only one entry can lead to improvement in productivity and quality by reducing discrepancies.

3. **Reduce or eliminate overkill**. Sometimes, a company's standard or recommended resources are designed to handle special, severe, or critical situations rather than the normal situation. Changing the standard to the appropriate level of resources for the normal situation reduces waste. Additional resources are used only when the situation warrants it.

4. **Reduce controls on the system**. Individuals and organizations use various types of controls to make sure a process or system does not stray too far from standards, requirements, or accepted practices. While useful for protection of the organization, these controls can increase costs, reduce productivity, and stifle improvement. Typical forms of controls include a layered management structure, approval signatures,

standardized forms, and reports. A regular review of all the organization's control procedures by everyone working in the system can result in identifying opportunities to reduce controls on the system without putting the organization at risk.

5. **Recycle or reuse.** Once a product is created and used for its intended purpose, it is natural to discard it and the by-products created by its use. However, if other uses can be found for the discarded product or by-products, the cost of producing the product can be spread out over its use and its reuse.

6. **Use substitution.** Waste can often be reduced by replacing some aspect of the product or process with a better alternative. One type of substitution is to include lower-cost components, materials, or methods that do not affect the performance of the process, service, or product (sometimes called *value engineering*). Another type of substitution is to switch to another process with fewer steps or less manual effort.

7. **Reduce classifications.** Classifications are often developed to differentiate elements of a system or to group items with common characteristics, but these classifications can lead to system complexity that increases costs or decreases quality. Classification should be reduced when the complexity caused by the classification is worse than the benefit gained.

8. **Remove intermediaries.** Intermediaries such as distributors, handlers, agents, and carriers may be part of a system. Consider eliminating these activities by linking production directly with the consumer. Some intermediaries add value to a process because of their specialized skills and knowledge. Often, however, eliminating these services can increase productivity without reducing value to the customer.

9. **Match the amount to the need.** Rather than using traditional standard units or sizes, organizations can adjust products and services to match the amount required for a particular situation. This practice reduces waste and carryover inventory. By studying how customers use the product, more convenient package sizes can be developed.

10. **Use sampling.** Reviews, checks, and measurements are made for a variety of reasons. Can these reasons be satisfied without checking or testing everything? Many times, the standard 100 percent inspection and testing results is a waste of resources and time. Formal sampling procedures are available that can often provide as good or even better information than 100 percent checking. This is discussed further in Chapter 13, "DMAIC Model: 'I' Is for Improve."

11. **Change targets or set points.** Sometimes problems go on for years because some piece of equipment is not designed or set up properly. Make sure that process settings are at desirable levels. Investigate places where waste is created, and consider adjustments to targets or set points to reduce the waste.

Improve Work Flow

Products and services are produced by processes. How does work flow in these processes? What is the plan to get work through a process? Are the various steps in the process arranged and prioritized to obtain quality outcomes at low costs? How can the work flow be changed so that the process is less reactive and more planned?

12. **Synchronize**. Production of products and services usually involves multiple stages. These stages operate at different times and at different speeds, resulting in an operation that is not smooth. Much time can be spent waiting for another stage to be reached. By focusing on the flow of the product (or customer) through the process, each of the stages can be synchronized.

13. **Schedule into multiple processes**. A system can be redesigned to include multiple versions of the same process focused on the specific requirements of the situation. Rather than a "one-size-fits-all" large process, multiple versions of the process are available, each tuned to the different types of needs of customers or users. Priorities can be established to allocate and schedule the inputs to maximize the performance of the system. The specific processes can then be greatly simplified since they only address a limited range of input requirements.

14. **Minimize handoffs**. Many systems require that elements such as a customer, a form, or a product be transferred to multiple people, offices, or workstations to complete the processing or service. The handoff from one stage to the next can increase time and costs and cause quality problems. The work flow can be rearranged to minimize any handoff in the process. The process can be redesigned so that any worker is only involved once in an iteration of a process.

15. **Move steps in the process close together**. The physical location of people and facilities can affect processing time and cause communication problems. If the physical location of adjacent steps in a process are moved close together, work can be directly passed from one step to the next. This eliminates the need for communication systems, such as mail, and physical transports, such as vehicles, pipelines, and conveyor belts.

16. **Find and remove bottlenecks**. A bottleneck or constraint is anything that restricts the throughput of a system. A constraint within an organization would be any resource for which the demand is greater than its available capacity. To increase the throughput of a system, the constraints must be identified, exploited if possible, and removed if necessary. Bottlenecks occur in many parts of daily life; they can usually be identified by looking at where people are waiting or where work is piling up.

17. **Use automation**. The flow of many processes can be improved by the intelligent use of automation. Consider automation to improve the work flow for any process to reduce costs, reduce cycle times, eliminate human slips, reduce repetitive manual tasks, and provide measurement.

18. **Smooth work flow**. Yearly, monthly, weekly, and daily changes in demand often cause work flow to fluctuate widely. Rather than trying to staff to handle the peak demands, steps can often be taken to better distribute the demand. This distribution results in a smooth work flow rather than in continual peaks and valleys.

19. **Do tasks in parallel**. Many systems are designed so that tasks are done in a series or linear sequence. The second task is not begun until the first task is completed. This is especially true when different groups in the organization are involved in the different

steps of a process. Sometimes, improvements in time and costs can be gained from designing the system to do some or all tasks in parallel.

20. **Consider people in the same system**. People in different systems are usually working toward different purposes, each trying to optimize their own system. Taking actions that help people to think of themselves as part of the same system can give them a common purpose and provide a basis for optimizing the larger system.

21. **Use multiple processing units**. To gain flexibility in controlling the work flow, try to include multiple workstations, machines, processing lines, and fillers in a system. This makes it possible to run smaller lots, serve special customers, minimize the impact of maintenance and downtime, and add flexibility to staffing. With multiple units, the primary product or service can be handled on one line to maximize efficiency and minimize setup time. The less-frequent products and services can be handled by the other units.

22. **Adjust to peak demand**. Sometimes it is not possible to balance the demands made on a system. In these cases, rather than keeping a fixed amount of resources (materials, workers, and so on), historical data can be used to predict peak demands. Then methods can be implemented to meet the temporarily increased demand.

Optimize Inventory

Inventory of any type can be a source of waste in organizations. Inventory requires capital investment, storage space, and people to handle and keep track of it. In manufacturing organizations, inventory includes raw material waiting to be processed, in-process inventory, and finished-goods inventory. For service organizations, the number of skilled workers available is often the key inventory issue. Extra inventory can result in higher costs with no improvement in performance for an organization. How can the costs associated with the maintenance of inventory be reduced? An understanding of where inventory is stored in a system is the first step in finding opportunities for improvement.

23. **Match inventory to predicted demand**. Excess inventory can result in higher costs with no improvement in performance for an organization. How can the proper amount of inventory to be maintained at any given time be determined? One approach to minimizing the costs associated with inventory is to use historical data to predict the demand. Using these predictions to optimize lead times and order quantities leads to replenishing inventory in an economical manner. This is often the best approach to optimizing inventory when the process involves lengthy production times.

24. **Use pull systems**. In a *pull system* of production, work at a particular step in the process is done only if the next step in the process is demanding the work. Enough product is ordered or made to replenish what was just used. This is in contrast to most traditional *push systems*, in which work is done as long as inputs are available. A pull system is designed to match production quantities with a downstream need. This approach can often result in lower inventories than a schedule-based production system. Pull systems are most beneficial in processes with short cycle times and

high yields. Some features of effective pull systems are small lot sizes and container quantities, fast setup times, and minimal rework and scrap.

25. **Reduce choice of features**. Many features are added to products and services to accommodate the desires and wants of different customers and different markets. Each of these features makes sense in the context of a particular customer at a particular time, but taken as a whole, they can have tremendous impact on inventory costs. A review of current demand for each feature and consideration of grouping the features can allow a reduction in inventory without loss of customer satisfaction.

26. **Reduce multiple brands of same items**. If an organization uses more than one brand of any particular item, inventory costs will usually be higher than necessary since a backup supply of each brand must be kept. Consider ways to reduce the number of brands while still providing the required service.

Change the Work Environment

Changes to the environment in which people work, study, and live can often provide leverage for improvements in performance. As organizations try to improve quality, reduce costs, or increase the value of their products and services, technical changes are developed, tested, and implemented. Many of these technical changes do not lead to improvement because the work environment is not ready to accept or support the changes. Changing the work environment itself can be a high-leverage opportunity for making other changes more effective.

27. **Give people access to information**. Traditionally, organizations have carefully controlled the information available to various groups of employees. Making relevant information available to employees allows them to suggest changes, make good decisions, and take actions that lead to improvements.

28. **Use proper measurements**. Measurement plays an important role in focusing people on particular aspects of a business. Developing appropriate measures, making better use of existing measures, and improving measurement systems can lead to improvement throughout the organization.

29. **Take care of basics**. Certain fundamentals must be considered to make any organization successful. Concepts like orderliness, cleanliness, discipline, and managing costs and prices are examples of such fundamentals. It is sometimes useful to take a fresh look at these basics to see whether the organization is still on track. If there are fundamental problems in the business, changes in other areas may not lead to improvements. Also, when people's basic needs are not being met, meaningful improvements cannot be expected in other areas. The Five-S movement, which was the beginning of quality control in Japanese workshops, got its name from the Japanese words for straighten up, put things in order, clean up, cleanliness, and discipline.

30. **Reduce demotivating aspects of the pay system**. Pay is rarely a positive motivator in an organization, but it can cause confusion and become a demotivator. Some pay systems can encourage competition rather than cooperation among employees. Another result of some pay systems is the reluctance to take risks or make changes.

Review the organization's system for pay to ensure that the current system does not cause problems in the organization.

31. **Conduct training**. Training is basic to quality performance and the ability to make changes for improvement. Many changes are not effective if people have not received the basic training required to do a job. Training should include the "why" as well as the "what" and the "how."

32. **Implement cross-training**. Cross-training means training people in an organization to do multiple jobs. Such training allows for flexibility and makes change easier. The investment required for the extra training pays off in productivity, product quality, and cycle times.

33. **Invest more resources in improvement**. In some organizations, people spend more than a full-time job getting their required tasks completed and fighting the fires created in their work. The only changes made are reactions to problems or changes mandated outside the organization. To break out of this mode, management must learn how to start investing time in developing, testing, and implementing changes that lead to improvements.

34. **Focus on core processes and purpose**. Core processes are the processes directly related to the purpose of the organization. They can be characterized as those activities that provide value directly to external customers. To reduce costs, consider reducing or eliminating activities that are not part of the core processes.

35. **Share risks**. Every business is faced with taking risks and reaping their accompanying potential rewards or losses. Many people become more interested in the performance of their organization when they can clearly see how their future is tied to the long-term performance of the organization. Developing systems that allow all employees to share in the risks can lead to an increased interest in performance. Types of plans for sharing risks and gains include profit sharing, gain sharing, bonuses, and pay for knowledge.

36. **Emphasize natural and logical consequences**. An alternative approach to traditional reward-and-punishment systems in organizations is to focus on natural and logical consequences. Natural consequences follow from the natural order of the physical world (for example, not eating leads to hunger), while logical consequences follow from the reality of the business or social world (for example, if you are late for a meeting, you will not have a chance to have input on some of the issues discussed). The idea of emphasizing natural and logical consequences is to get everyone to be responsible for their own behavior rather than to use power, judge others, and force submission. Rather than demanding conformance, the use of natural and logical consequences permits choice.

37. **Develop alliances/cooperative relationships**. Cooperative alliances optimize the interactions between the parts of a system and offer a better approach for integration of organizations.

Enhance the Producer/Customer Relationship

To benefit from improvements in quality of products and services, the customer must recognize and appreciate the improvements. Many ideas for improvement can come directly from a supplier or from the producer's customers. Many problems in organizations occur because the producer does not understand the customer's needs, or because customers are not clear about their expectations of suppliers. The interface between the producer/provider and the customer provides opportunities to learn and develop changes that lead to improvement.

38. **Listen to customers**. It is easy for people to get caught up in the internal functioning of the organization and forget why they are in business: to serve their customers. Time should be invested on a regular basis in processes that "listen" to the customers. Sometimes it is important to figure out how to communicate with customers farther down the supply chain, or even with the final consumer of the product or service. Talk to customers about their experiences in using the organization's products. Learn about improvement opportunities.

39. **Coach customers to use the product/service**. Customers often encounter quality problems and actually increase their costs because they do not understand all the intricacies of the product or service. Companies can increase the value of their products and services by developing ways to coach customers on how to use them.

40. **Focus on the outcome to a customer**. Make the outcome (the product or service) generated by your organization the focus of all activities. First, clearly understand the outcomes that customers expect from your organization. Then, to focus improvement efforts on a particular work activity, consider the question, How does this activity support the outcome to the customer? Make improvements in such areas as the quality, cost, efficiency, and cycle time of that activity. Organize people, departments, and processes in a way that best serves the customer, paying particular attention to the product/customer interfaces. This change concept can also be described as "begin with the end in mind."

41. **Use a coordinator**. A coordinator's primary job is to manage producer/customer linkages. For example, an expeditor is someone who focuses on ensuring adequate supplies of materials and equipment or who coordinates the flow of materials in an organization. Having someone coordinate the flow of materials, tools, parts, and processed goods for critical processes can help prevent problems and downtime. A coordinator can also be used to work with customers to provide extra services. One example is a case manager, who acts as a buffer between a complex process and the customer.

42. **Reach agreement on expectations**. Many times customer dissatisfaction occurs because the customers feel that they have not received the products or services they were led to expect as a result of advertising, special promotions, and promises by the sales group. Marketing processes should be coordinated with production capabilities. Clear expectations should be established before the product is produced or the service is delivered to the customer.

43. **Outsource for "free."** Sometimes it is possible to get suppliers to perform additional functions for the customer with little or no increase in the price to the customer. A task that is a major inconvenience or cost for the customer can be performed inexpensively and efficiently by the supplier. The supplier might be willing to do this task for "free" to secure ongoing business with the customer.

44. **Optimize level of inspection.** What level of inspection is appropriate for a process? All products eventually undergo some type of inspection, possibly by the user. Options for inspection at any given point in the supply chain are no inspection, 100 percent inspection, or reduction or increases to the current level of inspection. A study of the level of inspection can potentially lead to changes that increase quality of outcomes to the customers and/or decrease costs. Identifying the appropriate level of inspection for a process is discussed in Chapter 13.

45. **Work with suppliers.** Inputs to a process sometimes control the costs and quality of performance of a process. Working with suppliers to use their technical knowledge can often reduce the cost of using their products or services. Suppliers may even have ideas on how to make changes in a company's process that will surprise its customers.

Manage Time

Cut cycle time as a strategy for improving any organization. An organization can gain a competitive advantage by reducing the time to develop new products, waiting times for services, lead times for orders and deliveries, and cycle times for all functions and processes.

46. **Reduce setup or startup time.** Setup times can often be cut in half just by getting organized for the setup. Minimizing setup or startup time allows the organization to maintain lower levels of inventory and get more productivity out of its assets.

47. **Set up timing to use discounts.** The planning and timing of many activities can be coordinated to take advantage of savings and discounts that are available, resulting in a reduction of operating costs. An organization must have a system in place to take advantage of such opportunities. For example, available discounts on invoices offered by suppliers for paying bills within ten days of the invoice date require a system that can process an invoice and cut a check within the discount period. Opportunities to apply this concept require a flexible process and knowledge of the opportunity to take advantage of the timing.

48. **Optimize maintenance.** Time is lost and quality often deteriorates when production and service equipment break down. A preventive maintenance strategy attempts to keep people and machines in good condition instead of waiting until there is a breakdown. Through proper design and the study of historical data, an efficient maintenance program can be designed to keep equipment in production with a minimum of downtime for maintenance. Learning to observe and listen to equipment before it breaks down is also an important component of any plan to optimize maintenance.

49. **Extend specialists' time.** Organizations employ specialists who have specific skills or knowledge, but not all of their work duties utilize these skills or knowledge. Try to

remove assignments and job requirements that do not use the specialists' skills. Find ways to let specialists have a broader impact on the organization, especially when the specialist is a constraint to throughput in the organization.

50. **Reduce wait time**. Reduction in wait time can lead to improvements in many types of services. Ideas for change that can reduce the time that customers have to wait are especially useful. This refers not only to the time to perform a service for the customer, but the time it takes the customer to use or maintain a product.

Manage Variation

Many quality and cost problems in a process or product are due to variation. Reduction of variation improves the predictability of outcomes and may actually exceed customer expectations and help to reduce the frequency of poor results. Many procedures and activities are designed to deal with variation in systems. Consideration of Shewhart's concept of common and special causes opens up opportunities to improve these procedures. By focusing on variation, some ideas for changes can be developed. Three basic approaches can be taken: Reduce the variation, compensate for the variation, or exploit the variation.

51. **Standardization (create a formal process)**. The use of standards, or standardization, has a negative and bureaucratic connotation to many people. However, an appropriate amount of standardization can provide a foundation upon which improvement in quality and costs can be built. Standardization is one of the primary methods for reducing variation in a system. The use of standardization, or creating a more formal process, should be considered for the parts of a system that have big effects on the outcomes, or the leverage points.

52. **Stop tampering**. Tampering is defined as interfering so as to weaken or change for the worse. In many situations, changes are made on the basis of the last result observed or measured. Often these changes actually increase the variation in a process or product, as illustrated by the Funnel Experiment, discussed in Chapter 1, "You Don't Have to Suffer from the Sunday Night Blues!" The methods of statistical process control can be used to decide when it is appropriate to make changes based on recent results.

53. **Develop operational definitions**. Reduction of variation can begin with a common understanding of concepts commonly used in the transaction of business. The meaning of a concept is ultimately found in how that concept is applied. Simple concepts such as on time, clean, noisy, and secure, need operational definitions to reduce variation in communications and measurement.

54. **Improve predictions**. Plans, forecasts, and budgets are based on predictions. In many situations, predictions are built from the ground up each time a prediction is required, and historical data is not used. The study of variation from past predictions can lead to alternative ways to improve the predictions.

55. **Develop contingency plans**. Variation in everyday life often creates problems. Reducing the variation may eventually eliminate the problems, but how do people cope in the meantime? One way is to prepare backup plans, or contingencies, to deal with the

unexpected problems. When the variation is due to a special cause that can be identified, contingency plans can be ready when these special causes of variation occur.

56. **Sort product into grades**. Creative ways can be developed to take advantage of naturally occurring variation in products. Ways of sorting the product or service into different grades can be designed to minimize the variation within a grade and maximize the variation between grades. The different grades can then be marketed to different customer needs.

57. **Desensitize**. It is impossible to control some types of variation: between students in a class, among the ways customers try to use a product, in the physical condition of patients who enter the hospital. How can the impact on the outcome (education, function, and health) be minimized when this variation is present? It can be done by desensitizing or causing a nonreaction to some stimulus. This change concept focuses on desensitizing the effect of variation rather than reducing the incidence of variation.

58. **Exploit variation**. It is sometimes not clear how variation can be reduced or eliminated. Rather than just accepting or "dealing with" the variation, ways can be developed to exploit it. This change concept deals with some ways to turn the negative variation into a positive method to differentiate products or services.

Design Systems to Avoid Mistakes

Making mistakes is part of being human; they occur because of the interaction of people with a system. Some systems are more prone to mistakes than others. Mistakes can be reduced by redesigning the system to make their occurrence less likely. This type of system design or redesign is called *mistake proofing* or *error proofing*. Mistake proofing can be achieved by using technology, such as adding equipment to automate repetitive tasks, by using methods to make it more difficult to do something wrong, or by integrating these methods with technology. Methods for mistake proofing are not directed at changing people's behavior, but rather at changing the system to prevent slips. They aim to reduce mistakes from actions that are done almost subconsciously when performing a process or using a product.

59. **Use reminders**. Many mistakes are made as a result of forgetting to do something. Reminders can come in many different forms: a written notice or a phone call, a checklist of things to accomplish, an alarm on a clock, a standard form, or the documented steps to follow for a process.

60. **Use differentiation**. Mistakes can occur when people are dealing with things that look nearly the same. A person may copy a wrong number or grab a wrong part because of similarity or close proximity to other numbers or parts. Mistakes can also occur when actions are similar. A person may end up in the wrong place or use a piece of equipment in the wrong way because the directions or procedures are similar to others they might have used in a different situation. Familiarity that results from experience can sometimes increase the chance of committing mistakes of association. To reduce mistakes, steps should be taken to break patterns. This can be done by, for example, color coding, sizing, using different symbols, or separating similar items.

61. **Use constraints**. A constraint restricts the performance of certain actions. A door that blocks passage into an unsafe area is a constraint. Constraints are an important method for mistake proofing because they can limit the actions that result in mistakes. They do not just make information available in the external world, they also make it available within the product or system itself. To be effective, constraints should be visible and easy to understand. Constraints can be built into a process so that accidental stopping or an unwanted action that will result in a mistake can be prevented. Constraints can also be used to make sure that the steps performed in a process or when using a product are accomplished in the correct sequence.

62. **Use affordances**. An affordance provides insight, without the need for explanation, into how something should be used. In contrast to a constraint, which limits the actions possible, an affordance provides visual (or other sensory) prompting for the actions that should be performed. Once a person sees the fixtures on a door, he should be able to determine whether it opens in, opens out, or slides. There should not be a need to refer to labels or to use a trial-and-error approach. If a process or product can be designed to lead the user to perform the correct actions, fewer mistakes will occur.

Focus on the Product or Service

Most of the change concepts in the other categories address the way that a process is performed; however, many of the concepts also apply to improvements to a product or service. This category comprises eight change concepts that are particularly useful for developing changes to a product or service that does not naturally fit into any of the other groupings.

63. **Mass customize**. Most consumers of products and services would agree that quality increases as the product or service is customized to the customer's unique circumstances. Most consumers would also expect to pay more or wait longer for these customized offerings than for a mass-produced version. To mass customize means combining the uniqueness of customized products with the efficiency of mass production.

64. **Offer the product or service anytime**. Many products and services are available only at certain times. Such constraints almost always detract from their quality. How can these constraints be removed? In some cases a technology breakthrough, such as the ATM, is needed. In other cases, prediction plays an important role—for example, predicting what type of cars customers will order. However, in many situations the constraint is created because it is more convenient for the provider of the service than for the customer. Offering the product or service anytime is different from just reducing wait time. To achieve this goal often takes a totally new conceptualization of the product or service. For this reason, "anytime" is an important concept for expanding the expectations of customers.

65. **Offer the product or service anyplace**. An important dimension of quality for most products and services is convenience. To make a product or service more convenient, free it from the constraints of space. Make it available anyplace. For products, the constraint of space is often related to the size of the product. Making a product smaller or

lighter without adversely affecting any of its other attributes almost always improves the quality of the product. One of the most striking examples is the miniaturization of the computer to the point that it can now be carried in a briefcase and used virtually anywhere.

66. **Emphasize intangibles**. Opportunities for improvement can be found by embellishing a product or service with intangible features. Three ways to accomplish this are by miniaturizing, providing information (electronically or otherwise), and developing producer-customer relationships.

67. **Influence or take advantage of fashion trends**. The features and uniformity of a product or service define its quality. Uniformity is often assumed to exist in a product or service, while features can affect customer expectations. Features are frequently subject to fashion trends.

68. **Reduce the number of components**. Reducing handoffs was one of the change concepts for simplifying a process. Similarly, reducing the number of component parts is a way to simplify a product. Components in this context can mean component parts, ingredients, or multiple types of the same component. Reduction in the number of components can be achieved through design of the product so that one component performs the functions previously performed by more than one; or by standardizing the size, shape, or brand of similar components; or by packaging components into modules.

69. **Disguise defects and problems**. In some instances, especially in the short term, it may be more effective to hide the defect in a product or service than to remove it. However, the longer-term strategy is to remove the defect. Included in this category are actions taken to make the defect more palatable to the customer. This change concept does not include false advertising, in which claims about the product are made that are not true, nor does it include defects that are hidden at the time of sale only to emerge in later use of the product.

70. **Differentiate product using quality dimensions**. Customer satisfaction is improved as the match between process output and customer needs/wants is increased. The degree of matching is determined using customer research. Customer research can provide an understanding of customer's needs and wants.

How to Do Communication Plans

Communication plans are created by thinking about what human factors need to be addressed in your organization as relates to a process improvement project or initiative. Some possible questions to consider are

- Does everyone in the organization understand process improvement in general?
- Who do we need to communicate to?
- Who is committed (not just supportive) to the project from top management?

- What does the project mean to top management?
- What does the project mean for individual employees?
- What does the project do for the organization?
- What is expected from various stakeholders with respect to the project?
- What are the potential benefits and risks of the project; do we need to communicate them?
- What is the timeline for the project?
- What resources will be required for the project; especially human resources?
- How will people be affected by the process improvement project, and how can we alleviate their potential concerns?

Once you have answered these questions you need to plan different events/communications using channels most appropriate for the selected audience by filling in the columns on the communication plan presented in Table 6.19.

- **Event/communication**—What do you need to communicate?
- **Participants**—Who does the event/communication need to reach? Are there opinion leaders (those who have an active voice in a community and whose advice is sought by others) that we want to be involved to help facilitate project buy-in?
- **Medium**—How are you going to reach them? Possible ways are meetings, website, email, posters, information sessions, conference calls, private discussions, and so on.
- **Frequency**—How often will this event/communication take place?
- **When**—What is the specific date(s) and time it will happen?
- **Lead**—Who is taking the lead on the communication?
- **Scheduled**—Who is doing the scheduling and are they doing it on time?
- **Status**—TBD (to be done), Done (completed), WIP (work in progress)
- **Notes**—Are there any miscellaneous notes?

Table 6.19 Communication Plan Outline

#	Event/ Communication	Participants/ Audience	Medium	Frequency	When	Lead	Scheduled?	Status	Notes
1									
2									
3									
4									

#	Event/ Communication	Participants/ Audience	Medium	Frequency	When	Lead	Scheduled?	Status	Notes
5									
6									

See Table 6.7 for an example of a communication plan for reducing no shows in a clinic project.

Takeaways from This Chapter

- A flowchart is a tool used to define and document a process.
- Voice of the Customer analysis involves surveying stakeholders of a process to understand their requirements and needs.
- A SIPOC analysis is a simple tool for identifying the Suppliers and their Inputs into a process, the high level steps of a Process, the Outputs of the process, and the Customer segments interested in the outputs.
- An operational definition promotes understanding between people by putting communicable meaning into words.
- Measurement systems analysis studies are used to calculate the capability of a measurement system to determine whether the data can be used for meaningful analysis of a problematic process.
- Failure Modes and Effects Analysis (FMEA) is a tool used to identify, estimate, prioritize, and reduce risk among the Xs for a project.
- Check sheets are used for collecting or gathering data in a logical format, called rational subgrouping. The data collected can be used to construct a control chart, a Pareto diagram, or a histogram.
- Brainstorming is a process used to elicit a large number of ideas from a group of people in a brief amount of time. Team members use their collaborative thinking power to generate unlimited ideas and thoughts.
- An affinity diagram is a tool used by teams to organize and consolidate a substantial and unorganized amount of verbal, pictorial, and/or audio data relating to a problem.
- A cause and effect (C&E) diagram, also known as a fishbone diagram as its structure resembles the spine and head of a fish, is a tool that is used to organize the potential causes of a problem, select the most probable cause, and verify the cause and effect relationship between the problem and the most probable cause.
- Pareto diagrams are used to identify and prioritize issues that contribute to a problem we want to solve. Pareto analysis focuses on distinguishing the "vital few versus the trivial many."

- A Gantt chart is a simple scheduling tool. It is a bar chart that plots tasks and subtasks against time.

- Change concepts are approaches to change that have been found to be useful in developing solutions that lead to improvement in processes.

- A communication plan is created for a project to identify and appease human concerns regarding the project as many people will typically be affected by it.

References

Gitlow, H., A. Oppenheim, R. Oppenheim, and D. Levine (2015), *Quality Management*, 4th ed. (Naperville, IL: Hercher Publishing Company). This book is free online at hercherpublishing.com.

Gitlow, H. and D. Levine (2004), *Six Sigma for Green Belts and Champions: Foundations, DMAIC, Tools and Methods, Cases and Certification* (Upper Saddle River, NJ: Prentice-Hall).

Langley J., K. Nolan, T. Nolan, C. Norman, and L. Provost (1996), *The Improvement Guide* (San Francisco: Jossey-Bass).

Additional Readings

Pyzdek, Thomas (2003), *Quality Engineering Handbook*, 2nd ed. (New York: Marcel Dekker).

Rath and Strong (2000), *Six Sigma Pocket Guide* (Lexington, MA: Rath and Strong).

7

Overview of Process Improvement Methodologies

What Is the Objective of This Chapter?

The objective of this chapter is to introduce the different methods available to solve process improvement problems. We give you a high level overview of the methods and some simple examples. Later in the book we go into more detail about the methods, as well as when to use them and how to use them.

SDSA Cycle

The SDSA cycle is a series of four steps used to standardize a process and reduce variation in process outcomes (Gitlow et al., 2015; Gitlow and Levine, 2004). After forming a project team, determining a process to standardize, and agreeing on a way to measure success, the following steps are executed:

- **Standardize**—Each employee flowcharts the process under study, usually their collective job. They are then brought together and agree upon a best practice flowchart that takes the best aspect of their individual flowcharts and eliminates the worst aspects of their individual flowcharts. Basically they move from their individual current state flowcharts to a new and hopefully improved standardized future state flowchart that all employees using the flowchart agree upon and follow. Standardization can dramatically reduce variability in the outcomes of a process. It is also a great test of a workforce's ability to exhibit discipline in the way they do their jobs. Personal discipline to follow the standardized flowchart, or an improved flowchart, is *critical* to a successful Six Sigma organizational transformation.

- **Do**—Each employee uses the new, standardized flowchart, and they collectively gather data of their efforts so the results can be analyzed in the Study step.

- **Study**—Employees analyze the results of the experiment or test on the new standardized flowchart to see whether it has the intended outcome.

- **Act**—Employees decide whether to adopt the new standardized flowchart, and if the answer is yes, they lock it in with documentation and training, or modify the plan and repeat the SDSA cycle. If the answer is no, the team goes back to the beginning and tries again.

SDSA Example

A medical records department in a hospital must file 80% of all patient records within 30 days of the patient leaving the hospital; it is the law. Currently, the percentage of medical records filed within 30 days has an average of 22% and a standard deviation of 2%.

- **Standardize**—First, the three employees who work in the medical records department flowchart the medical records process as they see it. Second, they compare their flowcharts; see the flowcharts for employees A, B, and C in Figure 7.1. Third, they develop one best practice flowchart that includes all their individual flowcharts' strengths and avoids all of their individual flowcharts' weaknesses; see the flowchart called Standardized Flowchart in Figure 7.1.

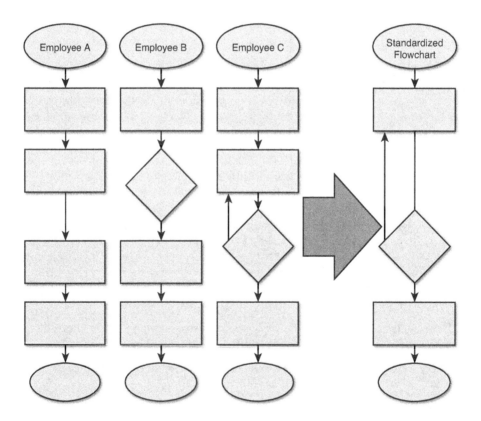

Figure 7.1 Employee flowcharts and standardized flowchart for medical records department

- **Do**—All employees involved in the process perform a pilot test using the new best practice flowchart, called the Standardized Flowchart in Figure 7.1.

- **Study**—The employees involved in the process compare the percentage of medical records filed within 30 days after the pilot test with the data from before the pilot test. We can see from the Current panel and the Standardize panel in the chart in Figure 7.2 that it has improved from 22% to about 45% through standardization!

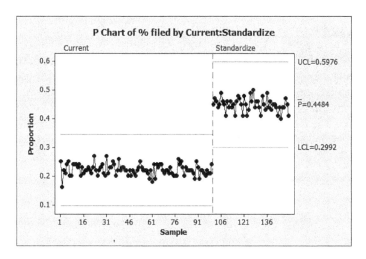

Figure 7.2 Current and standardized panel showing improvement

- **Act**—Finally, employees make the combined flowchart the best practice method that they all follow through documentation and training.

Anecdote about the SDSA Cycle

One of the authors always forgot which teeth he had flossed and consequently had to start over. It was very frustrating. So, he developed a standardized best practice method. He has four bridges in in his mouth all in the same place: the last two back teeth on the top and bottom on both sides of his mouth. So, he standardized how he flossed.

Standardize: First the upper left bridge, next the lower left bridge, then the lower right bridge, then the upper right bridge, then the upper teeth from left to right, and finally, the lower teeth from left to right.

Do: He followed the standardized method for one month.

Study: He studied the results and realized he had never gotten mixed up!

Act: He locked in the standardized method and now uses it every day.

PDSA Cycle

The PDSA cycle, also called the Deming cycle or the Shewhart cycle, consists of four steps used to continually improve a process by reducing variation, centering the mean on the desired nominal level, or eliminating waste from the process (Gitlow et al., 2015; Gitlow and Levine, 2004; Deming, 1994). After forming a project team, determining an aim for the team, and agreeing on a way to measure success (a metric), the following steps are executed:

- **Plan**—Team members figure out what changes can be made to the current best practice process (see standardized flowchart on the left of the Figure 7.3) to achieve improved results from the revised best practice method for the metric of interest (see revised practice flowchart on the right of Figure 7.3)—for example, deleting an unnecessary step in the process as shown in Figure 7.3. Basically, team members are moving from the current state flowchart to a new and hopefully improved future state flowchart by changing the flow of the process. In this book we discuss many methods for creating the revised best practice method. These methods are ideas on how to improve a process that are noted in the revised best practice flowchart.

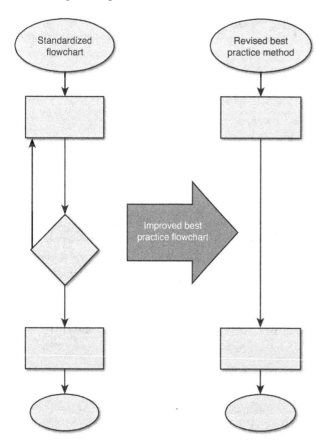

Figure 7.3 Standardized and revised best practice flowcharts

- **Do**—Next, team members conduct an experiment or test of the revised flowchart (Plan) on a small scale and collect data so the results can be analyzed in the Study step.
- **Study**—Team members, perhaps with the assistance of an expert, analyze the results of the experiment or determine whether it has the intended outcome. Note: Originally this step was known as "Check," but Dr. Deming changed it to "Study" as he felt check meant to stop someone in ice hockey.
- **Act**—Team members decide whether to (1) adopt the revised best practice flowchart and lock it in with documentation and training, (2) abandon the revised best practice method because it failed to yield the desired results, or (3) modify the revised best practice method and repeat the PDSA cycle.

These four steps are repeated over and over again in the quest for never ending improvement.

PDSA Example

Using the SDSA cycle we were able to improve the percentage of medical records filed within 30 days from 22% to about 45% by standardizing the process among employees, which is good, but not enough to get to the 80% required by law. The next course of action is to turn the PDSA cycle:

- **Plan**—After various conversations with process stakeholders, team members find that records require three signatures from the lab before being released from the lab for filing by the medical records department. Upon further study, it is determined that two of those signatures are unnecessary and just increase the cycle time it takes to file the medical records.
- **Do**—The lab agrees to eliminate the two signatures. Then team members run a pilot test on the new process without the two signatures from the lab.
- **Study**—Next, team members compare the percentage of medical records filed within 30 days after the new pilot test (panel 3 in Figure 7.4, called After Lab) with before the pilot test (panel 2 in Figure 7.4, called Standardize). We can see from Figure 7.4 that it has improved from 45% to about 65% through the elimination of the two signatures in the lab!
- **Act**—Finally, team members lock in and sustain the new process through documentation and training. The PDSA cycle continues forever, although perhaps with less data collection, to constantly improve the process absent capital investment. Once capital investment is required, additional financial considerations come into play. These financial considerations are not discussed in this book. However, many problems that are attempted to be fixed through capital investment can frequently be fixed using the PDSA cycle without any capital investment.

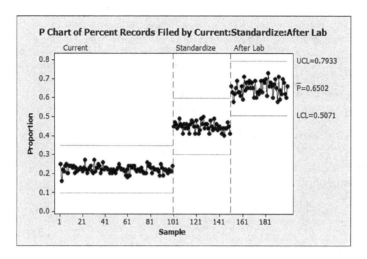

Figure 7.4 After Lab

> **NOTE**
>
> The team was about to run the PDSA cycle again to move from 65% to 80% of medical records filed within 30 days when the CEO who was also the Project Champion was fired, c'est la vie!

> **Anecdote about the PDSA Cycle**
>
> I was out to a Tai dinner with a friend who was 70 years old. He always ordered Spring Break Duck as his entrée. We have been going to this restaurant at least twice a week for years. One day I noticed that he always leaves the onions over on his plate. I said to myself, "Is it possible he doesn't know he can order the duck without onions?" It seemed to stretch the boundaries of credibility that he could be 70 and be so unaware of something as common as this. So I asked him, why don't you order the duck without onions. And my 70-year-old friend said to me: "Can I do that?" I was amazed that in 70 years he hadn't realized that he could change items on a restaurant menu. So this was an opportunity for a turn of the PDSA cycle.
>
> - **Plan**—The next time he eats Spring Break Duck, he will order it without onions and with extra pineapple.
> - **Do**—The next time he orders Spring Break Duck without onions and with extra pineapple.
> - **Study**—The waiter and the restaurant owner didn't kick him out of the restaurant for being obnoxious and overly demanding.

> - **Act**—He realized that he could order his food the ways he likes it, and that the restaurant personnel are happy to oblige him, because then he will be more likely to return to the restaurant.
>
> This story is amazing and true. PDSA can really improve our lives!

Kaizen/Rapid Improvement Events

Kaizen is Japanese for "improvement" or "change for the best" (Wikipedia: Kaizen); it refers to a philosophy or practices that focus upon continuous improvement of processes. Kaizen refers to activities that continually improve all functions and involves all employees. Kaizen aims to eliminate waste by improving standardized processes.

Kaizen is a daily process, the purpose of which goes beyond simple productivity improvement. It is also a process that, when done correctly, humanizes the workplace, eliminates overly hard work (called "*muri*" in Japanese), and teaches people how to perform experiments on their work using the PDSA cycle and how to learn to spot and eliminate waste in processes. In all, kaizen provides a humanized approach for workers to increase productivity; in other words, kaizen nurtures employees and promotes participation in process improvement activities.

Kaizen works as follows: The PDSA cycle is turned, and at the Do stage employees make small and rapid modifications to the Plan stage and determine their success in the Study phase of the PDSA cycle. The small and rapid changes to the Plan find any problems with the Plan, display the problems, clear the problems that stand in the way of the Plan, and acknowledge the correctness of the modification to the Plan. Next, employees Study the results of the effort by viewing the critical metric (CTQ or Y), and finally, Act. Kaizen or Rapid Improvement Events are typically conducted on processes where the root cause is known but the solution to eliminate the root cause is not.

Kaizen or Rapid Improvement Events generally consist of three stages that correspond to the PDSA cycle. An example of a typical Kaizen or Rapid Improvement Event (RIE) looks like this:

Pre-event (Plan)

- Identify the area to be *kaizened*.
- Identify team members.
- Plan the event (secure meeting space, reserve equipment, obtain supplies, order meals, prepare training, create certificates, and so on).
- Conduct Voice of the Customer (VoC) analysis and perform process observations.
- Collect baseline data.
- Create Kaizen/Rapid Improvement Event (RIE) charter.

- Obtain management commitment.
- Select the Project Champion.
- Prepare a schedule of meetings with the Project Champion in which he can hear the results of the kaizen blitz.

Event (Plan, Do, and Study)

Three to five day events consisting of

- Event kickoff.
- Conduct basic RIE training.
- Review the charter with all relevant personnel.
- Review the data.
- Map out current state process.
- Observe the current state process.
- Do root cause analysis on current state process.
- Brainstorm/prioritize/select improvements to the current state process.
- Develop an implementation plan for the future state process.
- Train staff on new future state process.
- Create new tools for the future state process.
- Implement future state process.
- Prepare and deliver final report to the Process Owner and the Project Champion.

Post-event (Act)

- Follow up on outstanding action items.
- Celebrate success.
- Track progress over time, perhaps on the organization's dashboard.
- Update the Project Champion.
- Review data periodically to make sure the process has not fallen back into its old bad habits.

Kaizen/Rapid Improvement Events Example

First case starts (FCS) in an operating room (OR) are defined as the percentage of first cases of the day in each OR that start on time. Start time is measured by when the patient is wheeled into the OR (wheels in). So any case that has "a wheels in time" greater than the scheduled time is defined as late.

XYZ hospital has a problem with first case starts in its 16 ORs; only 55% of cases start on time. Due to the high cost of OR time (reported to be in the neighborhood of $1,800/hour) and the opportunity cost of cases not scheduled (if cases are delayed long enough it means that other margin generating cases cannot be done), the CEO asks the process improvement team to get involved to improve FCS.

Since many of the root causes are thought to be known (physician late, patients not cleared), but no solution has been put in place, the team decides to conduct a Kaizen/Rapid Improvement Event.

Pre-Event

The Process Improvement Executive names a Project Leader, who forms the team, which consists of employees from related areas including operating room staff, nursing, surgeons, surgical services, anesthesiology, and central sterile staff. Over the next month the team completes all the pre-event work, such as completing the project charter, planning the event, conducing Voice of the Customer interviews, and collecting baseline data. At this point the team is ready for the event.

Event

To limit disruption to the OR, the team schedules the event for a Friday, Saturday, and Sunday. As an aside, each participant is given two vacation days to be used at a later time determined by management.

Over the next three days, the team maps out the current state process, creates a fishbone diagram to identify (confirm) potential root causes, studies data and identifies the main reasons for the delays, brainstorms potential solutions, maps out the new future state process, and creates an implementation plan.

They find that FCS are delayed for the following three reasons:

- Surgeons are late.
- Patients are not cleared properly.
- Missing items.

The solutions they come up with to address the preceding issues are the following:

- **Surgeons are late**—The surgeon on the team worked with his colleagues to come up with a plan. Since first case starts are coveted by surgeons, any surgeon who is late for two or more first cases may lose his first case start privileges for the next month. Also, surgeons are required to be in the hospital a minimum of 30 minutes prior to their scheduled start.

- **Patients not being cleared properly**—Each patient is required to be scheduled for pre-admission testing and clearance, and anesthesia is to be notified when the patient is cleared.
- **Missing items**—A standard checklist of items to be kept in each OR was created, and a member of the OR staff makes sure each item is in each room the night before.

Post-Event

Since the ORs were closed on the final day of the event, the team had to wait until the next week when the ORs opened back up to implement changes.

All staff is trained on the new process and despite the usual resistance from some employees, the changes are implemented. Minor changes are made along the way, but the new plan seemed to be working well.

The team collects a month's worth of data and presents it to the CEO, who is anxious to see how the team did. To her amazement, the team had gone from 55% of FCS on time to 89% of FCS on time.

She thanks the team and throws a party for everyone involved to celebrate!

All new processes are locked into place with documentation, and the team creates a plan to track the data to ensure that the improvements are sustained.

Kaizen and Rapid Improvement Events are not discussed further in this book. If you are interested in this topic, many excellent resources are available on the Web.

DMAIC Model: Overview

Based on the scientific method, the DMAIC model is like the PDSA cycle in that it is used to improve a current process, product, or service (Gitlow et al., 2015; Gitlow and Levine, 2004). You can think of the DMAIC model like the PDSA cycle on steroids in that it is much more detailed in the number of substeps within each of its five phases.

The easiest way to explain the DMAIC model is to think of it in terms of the CTQ (Y) is a function of one or more Xs. For example, suppose that it is believed that patient waiting time is affected by insurance type, physician, and availability of an examination room. This could be stated as follows:

> Patient waiting time is a function of insurance type, physician, and availability of an examination room.
>
> If we call Patient waiting time Y, insurance type (X_1), physician (X_2), and availability of an examination room (X_3), then we can write the relationship between the Y and the Xs as follows:
>
> $$Y = f(X_1, X_2, X_3).$$

where:

- Y is a measure of patient waiting time.
- f is the symbol of relationship.
- X_1 is insurance type.
- X_2 is physician.
- X_3 is availability of an examination room.

The five phases of the DMAIC model are *Define, Measure, Analyze, Improve,* and *Control.* They are aimed at finding revised methods that modify how the Xs are performed with the intention of moving the CTQ (Y) in a specified direction, reducing variation in the output (Y), or both.

Define Phase

The Define phase involves four steps; they are

1. Prepare a project charter. This is the rationale for doing the project.

2. Understand the relationships between Suppliers-Inputs-System-Outputs-Customers (called SIPOC analysis). SIPOC analysis allows team members to get an initial understanding of all the stakeholders of the project.

3. Analyze Voice of the Customer (VoC) data to identify the critical to quality (CTQs or Ys) important to customers. A VoC analysis provides data from each stakeholder group directly to team members about the needs and wants of each stakeholder group in the language of the stakeholders.

4. Develop an initial project objective. A project objective states the following: the problematic process to be studied, a metric for measuring the CTQ of the process, a direction for the CTQ (increase or decrease), a target for the metric that is rational not arbitrary, and a deadline for the completion of the project.

Measure Phase

The Measure phase involves five steps; they are

1. Create an operational definition for each CTQ. Operational definitions give meaning to terms so that all concerned parties agree on the definition of the metric being studied in the DMAIC model.

2. Develop a data collection plan for the CTQ.

3. Perform a measurement system that determines if there is accurate data about each CTQ.

4. Collect and establish a baseline for each CTQ.

5. Estimate the process capability for each CTQ.

Once all of this is done, you can collect data to determine the performance of the process. In other words, you can determine the current state performance of your CTQ or Y. This allows you to compare the current process performance, as measured by the CTQ, with the performance of the improved process.

Analyze Phase

The Analyze phase involves the following steps:

1. Create a detailed flowchart of the current state process. This allows all concerned parties to understand exactly how the process works in its current form.

2. Identify the Xs, which are factors that cause each of your CTQs to be problematic. Xs can be identified in many ways, such as analyzing your current state flowchart, talking to process experts, surveying the literature, analyzing data, using the list of 70 change concepts, and benchmarking, to name a few sources of Xs.

3. Reduce the number of potential Xs that can impact the CTQ by using a Failure Modes and Effect Analysis (FMEA). FMEA is a tool used to identify the Xs that are critical to reducing the variation of the CTQ, moving the CTQ toward the desired nominal level, or eliminating waste to get better results for the CTQ.

4. Develop an operational definition for each X (see Measure phase).

5. Develop a data collection plan for the Xs (see Measure phase).

6. Create a measurement system for each of the Xs (if necessary; see Measure phase).

7. Collect data and testing theories to determine which of the Xs cause your CTQs to be problematic.

8. Develop hypotheses/takeaways about the relationships between the critical Xs and the CTQ(s).

Testing of theories between CTQs and Xs involves determining how and why each X causes the CTQ to be problematic. For example, you can develop hypotheses like the following (if you have 5 Xs):

$$Y = f(X_1), \text{ or } Y = f(X_2, X_3, X_5)$$

A test would look like this:

If cycle time for a nurse to answer a call bell in a hospital is the CTQ (or Y), and X_1 is the nurse, then we can see from Figure 7.5 that nurse A takes longer to answer a call bell than nurse B. In this hospital, X1 is a potentially important X for improving the CTQ.

Suppose that the cycle time to answer a call bell in a second hospital is recorded for Nurse X and Nurse Y for 100 consecutive patients. From the distributions of call bell response times in Figure 7.6, we can easily see that the nurses have the same call bell response time distribution.

Consequently, we develop a hypothesis that "X_{Nurse}" is NOT a potentially important X to explain the behavior of call bell response time (CTQ or Y) in the second hospital, unless it is correlated to another X that is significant and interacts with X_{Nurse}.

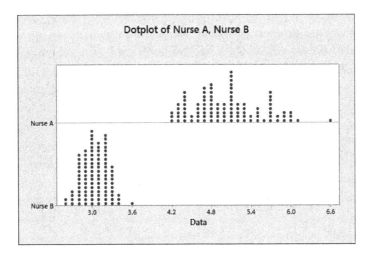

Figure 7.5 Dot plot of Nurse A versus Nurse B

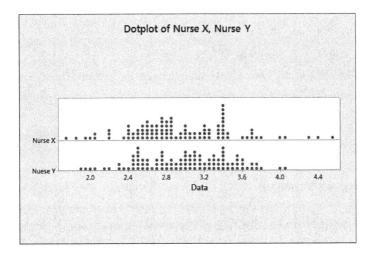

Figure 7.6 Dot plot of Nurse X versus Nurse Y

Improve Phase

The Improve phase involves the following steps:

1. Generate potential solutions to optimize the mean, variation, and shape of the distribution of the CTQ. This can be done many ways; for example, experimental designs, FMEA, brainstorming, using the 70 change concepts (discussed previously in this book), simulation, or benchmarking. (**Note:** We explain each of these methods when we get into the Improve phase in more detail later in the book.)
2. Select solutions.
3. Create a flowchart for the new optimized process.
4. Identify and mitigate risk for the new process.
5. Run a pilot test of the new improved process.
6. Collect and analyze data from the pilot test.
7. Implement the process full scale across the organization.

Control Phase

The Control phase involves the following steps:

1. Reduce the effects of collateral damage to related processes.
2. Standardize improvements.
3. Develop a control plan for the Process Owner.
4. Identify and document the benefits and costs of the project.
5. Input the project into the Six Sigma database.
6. Diffuse the improvements throughout the organization.
7. Champion, Process Owner, and Black Belt sign off on the project.

DMAIC Model Example

ABC Health System has a problem with patients not showing up for their scheduled appointments in one of its behavioral health clinics. This causes idle time for the physicians and staff and represents lost revenue for the system. The process improvement team undertakes a Six Sigma DMAIC project to solve the problem.

Define

In the Define phase the team completes their business case including financial impact to the department, speaks to different process stakeholders by doing a VoC analysis, and identifies their process Y (CTQ); that is, "percentage of no shows by month." Their final project objective is to reduce patient no shows in the behavioral health clinic from 30% per month to as low as possible per month by January 1, 2015.

Measure

In the Measure phase the team collects data on the CTQ (Y) to get a baseline of current performance and finds that 30% of patients do not show up to their appointments on average in the largest behavioral health clinic.

Analyze

In the Analyze phase the team uses a variety of methods to identify the Xs or the causes for patients not showing up to their appointments and finds that there are three main factors responsible for the patient no shows:

- Age of the patient and time of appointment
- No reminder call
- Lack of penalty for no shows

Specifically the team finds that younger patients no show more often if their appointments are early in the morning. They also find that patients who receive an appointment reminder call have a higher likelihood of showing up for their appointment. Finally, during the VoC interviews the team finds that because there is no penalty for not showing up, the patients had a laissez faire attitude about keeping their appointments.

Improve

The team modifies the process as follows to address the critical Xs found in the Analyze phase:

- Schedule older patients earlier in the day and younger patients later in the day.
- Have reminder calls made to the patients two days prior to their appointment.
- Institute a $50 penalty for patients who do not cancel at least 48 hours in advance.

After conducting a pilot test with these three process changes the team saw no shows drop to an average of 5% over the next three months.

Control

To sustain their improvements the team locks in the process changes by documenting the newly improved process and trains all employees in the new process. They then celebrate their success and hand the process over to the Process Owner, which includes reviewing no show data monthly to ensure the new process is still being followed. Additionally, the Process Owner continues to turn the PDSA cycle to drive the percentage of no shows even lower than the 5% mark.

DMADV Model: Overview

Sometimes you hit a wall with a current process in terms of making further improvements. If this happens you need to redesign the process, or sometimes you have no process at all so you need to create one. In both of these cases, you use what is called the DMADV model (Gitlow, et al., 2006). By using the DMADV model you are designing the process, product, or service with customer requirements in mind at every step to ensure a high quality outcome. The DMAIC model is about improving a process, product, or service, while the DMADV model is about inventing and/or innovating a process, product, or service.

The *DMADV model* is the *Design for Six Sigma (DFSS)* model used to create major new features of existing products, services, or processes, or to create entirely new products, services, or processes. It has five phases: Define, Measure, Analyze, Design, and Verify/Validate.

Define Phase

The Define phase of the DMADV model has five components; they are

- Establishing the background and business case for the project.
- Assessing the risks and benefits of the project.
- Forming the Six Sigma DMADV team.
- Developing the project plan.
- Writing the project objective.

Measure Phase

The Measure phase of a Design for Six Sigma project has three steps; they are

1. Segmenting the market, designing and conducting Voice of the Stakeholder surveys.
2. Using the survey results as inputs to find Critical to Quality characteristics (CTQs or Ys).
3. Creating a matrix to understand the relationships between the needs and wants of stakeholders and the features of the product, service, or process design.

Analyze Phase

The Analyze phase contains four steps; they are

1. Generate several possible designs at a high level; no details are developed in this step.
2. Compare all the possible designs and select the best one based on relevant criteria.
3. Perform a risk analysis on the best high level design.
4. Create a high level model of the best design.

The aim of these four steps in the Analyze phase is to develop preliminary designs from which a best design can be identified based on relevant criteria.

Design Phase

The Design phase of a Design for Six Sigma project has four steps; they are

1. Construct a detailed design of the "best" design from the Analyze phase.
2. Develop and estimate the capabilities of the critical features of the best design. These features are called Critical to Process (CTP) elements. CTPs are the components that make up the detailed design. They include detailed drawings of each component, as well as nominal values and specification limits.
3. Do a trial run of the new design.
4. Prepare a verification plan to enable a smooth transition among all affected departments. A verification plan ensures that the final design works well for all relevant stakeholders.

Verify/Validate Phase

The intent of the Verify/Validate phase is to

- Facilitate buy-in of Process Owners.
- Design a control and transition plan so that the design can hit the ground running.
- Conclude the DMADV project and celebrate the success.
- Turn the new product, service, or process over to the Process Owner.

The final part of the Verify phase is to maintain communication between the Champion and the Process Owner. These lines of communication alleviate any confusion or other unforeseen problems that will inevitably develop. It ensures that the conceptual design is not compromised by outside forces and neglect. All phases of the DMADV model end with a tollgate review by the Process Owner and the Project Champion.

DMADV Model Example

ABC Hospital has seen a substantial increase in the number of transfer admissions coming into the hospital. Transfer admissions are admissions of patients coming from other hospitals that cannot provide the specialized services that ABC Hospital can. However, due to delays in the process and a lack of standardization, referring hospitals have started to send their patients elsewhere. Currently transfers are handled by nursing administration, but after looking at industry best practices, the C-suite of the hospital decides to create a Transfer Center within the hospital. The Transfer Center would have employees whose sole job is to ensure a standardized process to screen and admit transfers from other hospitals. To design the new

Transfer Center to meet customer expectations in every way, the CEO commissioned the process improvement team to use the DMADV model to make it happen.

Define

In the Define phase the team completes their business case including financial impact to the department, conducts a risk assessment on the project, and forms the team. Their final project objective is to create a Transfer Center that increases transfers to the hospital from 400 per year to as high as possible per month by January 1, 2015. At a contribution margin of $3,000 per transfer, the financial impact is significant.

Measure

In the Measure phase the team segments the market and conducts VoC interviews with various stakeholders, including physicians, nurses, hospital staff, insurance providers, and referring hospitals. The CTQs for the team are number of transfers and number of complaints from referring hospitals, patients, and ABC Hospital staff. The team also creates a matrix to understand the relationships between the needs and wants of stakeholders and the features of the product, service, or process design.

Analyze

In the Analyze phase the team uses a variety of methods to identify several low level design options; specifically the team looks at

- Various IT and data collection systems (four systems were looked at)
- Number of staff (using three, four, or five employees was discussed)
- Skillset of staff (having a registered nurse versus not having a registered nurse)
- Location of Transfer Center within the hospital (near the Emergency Department or near the Nursing Administration office)
- Pricing packages for international patients (three options were discussed)
- Hours (24/7 operation versus usual business hours)

The team then uses a matrix to select the best design, which includes

- Using an add-on from the vendor who is providing the new electronic medical record system.
- Three staff is deemed the proper amount to handle the projected increased volume.
- A registered nurse is critical to properly screen patients.
- Being close to nursing administration is important as that is where bed assignment occurs.
- Pricing their services similarly to their closest competitors.

- Providing service 24/7 as their competitors do not offer 24/7 service, and management decided that the incremental cost is worth the differentiation from the competition that it provides.

Design

In the Design phase, the team creates a detailed design for the best design from the Analyze phase. An implementation plan is then created to execute on the detailed design.

Verify

The Verify phase includes meeting with the Process Owners and Project Champion to review the control and transition plan and then celebrating the completion of the project!

Part of the Verify phase is creating and tracking the CTQs for the project and calculating the financial impact.

After a year, the number of transfers to the hospital increased from 400 per year to 1,150, an increase of 287%! At a contribution margin of $3,000 per patient, the financial impact was $2.25 million! The number of complaints was reduced to next to none.

We do not cover the DMADV model in this book as it requires a book of its own. We simply want to expose you to the methodology at a high level. For details of the DMADV model, see Gitlow, Levine, and Popovich (2006).

Lean Thinking: Overview

Lean thinking is a process orientation that focuses on reducing waste by eliminating non-value added steps in a process. Lean thinking has several tools and methods that we briefly discuss in this chapter; they are the 5S methods, Total Productive Maintenance (TPM), quick changeovers, poka-yoke, and value stream mapping. If you are interested in learning about lean thinking in more depth, see the references at the end of this chapter. It is important to realize that all the lean tools and methods briefly discussed here can be used as potential change concepts to manipulate the Xs to optimize the CTQs in the Six Sigma DMAIC model.

The 5S Methods

The first step in process improvement is frequently accomplished using the 5S methods (Gitlow, 2009). The 5S methods are simple techniques for highlighting and eliminating waste, inconsistency, and unreasonableness from the workplace. The 5S methods can be used to accomplish these tasks by creating a neat and tidy workplace; everything has a place and everything is in its place.

Additionally, the 5S methods and standardization (SDSA) can be used to determine whether the workforce in an organization has the personal discipline to maintain over time a neat and tidy standardized workplace. Think of your junk closet at home. One day you clean it and throw out a ton of garbage. Two months later it is back to where it was before your 5Sed the closet. No discipline! Discipline is difficult to develop in a workforce; it takes years to really hone a personal discipline culture.

Table 7.1 briefly describes each of the 5S methods, followed by some explanations.

Table 7.1 Description of the 5S Methods

5Ss	Description
Sort	Eliminate unnecessary things (or put them away) and make necessary things visible.
Systematize	Order essential things so that they can be quickly and easily accessed and put away.
Spic and Span	Clean machines, equipment, and the work environment.
Standardize	Develop "best practices" to make the preceding 3Ss habits.
Self-discipline	Spread and maintain the 5S culture throughout the organization. Get everyone to use the preceding 4Ss in work every day.

1. **Sort** means organizing things so they are easy to find. Sort divides things that are needed from things that are not needed, thus, creating a workplace that has only what is needed and not a lot of junk that makes finding what is needed a time consuming task.
2. **Systematize** means tidily placing "things" that are needed in their proper places so anyone can access or put them away.
3. **Spic and Span** is an attitude that considers a dirty and untidy workplace intolerable; it is proactive cleaning and maintenance. Make cleaning a cultural cornerstone of your organization. Spic and Span is analogous to personal hygiene for people.
4. **Standardize** is the development of an integrated system of "best practice" methods for sorting, systematizing, and spic and span. Standardize is used to make the first 3Ss a habit for each individual using it.
5. **Self-discipline** is about spreading and maintaining the 5S culture throughout an organization. Everyone should be involved in the 5S culture.

5S Methods Example

We demonstrate the 5S methods with an example that resonates with many of us, a messy closet! Many people's clothes closets at home are a bit of a disaster in that there is so much clutter in them that they end up spending more time finding something than they spend getting dressed in the morning. The 5Ss are a perfect way to Lean out that problematic closet.

Start with **Sort**. Go through your closet and pull out anything that you haven't worn in the past year and put it in a pile. Unless it is seasonal clothing, odds are if you haven't worn it in the past year it is simply taking up valuable closet space and making it hard for you to find what you need. Donate it to someone in need or give it to a friend and let it take up space in their closet! You might even want to remove seasonal items and put them somewhere else until that season approaches.

Your next task is then to **Systematize** and organize your closet so things are in logical places and are easy to find. Perhaps put your shirts on one side, your dress clothes on another, your jeans on another, and have a specific spot to hang your belts. If you really want to get sassy you can organize them by color as well.

Then you are going to **Spic and Span** or clean your closet. This can be accomplished by throwing away old or broken hangers, dusting and cleaning the area, vacuuming the floor, and throwing away anything that has accumulated since the last time (probably never) that you cleaned your closet.

Next is **Standardize;** this makes the preceding 3Ss a habit.

Finally determine how you will **Self-discipline** your 5S efforts. This involves spreading the 5S throughout the entire organization (family).

Total Productive Maintenance (TPM)

Total Productive Maintenance (TPM) is useful for maintaining plant and equipment with total involvement from all employees (Venkatesh, 2005; Japan Institute of Plant Maintenance, 1997). Its objectives are to dramatically increase production and employee morale by

1. Decreasing waste
2. Reducing costs
3. Decreasing batch sizes
4. Increasing production velocity
5. Increasing quality

TPM is built on a base of the 5S methods. Problems cannot be clearly seen when the workplace is unorganized. Making problems visible is the first step to eliminating them. TPM has seven component parts. They are *jishu hozen* (autonomous maintenance); *kaizen*; planned maintenance; quality maintenance; training; office TPM; and safety, health, and environment TPM.

TPM Example

An example on preventative maintenance that is near and dear to all of us is our bodies. This is something that not only helps each of us individually but also helps us as a nation reduce our skyrocketing healthcare costs.

- **Autonomous maintenance**—These are things that the operator (or human being) does without the need of any specific skill. In terms of the human body this consists of personal hygiene, eating, getting enough sleep, reducing stress, and so on.
- **Planned maintenance**—This refers to actions that can be taken on a routine or recurring basis. Some examples related to the human body would be annual physicals,

yearly skin cancer screenings, colonoscopies after the age of 50, annual mammography, and so on.

- **Training**—These are skills that the operator (human) does not have currently that would help prevent equipment breakdowns in the future. In terms of the human body some examples would be stress management classes, nutrition education, learning how to meditate, learning how to monitor blood pressure or glucose levels, and so on.

- **Health maintenance**—Finally, equipment design changes are fundamental design changes to the equipment (body) itself to prevent breakdowns in the future. Some examples related to improving the human body would be to begin a weight training program to improve strength of the body, walking daily to improve your cardiovascular system to prevent heart issues, stretching daily to improve flexibility and prevent muscle pulls, removing tonsils to prevent recurring strep throat infections.

Quick Changeover (Single Minute Exchange of Dies—SMED)

Quick changeover (Productivity Development Team, 1996), or Single Minute Exchange of Dies (SMED), is a technique that team members can use to analyze and then reduce

- Setup time for equipment (including tools and dies) and people (for example, shift to shift setup for cashiers in a supermarket)
- Resources required for a changeover
- Materials necessary for a changeover

Quick changeovers are accomplished through the SMED system. Single minute states that the time it takes to change a process from producing product A to product B, and so on, is less than 10 minutes. Quick changeovers create the opportunity to institute small batch sizes, or even one-piece flows, due to the short switch over times from unit A to unit B. Ideally, one unit batch sizes occur when switchover time from product A to product B is zero, or very small. This allows organizations to produce to customer demand, not to produce a batch of A, then due to a long switchover time, produce a large batch of B. This creates large and expensive inventories.

SMED Example

Auto races are competitive in that after 500 miles cars may be separated by only 50 feet. As all cars need to make "pit stops" to change tires and refuel, performing this type of quick changeover can be a real competitive advantage to a team and mean the difference between first place and tenth place.

It takes most people 15 minutes to change one tire, while a NASCAR pit crew can change a tire in less than 15 seconds! Why? They are using SMED principles such as

- Performing as many steps as possible before the pit stop begins (having all the tools and parts ready for when the car stops, redesigning the wheel so that all nuts are attached to the tire except one, necessitating only tightening of the nuts)

- Using a coordinated team to perform multiple steps in parallel (having a team working on changing the tires while a separate team works on refueling)
- Creating a highly optimized and standardized process with roles and responsibilities to execute while the car is stopped as well as before and after it is stopped and practicing to make sure everyone can execute their responsibility

Poka-Yoke

Poka-yoke (pronounced *POH-kah YOH-kay*) is Japanese for "mistake-proofing device" (Shingo, 1986). A mistake-proofing device is used to prevent the causes of defects and/or defective output (called errors), or to inexpensively inspect each item produced to determine whether it is conforming or defective. Poka-yoke devices promote Zero Quality Control (ZQC). ZQC means the production or delivery of zero errors, scrap, downtime, or rework.

Poka-Yoke Examples

- Sponge bags to let surgical teams count sponges to prevent sponges from being left inside patients after surgery.
- Bar coding on patient bracelets to prevent medication errors.
- Rumble strips—those little carved depressions in the road that provide the driver with tactile and audible warning that the car is not in its lane
- Beeps in cars if key is left in ignition so you don't lock your keys in your car
- Ceiling paint that goes on pink but dries white to let know if you have missed a spot
- Gas cap tethers, which keep you from leaving your gas cap behind after refueling your car

An Anecdote for Poka-Yoke

There are many humorous examples of poka-yoke. One such example concerns keeping urinals clean in public places. This largely amounts to keeping the floors clean and dry, which is often a problem because of the poor aim of the male users of the facilities. One poka-yoke device is to fill the urinal with ice. We assume that most of the male readers of this text have encountered this situation. One purpose of the ice is to bring out the hunter gene that lies deep in most males from prehistoric times. Frequently what happens is that the male going to the bathroom takes aim at one particular ice cube with great concentration and attempts to completely melt it, thereby satisfying his need for killing the cube and having a successful hunting experience. By the way, this has been shown to reduce spillage around the urinal by over 80%.

Value Streams

Value stream (Rother and Shook, 2003) is a term that describes all the value added and non-value added process steps and decisions necessary to move a product or service from supplier to customer. These steps include design and redesign, raw material flows, subcomponent flows, information flows, production and service flows, and people flows, to name a few steps. Value stream maps can be created for different points in time. A current state value stream map follows a product or service from supplier to customer to represent current conditions. A future state value stream map incorporates the opportunities for improvement identified in the current state value stream map to improve the process at a future point in time using the previous lean tools and methods (see Figures 7.7a and 7.7b).

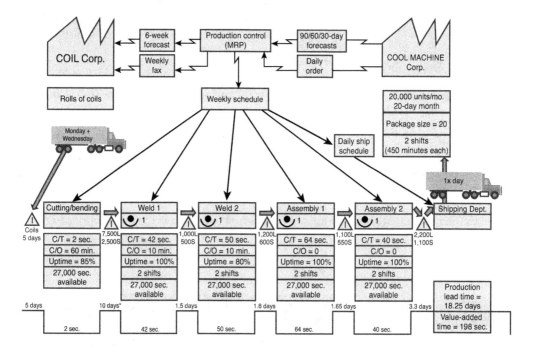

* 7,500 + 2,500 = 10,000
10,000 units/1,000 demand per day = 10 days

Figure 7.7a Example of current and future state value stream maps

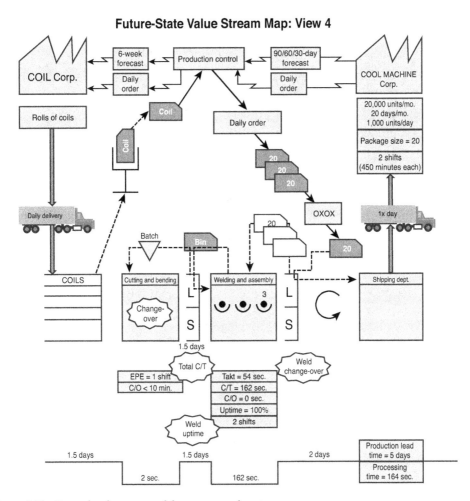

Figure 7.7b Example of current and future state value stream maps

We do not discuss lean tools and methods any further in this book. If you are interested in lean tools and methods, a wealth of literature is available on the Web. Remember, each lean tool can be used to find Xs for improving a CTQ!

Takeaways from This Chapter

- The SDSA cycle is a series of four steps used to standardize a process and reduce variation and waste in process outcomes.
- The PDSA cycle consists of four steps used to continually improve a process by reducing variation, moving the process toward nominal, and eliminating waste.

- A kaizen or Rapid Improvement Event (RIE) is a methodology that consists of a three to five day event that empowers front-line staff and utilizes their process knowledge to analyze and make immediate improvements to problematic processes.

- Based on the scientific method, the DMAIC model is a five-phase method used to improve a current process, product, or service.

- The DMADV model is a five-phase method used to design a new or redesign a current process, product, or service.

- Lean thinking is focused on promoting the reduction of waste by eliminating non-value added steps in a process.

- The 5Ss are a lean methodology focused on highlighting and eliminating waste, inconsistency, and unreasonableness from the workplace. It is used to create a neat and tidy workplace.

- TPM is a theory that is useful for maintaining plants and equipment with total involvement from all employees.

- Quick changeover (SMED) is a technique that team members can use to analyze and then reduce setup time for equipment and people, resources required for a changeover, and materials necessary for a changeover.

- Poka-yoke is Japanese for *mistake-proofing device*. A mistake-proofing device is used to prevent the causes of defects and/or defective output (called errors), or to inexpensively inspect each item produced to determine whether it is conforming or defective.

- Value stream is a term that describes all the value added and non-value added process steps and decisions necessary to move a product or service from supplier to customer.

References

Deming, W. E. (1994), *The New Economics for Industry, Government, Education*, 2nd ed. (Cambridge, MA: Massachusetts Institute of Technology).

Gitlow, H. (2009), *A Guide to Lean Six Sigma* (Boca Raton, FL: CRC Press - A Division of Taylor and Francis).

Gitlow, H., D. Levine, and E. Popovich (2006), *Design for Six Sigma for Green Belts and Champions: Foundations, DMADV, Tools and Methods, Cases and Certification* (Upper Saddle River, NJ: Prentice-Hall).

Gitlow, H., A. Oppenheim, R. Oppenheim, and D. Levine (2015), *Quality Management: Tools and Methods for Improvement*, 4th ed. (Naperville, IL: Hercher Publishing Company). This book is free online at hercherpublishing.com.

Gitlow, H. and D. Levine, D. (2004), *Six Sigma for Green Belts and Champions: Foundations, DMAIC, Tools and Methods, Cases and Certification* (Upper Saddle River, NJ: Prentice-Hall).

Japan Institute of Plant Maintenance (1997 English Edition), *Focused Equipment Improvement for TPM Teams* (New York: Productivity Press).

Productivity Development Team (1996), *Quick Changeovers for Operators: The SMED System* (New York: Productivity Press).

Rother, M. and J. Shook (2003), *Learning How to See* (Cambridge, MA: The Lean Enterprise).

Shingo, S. (1986), *Zero Quality Control: Source Inspection and the Poka-Yoke System* (New York: Productivity Press).

Venkatesh, J. (2005), *An Introduction to Total Productive Maintenance (TPM)*, The Plant Maintenance Resource Center, 1996-2005. Revised: Sunday, 27-Feb-2005.

8

Project Identification and Prioritization: Building a Project Pipeline

What Is the Objective of This Chapter?

The objective of this chapter is to discuss the process used to create the project pipeline at your organization. We walk you through project identification, project screening and scoping, project prioritization and selecting projects, and finally managing the project pipeline. The process of creating and managing the pipeline is typically led by a Process Improvement Executive and senior-level Executive Champions.

Project Identification

The first step in the process is project identification. We have found that potential process improvement projects can come to the Process Improvement Executive, senior level executives, employees, and customers, to name a few sources, in a variety of ways.

There are four basic categories of ways to identify process improvement projects; they are summarized in Table 8.1. Each of the methods in the cells of Table 8.1 is explained in this chapter.

Table 8.1 Project Identification Matrix

	Proactive	**Reactive**
Internal	Strategic/tactical plans Voice of the Employee (VoE) interviews Employee focus groups or forums Employee surveys	Employee feedback (suggestions and complaints)
External	Voice of the Customer (VoC) interviews Customer focus groups Customer surveys Regulatory issues dealt with proactively	Customer feedback (suggestions and complaints) Regulatory issues dealt with reactively

Internal Proactive

The first category for project identification we discuss is internal proactive sources. *Internal* refers to the fact that the data for project identification comes from within the organization; for example, Voice of the Employee interviews. *Proactive* refers to the fact that employees in the organization actively collect the data for project identification; it does not come passively to the organization.

Strategic/Tactical Plans

Strategic planning is a process frequently used within organizations to identify the objectives necessary to pursue the mission; this includes setting the mission and objectives, determining actions to achieve the objectives, and allocating resources to achieve the objectives. A strategy describes how the objectives will be achieved given the budget. Many times those actions involve potential process improvement projects.

Most projects evolve out of an organization's strategic plan. Each section of the organization submits, usually on a yearly basis, its strategic plans for the next year. Part of the plan is a gap analysis that compares the current state with the desired state objectives. If the gap between the current state and desired state is larger than the organization would like, a tactical thrust is initiated to close that gap. The tactical thrust in this case is a potential process improvement project that depending on the gap could be a 5S project, an SDSA project, a PDSA project, a Six Sigma DMAIC project, a Six Sigma DMADV project, or a Lean project.

Potential strategic/tactical projects may also be identified through internal metrics found on

- **Organizational dashboards**—A dashboard is a set of interlocking and cascading objectives with appropriate metrics (called *key performance indicators [KPIs]* or CTQs) that go from the top to the bottom of the organization; the objectives and metrics define employees' jobs. Many dashboards have adopted the four types of objectives: employee, process, customer, and financial.

- **Balanced scorecards**—A balanced scorecard is similar to a dashboard, except it always focuses on four types of objectives: financial, customer, process, or employee learning and growth. The basic idea is happy employees improve the processes, which creates customer satisfaction, and ultimately, results in good financial performance.

Voice of the Employee (VoE) Interviews—Finding the Problems

Another way of identifying projects is by having a process improvement professional meet individually with senior executives or managers to identify potential projects by asking any of the following questions:

- What concerns do you have within your organization/area/department/ division?

- What is your actual performance versus your desired performance? (to identify performance gaps)

- How are you not meeting your customers' expectations? What comments or feedback have you received from your customers?

- What are your competitive threats?
- How is the market changing?
- What are your performance results from past years? Are they stable, predictable, and acceptable?
- What are the areas in which you have less dollars (revenue, profit, and so on) than expected in your strategic plan?
- Are you over budget? Is your budget based on rational expectations? If you are exceeding a rational budget where are you exceeding it? What are you going to do about it? Remember the effects of treating common variation as special variation.

After all these meetings you should have a list of problems and issues. If there are many vague and similar issues you may want to consolidate them using an affinity diagram. You may have to meet with the senior leaders again to dig deeper into the issues to get at their root causes so you understand the project that is needed to resolve the issue(s) caused by some problematic process.

Employee Focus Groups

An employee focus group is a type of qualitative research. They typically consist of 6 to 12 employees and are facilitated by a moderator. The members of the focus group share an interest in the well-being of the organization. The purpose of a focus group is to dive deep into a few high priority issues to learn about employees' perceptions, feelings, opinions, beliefs, needs, and wants, which may lead to ideas for process improvement projects.

Focus groups that are well run help uncover feelings and emotions that are not visible in surveys due to the depth of the questioning on a selected few high priority topics. Employees that are on the line often see issues not visible to management. Hence, focus groups can be a valuable tool to identify issues and ideas for potential process improvement projects.

However, if management does not react quickly to the employees' suggestions, the focus groups become a joke! At a minimum, management has to listen to the suggestions and get back to the employees on what they plan to do or not do, but true effectiveness comes when management acts on the suggestions.

Employee Forums

Many organizations conduct quarterly, semi-annual, or annual employee forums to give their employees a chance to

- Present feedback to management on their areas of the organization.
- Communicate issues and concerns on their areas of the organization in particular, or the entire organization in general.
- Ask questions about the organization.
- Give their views and opinions about the organization.

Employee forums are another great way to identify potential process improvement projects. Front-line employees who attend the forums may have their finger on the pulse of the company to a greater degree than management with their areas, and may be exposed to issues that management wouldn't ordinarily be aware of. Again, if management does not react quickly to the employees' suggestions, the forums become a joke! Again, at a minimum, management has to listen to the suggestions and get back to the employees with their plans for acting on, or not acting on, their suggestions, but true effectiveness comes when management acts on the valuable suggestions.

Employee Surveys

Many organizations distribute employee engagement or satisfaction surveys on an annual basis. Often these surveys give employees an opportunity to give suggestions on how the organization may improve. This can be a great way to identify potential process improvement projects.

The problem with employee surveys is that the information is qualitative in nature and is difficult to analyze. A potential solution is to take the qualitative employee feedback and use techniques such as the affinity diagram and the cause and effect diagram to organize the data into meaningful opportunities for improvement of problematic processes. Once again, if management does not react quickly to the employees' suggestions, the surveys become a joke!

Internal Reactive

The second category for project identification we discuss is internal reactive sources. *Internal* is defined as before—for example, employee feedback. *Reactive* refers to the fact that the data comes to the organization as a direct result of doing business; again, like employee feedback.

Employee Feedback (Suggestions or Complaints)

Another way to identify potential process improvement projects is through unsolicited employee feedback in the form of complaints or suggestions, which may come in various forms:

- Through email
- Via suggestion boxes
- Complaints filed through human resources
- From coworkers, employees, or managers

Valuable information can be obtained from employee feedback so it must be encouraged and responded to for it to continue. One way to deal with employee feedback is through a closed feedback loop as shown in Figure 8.1. First, collect feedback from the preceding sources and analyze it on a regular basis. Second, investigate whether the feedback (data) indicates that a problem warrants further action. Make sure that you communicate to employees that you listened to their feedback. Finally, if you have made changes based on the employee feedback, refine the changes through continuous improvement; see Figure 8.1.

It is important to ensure that employees are satisfied with the outcome(s) of their input. So following up with employees who have given feedback may be an investment you want to make. Often all that employees want is to be listened to and thanked for their feedback. Remember, if management does not react quickly to the employees' suggestions, the feedback becomes a joke.

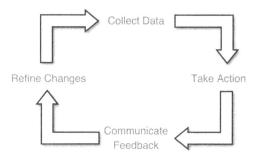

Figure 8.1 The closed-loop feedback process (source: www.mindtools.com)

An Anecdote about Collecting Customer Information

Years ago when you went to buy a new or used car, you would make a deal with the salesperson and then he would have to go back to the sales manager for approval. What you didn't know was that some car retailers had a hidden microphone under the desk where you were sitting. The sales manager and the salesman were listening to your conversation about the lowest price you would pay for the car, and of course used this information against you. The moral to this anecdote is that data collection is not always done ethically. You need to be careful.

External Proactive

The third category for project identification we discuss is external proactive sources. *External* refers to the fact that the data for project identification comes from outside of the organization; for example, Voice of the Customer interviews. *Proactive* refers to the fact that employees in the organization actively collect the data for project identification; it does not come passively to the organization.

Voice of the Customer (VoC) Interviews

Another way of identifying projects is by having a process improvement professional meet individually with your customers (stakeholders) to identify potential projects by asking the following questions:

- How do you feel about our product/process/service?
- What do you like about our product/process/service? How can we do better?
- What areas can we improve upon with respect to our product/process/service?
- How are we not meeting your expectations?
- How do you feel about our products/services/processes?
- What images and emotions come to mind when you think about our products/processes/services?

Customer Focus Groups

Like employee focus groups discussed previously, a customer focus group is a type of qualitative research typically consisting of 6 to 12 customers. The group is facilitated by a moderator, and its members are customers who share an interest in the well-being of the organization. The purpose of a focus group is to dive deeply into customers' (stakeholders') perceptions, feelings, opinions, beliefs, needs, and wants, which may lead to ideas for process improvement projects.

Focus groups that are well run help uncover feelings and emotions that are not visible in surveys. Hence, focus groups can be a valuable tool to identify issues and ideas for potential process improvement projects.

Customer Surveys

Customer surveys can either be proactive or reactive. Both proactive and reactive surveys may be done by organizations or third-party vendors (for example, Hospital Consumer Assessment of Health Care Providers and Systems [HCAHPS] inpatient surveys in healthcare) in various different ways:

- Phone surveys
- Email surveys
- Mail surveys
- Personal interviews

Surveys must be done properly or they yield garbage information; for more information on how to conduct a proper survey refer to any basic marketing research book.

External Reactive

The fourth category for project identification we discuss is external reactive sources. *External* is defined as before; for example, customer feedback. *Reactive* is defined as before; for example, reacting to regulatory compliance warnings.

Customer Feedback (Suggestions or Complaints)

Another way to identify potential process improvement projects is through unsolicited customer feedback in the form of complaints or suggestions, which may come in various forms:

- Through email
- Via suggestion boxes
- From complaints filed through the customer service department
- From customers directly at the point of service

Valuable information can be obtained from customer feedback, so it must be encouraged and responded to for it to continue. One way to deal with customer feedback is through a closed feedback loop as shown previously in the closed-loop feedback process in Figure 8.1.

It is important to manage feedback by ensuring customers are satisfied with the outcome of their feedback. So following up with customers who have given feedback may be an investment you want to make. Often all that customers want is to be listened to and thanked for their feedback. Again, nothing gets the customers' attention like acting on their feedback.

Regulatory Compliance Issues

Regulatory compliance describes an effort organizations make to ensure that they are aware of and take the necessary steps to ensure they comply with various laws and regulations. For example, healthcare organizations are typically faced with many compliance requirements aimed at patient safety and information security, service delivery, operational practices, and electronic medical record management. Such requirements include various regulatory bodies and industry standards such as

- HIPAA (Health Insurance Portability and Accountability Act)
- JCAHO (Joint Commission on Accreditation of Healthcare Organizations)
- SOX (Sarbanes Oxley Act)
- CMS (Centers for Medicare and Medicaid Services)

The preceding list is merely some common regulatory bodies; there are many more federal and state regulations for patient safety that healthcare organizations need to comply with, as well as other organizations, such food and drug companies and utilities. Depending on the violation, the repercussions for noncompliance vary among the regulatory bodies—from warnings to fines to exclusion from participating in some programs to being shut down. Civil and criminal penalties also can be imposed for violations of certain regulations. Many industries have regulatory bodies that *must* receive proper attention—or else!

Process improvement projects related to regulatory compliance can therefore be identified both proactively and reactively.

Proactive

Organizations with a strong regulatory compliance department can stay ahead of the curve by staying up to date on the latest rules and regulations, and building objectives and metrics to track them into their organizational dashboards or balanced scorecards. By establishing and monitoring key metrics they can identify potential problems immediately and execute process improvement projects to eliminate them.

Reactive

Periodically, healthcare organizations are subject to either scheduled or unscheduled surveys and/or visits from various regulatory bodies, for example, the Joint Commission in healthcare organizations. Often they find areas where the organization is not in compliance. To be compliant, the organization must execute a task, or if the issue is more complicated, the organization must undertake a process improvement project.

> **NOTE**
>
> Suggestions on issues and potential problems from interviews, focus groups, surveys, employee forums, and the like may result in a plethora of vague qualitative data. One way to make sense of all these great suggestions is to consolidate them using an affinity diagram and a cause and effect diagram.

Using a Dashboard for Finding Projects

A managerial dashboard is a tool used by management to clarify and assign accountability for the "critical few" key objectives, key indicators, and projects/tasks needed to steer an organization toward its mission statement (Gitlow et al., 2015; Gitlow and Levine, 2004). They do this by creating an interlocking and cascading set of objectives with metrics throughout the organization, from top to bottom. The objectives clarify employees' jobs. Managerial dashboards have both strategic and tactical benefits.

Structure of a Managerial Dashboard

The president's objectives and indicators emanate from the mission statement (see row 1 and columns 1 and 2 of Table 8.2). Direct reports identify their area objectives and area indicators by studying the president's key indicators (column 2 of Table 8.2) that relate to their area of responsibility. The outcome of these studies is to identify the key area objectives and area indicators (see columns 3 and 4 of Table 8.2) required to improve the president's key Indicator(s) (see column 2) to achieve a desirable state for presidential key objective(s) (see column 1). This process is cascaded throughout the entire organization until processes are identified that must be improved or innovated with potential process improvement projects or tasks (see column 5 of Table 8.2).

Table 8.2 Generic Managerial Dashboard

Mission Statement: A mission statement is a declaration of the reason for the existence of an organization. It should be short and memorable, as well as noble and motivational.				
President		**Direct Reports**		**Potential Process Improvement Projects or Tasks**
Presidential Objectives	Presidential Indicators	Area Objectives	Area Indicators	
Presidential objectives that must be achieved to attain the mission statement.	A key indicator is a measurement that monitors the status of a key objective. One or more presidential indicators show progress toward each presidential objective.	Area objectives are established to move each presidential indicator in the proper direction.	One or more area indicators show progress toward each area objective.	Process improvement projects or tasks are used to improve or innovate processes to move indicators in the proper direction.

Example of a Managerial Dashboard

Table 8.3 shows an example of a managerial dashboard for the ABC Hospital. The mission of ABC Hospital is to be the best university teaching hospital in the universe! It has a classic type of organizational structure led by a Chief Executive Officer, Chief Operating Officer, Chief Financial Officer, Chief Medical Officer, and Chief Nursing Officer. One of the areas that the Chief Operating Officer is responsible for is Surgical Services. Table 8.3 shows an example of a dashboard between the Chief Operating Officer and the Director of Surgical Services.

Table 8.3 Partial Managerial Dashboard for Hospital ABC

Mission Statement: To be the best university teaching hospital in the universe!				
Chief Operating Officer (COO)		**Director of Surgical Services (DSS)**		**Potential Process Improvement Projects**
COO Objectives	COO Indicators	DSS Objectives	DSS Indicators	
To have operating rooms that are efficient and cost effective	% of first case surgeries that start on time by day	To improve first case surgery start times	% of first start surgeries that are on time by surgeon by day	First case start project
	% of surgeries cancelled day of surgery by day	To decrease cancellations	% of cancellations by insurance provider by age of patient by day	Note: Data indicates that cancellations are very low; therefore this is not a candidate for a project.

Managing with a Dashboard

Top management uses a dashboard at monthly operations review meetings for several purposes. First, managers use dashboards to clarify key objectives and accountability among all personnel and areas. Second, managers use dashboards to promote statistical thinking by monitoring key indicators using control charts. For example, is the sales volume for last month due to a special or common cause of variation in the selling process? Third, a manager uses dashboards to develop and test hypotheses concerning potential changes to processes. A hypothesis test analyzes the effect of a change concept on a key indicator, and hence, on its objective. Fourth, a manager uses dashboards to ensure the routine and regular updating of key indicators and to prevent processes from sliding back into their old bad habits.

Managers can use (all or some of) the following questions when conducting a monthly review meeting to get the most out of their dashboard:

- Are the key objectives and key indicators on the dashboard the "best" set of objectives and indicators to attain the mission statement?

- Is the dashboard balanced in respect to employee, process, customer, and financial objectives? Do any areas have too much (or too little) representation on the dashboard?

- What products and/or services are most critical to your organization achieving its mission statement? List the top five or ten products and/or services.

- Are objectives being met in a timely fashion?

- What methods are used to manage, perform, and improve the processes that underlie key objectives?

- Which key indicators on the dashboard are used to measure customer satisfaction and dissatisfaction? Are these measures operationally defined? Are these measures adequate?

- What process is used to motivate employees to work on improvement projects? Hint: Managers redefine work to include doing work and improving work using the PDSA cycle. As things improve, employees experience the rush of intrinsic motivation from an improved and well done job.

- Does your organization have the ability to identify the return on investment from its dashboard? How is return on investment measured?

Project Screening and Scoping

The Process Improvement Executive has a list of potential projects; now what? The next steps in the process are to screen and scope potential projects as follows:

- Determine quickly which ones are worth investigating further. There are questions you can ask that eliminate projects right away so as not to waste your time and effort.

- Estimate the potential benefits of each project. If there is no benefit, why do it?
- Estimate the potential costs of each project. If the cost are too high in respect to the benefits, why do it?
- Determine which methodology would most likely be used for each potential project, for example, SDSA, PDSA, DMAIC, and so on.
- Create a high-level project charter and problem statement that can be used to prioritize the remaining potential projects.

Questions to Ask to Ensure Project Is Viable

A few questions can be asked right away to disqualify a potential process improvement project before it wastes valuable time and resources. Some of these questions are as follows:

- **Is the root cause and solution already known?** If the root cause and the solution are already known, there is no point starting a process improvement project because you already know what you have to do. So like Nike says, "Just Do It!"
- **Does the project have firm Champion commitment?** If you do not have firm Champion commitment, stop right now and run in the other direction! The number one key success factor is the commitment of top management (your Champion); if you don't have it, find another project or another Champion.
- **Is data readily available or collectable?** The process improvement methodologies we are teaching you in this book all necessitate data both to determine a baseline for the current state of the process, as well as to help identify root causes. If data is not readily available, or at least collectible, you should think long and hard about taking on that project.
- **Is the scope narrow? Can it be broken into smaller projects?** The last thing you want is a "solve world hunger" type project; they are almost impossible to execute and you will lose momentum quickly as their scope is way too big. An alternative is to break it down into smaller projects; if this is not possible, you probably want to stay away!
- **Are there organizational constraints that make this project risky?** Will key people in the organization be committed to keep the project on time and moving forward? Are there political issues that would side track the project? Are key people always in reactive mode and constantly putting out fires? If there is a potential for team members to frequently be pulled away to work on "more pressing issues," or they are overwhelmed with the "crisis du jour" (fires), you may want to reconsider this project. If either of these situations exist, the project Champion must really be committed to eliminate these barriers to a successful project! Additionally, schedule slippage is something you need to avoid as any loss in momentum can derail your project for good.
- **Are there environmental issues that you need to consider?** Is there something going on outside the scope of your organization that will change the landscape in your industry and eliminate the need for this project?

Estimating Project Benefits

Process improvement projects have both soft benefits and hard benefits (Gitlow et al., 2015; Gitlow and Levine, 2004). Examples of soft benefits include improving quality and morale, and decreasing cycle time. Examples of hard (financial) benefits include increasing revenues or decreasing costs. The Process Improvement Executive "guesstimates" the dollar impact of the process improvement project so management has an idea of its financial impact during the project prioritization process. This "guesstimate" will be refined through iterative learning as the project proceeds.

There are two taxonomies for classifying the potential cost related benefits that may be realized from a process improvement project.

Taxonomy 1: Cost Reduction Versus Cost Avoidance

Cost reduction includes costs that fall to the bottom of the profit and loss statement. A cost reduction can be used to offset price, increase profit, or can be reinvested elsewhere by management. Cost reductions are calculated by comparing the most recent accounting period's actual costs with the previous accounting period's actual costs. Cost avoidance includes those costs that can be reduced, if management chooses to do so, but until action is taken no real costs are saved. Examples include reducing labor hours needed to produce some fixed volume of work. Unless an increased volume of work is completed with the same headcount, no real savings are realized. The impact of cost avoidance is not visible on the profit and loss statement and is difficult to define, but is still important in meeting organizational goals.

Taxonomy 2: Tangible Costs Versus Intangible Costs

Tangible costs are easily identified—for example, the costs of rejects, warranty, inspection, scrap, and rework.

Intangible costs are costs that are difficult to measure—for example, the costs of long cycle times, many and long setups, expediting costs, low productivity, engineering change orders, low employee morale, turnover, low customer loyalty, lengthy installations, excess inventory, late delivery, overtime, lost sales, and customer dissatisfaction. It is important to realize the some of the most important benefits are unknown and unknowable. Hence, the "guesstimate" of benefits in the Define phase of a Six Sigma project often identifies a minimum estimate of intangible benefits.

The Process Improvement Executive develops a formula to "guesstimate" the potential benefits that the organization may realize due to the process improvement project. For example, here is a possible formula:

       Cost reductions   _____

       PLUS Cost Avoidance   _____

       PLUS Additional Revenue   _____

       LESS Implementation Costs   _____

       EQUALS Financial Benefits   _____

Project Methodology Selection—Which Methodology Should I Use?

Before prioritizing a project it is usually a good idea to determine the type of methodology you will be using; this allows you to project the time the project will take as well as the resources needed. Many times it is tough to know what methodology you will use until you have done further investigation, and sometimes the methodology you thought you would use changes once you get started.

You always want to choose the methodology that helps you execute your project the quickest to realize your objectives, while utilizing the least amount of resources. This is a fairly subjective decision that is best made by your Process Improvement Executive who can draw upon years of experience and leverage his understanding of different organizational factors, such as resources available, time needed, project complexity, political environment, and so on.

The authors created the flowchart in Figure 8.2 to help you figure out which methodology to use on a given project.

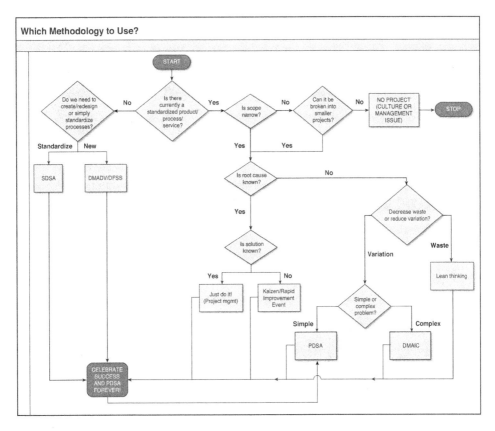

Figure 8.2 Which methodology to use?

Chapter 8 Project Identification and Prioritization: Building a Project Pipeline   243

The first question we ask is: Is there currently a standardized product/process/service?

- If there is *not* a current standardized product/process/service, then the project is one of the following:
  - **An SDSA project**—If everyone is doing things differently (they each have their own flowchart/process), team members need to get everyone together and take the best aspects from their individual flowcharts—that eliminates the weaknesses from their individual flowcharts—and create one best practice flowchart that everyone follows. This requires personal discipline. Using the 5Ss helps to develop personal discipline in the workforce.
  - **DMADV/DFSS project**—If there is no process or it is too screwed up to continue, we have to design one from scratch using the Design for Six Sigma (DMADV) methodology.
- If there is a current standardized product/process/service then:
  - If the scope is not narrow and it cannot be broken down into smaller projects, it is likely an organizational culture or management issue and we advise you to stay away!
  - If the scope is narrow, the root cause is known, and the solution is known, we call this a Just Do It project. All you need to do is create a list of action items and project manage it to success.
  - If the scope is narrow, the root cause is known, and the solution is not known, we conduct a Kaizen/Rapid Improvement Event over three to five days to come up with and implement a solution.
  - If the scope is narrow and the root cause is not known, the question is do we want to decrease variation/change a problematic CTQ (Y), or do we want to remove waste in a problematic CTQ?
    - If we want to decrease variation and the problem is not too complex, we suggest using the PDSA cycle. If the problem is more complex, we suggest using the Six Sigma DMAIC model.
    - If we want to remove waste, we suggest using one of the Lean thinking methodologies.

NOTE

Complex? A problem may be considered complex if it has some or all of the following characteristics:

- More than one CTQ
- Conflicting CTQs (one gets better and the other gets worse)

- Involves organizational politics
- Requires an Information Technology solution
- Requires capital investment

Estimating Time to Complete Project

At this point we like to make a rough estimate on how long the project will take so that we can prioritize and select projects. Many different methodologies can be used to estimate the time needed to complete a project. Many great project management books cover these methodologies in detail, so we are merely going to touch on a few of the ones you can use:

- **Top down (ambiguous) estimating**—Develop an estimated timeline based on past experience and past projects. At this point when we are only looking for a rough estimate to prioritize the project, this is the method typically used. The other methods listed here can be used after the project actually starts.
- **Bottom up estimating**—Breaking down big tasks into smaller ones and then estimating the time it will take to complete each one.
- **Three-point estimating**—With three-point estimating you come up with an expected estimate, a pessimistic estimate, and an optimistic estimate for each activity.
- **Project end date estimating**—Sometimes you will be given a hard deadline by when a project must be completed. In this case you consider the needed deliverables and resources available to create a timeline working backward.
- **Expert judgment estimating**—Many times you will have an expert who has worked on a similar project previously. Based on their experience they will have an idea of how long this type of project will take.

The Process Improvement Executive will estimate the time it takes to complete the project likely using top down estimating, and then once the project begins, you may want to use one of the preceding methods to narrow it down even further. At this stage, keep in mind some things to minimize the time it takes to complete a project, such as:

- Choose projects that have reliable and easily accessible data.
- Choose projects in areas where teams have total management support (Champion) and are allowed to meet as often as necessary.
- Potential project stakeholders are available for interviews (Interviewing stakeholders as part of the VoC analysis is time consuming. If your stakeholders have limited availability to participate it will slow you down.)
- Commitment by the Champion to change. At some point you will want to test change. You want to make sure your Champion will cut through any delays; so make sure you pick projects where this won't be an issue.

There are other actions you can take to speed up projects that we discuss later when covering different methodologies. These are just the ones you want to be cognizant of when prioritizing and selecting projects:

- The project does not involve politics.
- The project requires minimal resources.
- The project is under the Process Owner's and Champion's control, and they are "all in" on the project.
- The project directly affects the end user.

Creating a High Level Project Charter

The last step of project scoping is creating a high level project charter and problem statement for each project that you want to add to the pipeline. This will be carried forward to the next section where it is used to prioritize potential process improvement projects.

The project charter will be refined and expanded in detail during the initial part of the project. This is just to give the steering committee a starting point when prioritizing projects. Table 8.4 shows a format you can use for a high level project charter.

Table 8.4 Format for High Level Project Charter

Project Name:	Identify the issue that is the potential process improvement (PI) opportunity.
Department/Area Name:	What department or area will be impacted by the project?
High Level Problem Description:	Provide a concise description of the issue that needs to be addressed by the project team.
Project Type:	Methodology expected to be used (PDSA, DMAIC, etc.).
Time to complete:	Estimated time to complete project using a Gantt chart.
Problematic objective:	Problematic objective from the dashboard or other source.
Problematic indicator:	Metric that measures the status of the problematic objective.
Project Scope:	Process steps (flowchart) or functional boundaries of the proposed project. Where does the project start and where does it stop?
Potential Financial Impact:	Projected result of the project and anticipated savings or revenues to be realized.
Current Baseline Performance:	Actual current performance of problematic indicator.

Project Champion:	Senior executive who reviews projects, removes impediments for the team, and can secure adequate resources and support for the project team.
Process Owner:	The manager of the process under study who has the authority to change the process.
Finance Champion:	Member of the Finance department who can help the team estimate financial impact of the project.
Project Leader:	Process improvement expert who will be the project leader.
Potential Team Members:	Employees who will work on the project part-time.

Problem Statement

The problem statement is a paragraph that states the issue at hand in succinct terms so that you can understand the problem and its impact for the organization as a whole. A typical problem statement should be organized as follows:

Our organization is experiencing an issue with _____. The department/area where this issue is most problematic is _____. This issue has been in existence for _____. The impact of the issue on the organization is _____. It is affecting the organization by _____. The financial impact of the issue can be estimated at _____. Customers affected by the issue are _____.

Prioritizing and Selecting Projects

Now that you have identified projects, screened out projects that are not worth doing, and scoped the ones that you will consider doing, you have essentially created a project pipeline. However, just because a project makes it to the project pipeline does not mean it is a project you will take on. Due to limited time and resources, often you have to further prioritize projects so that you can maximize both your time and resources. One such methodology is to prioritize projects using a project prioritization matrix that essentially compares different projects in your pipeline and ranks them against each other based on certain criteria.

Depending on the organization, prioritization is done by senior executives who have process improvement training, as well as Process Improvement Executives. They then present the list of potential projects to a steering committee who makes the final decision on which projects to execute.

The first step in the process is to select the criteria for which you compare and rank the projects in your pipeline. There are many possible criteria you may use to compare projects. You can come up with your own, but some criteria we have seen used are the following:

- Financial impact (return on investment [ROI])
- Data availability
- Data quality
- Customer satisfaction
- Employee satisfaction
- Other stakeholder satisfaction (for example, investors, regulators, building, and grounds)
- Ease of implementation
- Alignment with mission
- Time to complete project (consider methodology and available resources)
- Ability to implement solution
- Probability of quick results
- Low investment cost
- Minimum collateral damage on other processes
- Scalable and replicable across organization
- Resources available
- Competitive advantage the project may create

Prioritizing Projects Using a Project Prioritization Matrix

The project prioritization matrix seen in Table 8.5 is a matrix we use to prioritize projects in our pipeline. The first column contains the evaluation criteria that we use to compare potential projects. The second column contains the weights that are usually assigned by the Finance department. The columns to the right contain the different potential projects in our pipeline.

Table 8.5 Project Prioritization Matrix

Evaluation Criteria	Weight	Project A	Project B	Project C	Project D
Criteria 1	Weight 1				
Criteria 2	Weight 2				
Criteria 3	Weight 3				
Criteria 4	Weight 4				
Criteria 5	Weight 5				
Criteria 6	Weight 6				
Weighted average of potential projects					

The cell values are assigned by top management, and they evaluate the strength of the relationship between each criteria and each potential project.

They are defined as follows:

  0 = no relationship between criteria and project

  1 = weak relationship between criteria and project

  3 = moderate relationship between criteria and project

  9 = strong relationship between criteria and project

Each cell value is then multiplied by its respective weight and summed so that each potential project has a weighted average. The projects are then prioritized by weight, which ranks them in terms of importance to the organization based on the evaluation criteria. Obviously this is not an exact science and is subjective in nature, so it is still possible that management will make an executive decision to go with a project that is lower on the list due to reasons they deem important.

In our example, we compare the four potential projects in our pipeline in Table 8.6 based on the criteria we selected and on the weights that came out of our evaluation criteria matrix.

Table 8.6 Project Prioritization Matrix Example

Evaluation Criteria	Weight	Project A	Project B	Project C	Project D
Financial impact	0.39	9	3	9	3
Customer satisfaction	0.28	9	3	3	9
Employee satisfaction	0.16	3	9	1	9
Probability of quick results	0.12	9	1	3	9
Data availability and quality	0.03	9	3	3	1
Ability to implement solution	0.02	3	1	1	3
Weighted average of potential		7.92	3.68	4.98	6.3

According to our project prioritization matrix the projects are prioritized as follows:

1. Project A   (7.92)
2. Project D   (6.3)
3. Project C   (4.98)
4. Project B   (3.6)

So we select project A (7.92) as the highest priority project. We keep going down the list of projects until resources are depleted.

Final Project Selection

The organization typically creates a process improvement steering committee that balances and manages the project pipeline. This steering committee is made up of senior executives who have been through process improvement training, process improvement personnel, and various key Champions and stakeholders throughout the organization. The committee reviews the pipeline, including high-level project charters and problem statements, as part of their monthly meeting and allocates resources they deem most important to the mission of the organization.

Executing and Tracking Projects

This section of the chapter discusses executing and tracking process improvement projects. The first part of this section explains the infrastructure needed to allocate resources to projects, while the second section discusses the role and significance of presidential reviews and monthly steering committee meetings to successful projects.

Allocating Resources to Execute the Projects

Once the projects are selected, the Process Improvement Executive identifies a project leader (process improvement professional), as well as other change agents with the needed skill sets, including process improvement professionals, process experts, and process stakeholders. Champions and process improvement experts rely on their understanding of the organization and their employees to assign the proper people to the right projects to leverage strengths, weaknesses, and time and resource availability to execute projects.

Once the project leader is assigned she rallies the troops with a kickoff meeting that includes

- Background on the project
- Roles and responsibilities for the project
- Methodology to be used by the project

- Project plan
- Communication plan
- Risk abatement plan
- Project charter

Monthly Steering Committee (Presidential) Reviews

Many organizations have a monthly steering committee review to hold people accountable, to track progress and benefits, as well as to review the pipeline and any new projects. The Project Champions of the various projects present updates on their projects to the steering committee. This is *crucial* as it ensures that top management stays committed to the projects in their areas. The project teams create the presentation and materials, and can help answer detailed questions, but it must be the Champion who does the talking. The steering committee also reviews and prioritizes the pipeline if resources are available to execute new projects.

Takeaways from This Chapter

- Creating a project pipeline consists of four main steps: project identification, project screening and scoping, prioritizing projects, and managing the pipeline.
- There are four basic categories of ways to identify process improvement projects: internal proactive, internal reactive, external proactive, and external reactive.
- Another way to identify projects presented by the authors is using managerial dashboards or balanced scorecards.
- Project scoping consists of asking questions to ensure a project is viable, calculating potential financial benefits, figuring out which methodology would potentially be used to execute the project, estimating the time the project will take, and finally creating a high level project charter.
- Next, the team prioritizes projects by selecting and ranking criteria to compare potential projects using a project prioritization matrix.
- The steering committee then selects the projects the team will be working on.
- The Champion and Process Improvement Executive then allocate resources to execute the projects.
- Finally a monthly steering committee meeting is held to hold people accountable, to track progress and benefits, as well as to review the pipeline and any new projects.

References

Gitlow, H., A. Oppenheim, R. Oppenheim, and D. Levine (2015), *Quality Management: Tools and Methods for Improvement*, 4th ed. (Naperville, IL: Hercher Publishing Company). This book is free online at hercherpublishing.com.

Gitlow, H. and D. Levine (2004), *Six Sigma for Green Belts and Champions: Foundations, DMAIC, Tools and Methods, Cases and Certification,* (Upper Saddle River, NJ: Prentice-Hall).

9

Overview of Six Sigma Management

What Is the Objective of This Chapter?

The objective of this chapter is for you to learn what you need to know about the fundamentals of Six Sigma including its non-technical and technical definitions, where it came from, its benefits, the key ingredient for success with Six Sigma, roles and responsibilities, and finally some terminology you need to know to implement Six Sigma at your organization. Six Sigma can be defined with both non-technical and technical definitions.

Non-Technical Definition of Six Sigma Management

You can think of the non-technical definition of Six Sigma management as your elevator pitch (60 second response). Six Sigma management can be defined as the relentless and rigorous pursuit of the reduction of variation in all critical processes to achieve continuous and breakthrough improvements that impact the bottom-line and/or top-line of the organization and increase customer satisfaction (Gitlow et al., 2015).

Technical Definition of Six Sigma

The technical definition is a bit more complicated to understand; hence we are placing it in Appendix 9.1 at the end of this chapter. For those readers with some background in statistics, read the appendix. For those readers without an understanding of basic statistics, skip it for now. Essentially, the technical definition states that a process should be improved so that it generates no more than 3.4 defects per million opportunities (Gitlow et al., 2015).

Where Did Six Sigma Come From?

Six Sigma was originally developed by Bill Smith at Motorola in 1985. He created the Six Sigma methodology to increase profitability while reducing defects by introducing the concept of *latent defect*, which revolved around reducing variation in processes that would then reduce defects to improve customer satisfaction and save money. Jack Welch at General Electric and Larry Bossidy at Allied Signal popularized the Six Sigma approach in the early and mid-1990s by attributing their increase in market capitalization to the results attained by their drive for Six Sigma quality (Gitlow and Levine, 2004).

Benefits of Six Sigma Management

There are two types of benefits from Six Sigma management: benefits to the organization and benefits to stakeholders. Benefits to an organization are gained through the continuous reduction of variation and centering of processes on their desired (nominal) levels (Gitlow et al., 2015). The benefits to the organization are the following:

- Improved process flows.
- Reduced total defects.
- Improved communication (provides a common language to everyone involved with a process).
- Reduced cycle times.
- Enhanced knowledge (and enhanced ability to manage that knowledge).
- Higher levels of customer and employee satisfaction.
- Increased productivity.
- Decreased work-in progress (WIP).
- Decreased inventory.
- Improved capacity and output.
- Increased quality and reliability.
- Decreased unit costs.
- Increased price flexibility.
- Decreased time to market.
- Faster delivery time.
- Conversion of improvements into hard currency.

Benefits to stakeholders are a by-product of the organizational benefits. The benefits to stakeholders include

- Stockholders receive more profit due to decreased costs and increased revenues.
- Customers are delighted with products and services.
- Employees experience higher morale and more satisfaction from joy in work.
- Suppliers enjoy a secure source of business.

Key Ingredient for Success with Six Sigma Management

The key ingredient for a successful Six Sigma management process is the commitment (not only support) of top management. Executives must have a burning desire to transform their organizations into Six Sigma enterprises. This means total commitment from the top to the bottom of the organization. Additionally, another ingredient for a successful Six Sigma management process is a labor force capable of following, and constantly improving, best practice methods for the processes they are part of. This requires personal discipline. My favorite example of employees desiring to exhibit personal discipline in following the best practice method is in training infantry on how to fight and protect themselves in a war zone. Why would someone not follow the best practice method that has been honed over thousands of years to keep themselves alive?

Six Sigma Roles and Responsibilities

Several jobs in an organization are critical to the Six Sigma management process. They are Senior Executive (CEO or President), Executive Committee (Senior Vice Presidents), Champion, Master Black Belt, Black Belt, Green Belt, and Process Owner. The roles and responsibilities of each of these jobs are described as follows (Gitlow et al., 2015; Gitlow and Levine, 2004).

Senior Executive

The Senior Executive provides the impetus, direction, and alignment necessary for Six Sigma's ultimate success. The Senior Executive should do the following:

- Study Six Sigma management.
- Lead the executive committee in linking objectives and metrics to Six Sigma projects (dashboards).
- Participate on appropriate Six Sigma project teams.
- Maintain an overview of the system to avoid suboptimization.
- Maintain a long-term view.
- Act as a liaison to Wall Street, explaining the long-term advantages of Six Sigma management, if appropriate.
- Constantly and consistently, publicly and privately, champion Six Sigma management.
- Improve the Six Sigma process.
- Conduct project reviews at the monthly operations review meeting.

The most successful, highly publicized Six Sigma efforts have one thing in common—unwavering, clear, and committed leadership from top management. There is no doubt in anyone's mind that Six Sigma is "the way we do business." Although it may be possible to initiate Six Sigma concepts and processes at lower levels, dramatic success is not possible until the Senior Executive becomes engaged and takes a leadership role.

Executive Steering Committee

The members of the executive committee are the top management of an organization; they report directly to the CEO. They should operate at the same level of commitment for Six Sigma management as the Senior Executive. They should *not* be allowed to follow their own management style in their areas even if they deliver good results. The reason for this is that creating a Six Sigma organization requires standardization of commitment to the Six Sigma management style from the entire C-suite. The members of the executive committee should do the following:

- Study Six Sigma management.
- Deploy Six Sigma throughout the organization.
- Prioritize and manage the Six Sigma project portfolio.
- Assign Champions, Master Black Belts, Black Belts, and Green Belts to Six Sigma projects.
- Conduct reviews of Six Sigma projects with the Senior Executive and within their own areas of control.
- Improve the Six Sigma process.
- Remove barriers to Six Sigma management or projects.
- Provide resources for the Six Sigma management process and projects.
- Improve the Six Sigma process.

Project Champion

Champions take an active sponsorship and leadership role in conducting and implementing Six Sigma projects. A Champion should be a member of the executive committee, or at least a trusted direct report of a member of the executive committee. She should have enough influence to remove obstacles or provide resources without having to go higher in the organization. Champions work closely with the executive committee, the project leader (called a Black Belt) assigned to their project, and the Master Black Belt (supervisor of Black Belts) overseeing their project. Champions have the following responsibilities:

- Commit to the high priority projects identified by the top management team.
- Develop and negotiate project objectives and charters with the executive committee.
- Select a Black Belt (or a Green Belt for a simple project) to lead the project team.
- Remove any political barriers or resource constraints to their Six Sigma project (run interference).
- Provide an ongoing communication link between their project team(s) and the executive committee.
- Help team members manage their resources and stay within the budget.

- Review the progress of their project in respect to the project's timetable.
- Keep the team focused on the project by providing direction and guidance.
- Ensure that Six Sigma methods and tools are being used in the project.
- Improve the Six Sigma process.

Process Owner

A Process Owner is the manager of a process. He has responsibility for the process and has the authority to change the process on his signature. The Process Owner should be identified and involved immediately in all Six Sigma projects relating to his area. A Process Owner has the following responsibilities:

- Be totally committed to the project's success.
- Be accountable for the monitoring and managing the output of his process.
- Empower the employees who work in the process to follow and improve the best practice method for the process.
- Focus the project team on the project charter.
- Assist the project team in remaining on schedule.
- Allocate the resources necessary for the project (people, space, and so on).
- Accept, manage, and sustain the improved process after completion of the Six Sigma project. (PDSA the process forever and prevent backsliding of the process.)
- Ensure that process objectives and indicators are linked to the organization's mission through the dashboard.
- Understand how the process works, the capability of the process, and the relationship of the process to other processes in the organization.
- Understand the relationship between the capability of a process and the demands management may put on the process in the form of targets, goals, deadlines, and so on.

Master Black Belt

A Master Black Belt takes on a leadership role as the keeper of the Six Sigma process and culture. She advises executives or business unit managers and leverages her skills with projects led by Black Belts and Green Belts.

Master Black Belts report directly to Senior Executives or business unit managers. There can be a dotted line relationship for a Black Belt between a Master Black Belt and an executive. The executive evaluates the results of the projects, while the Master Black Belt evaluates the skills set used to get the results in respect to the employees working on the project. Both the Master Black Belt and the executive work to improve the Black Belt's and/or Green Belt's skill set.

A Master Black Belt has successfully led many teams through complex Six Sigma projects. She is a proven change agent, leader, facilitator, and technical expert in Six Sigma management. Master Black Belt is a career path. It is always best for an organization to grow its own Master Black Belts. Unfortunately, sometimes it is impossible for an organization to grow its own Master Black Belts because it takes years of study, practice, and tutelage under a master to become a Master Black Belt. Ideally, Master Black Belts are selected from the Black Belts within an organization; however, sometimes circumstances require hiring Master Black Belts external to the organization. If this is the case, great care must be taken that the external Master Black Belt fits in with the culture of the organization.

Master Black Belts have the following responsibilities:

- Counsel Senior Executives and business unit managers on Six Sigma management.
- Help identify and prioritize key project areas in keeping with the mission.
- Continually improve and innovate the organization's Six Sigma process.
- Apply Six Sigma across both operations and transactions-based processes such as Sales, HR, IT, Facility Management, Call Centers, Finance, and so on.
- Coordinate Six Sigma projects from the dashboard.
- Teach Black Belts and Green Belts Six Sigma theory, tools, and methods.
- Mentor Black Belts and Green Belts.
- Improve the Six Sigma process throughout the organization.

Senior Master Black Belts have at least ten years of ongoing leadership experience and have worked extensively with mentoring the organizational leaders on Six Sigma management.

Black Belt

A Black Belt is a full-time change agent and improvement leader who may not be an expert in the process under study. The ideal candidate for a Black Belt is an individual who possesses the following characteristics:

- Has technical and managerial process improvement/innovation skills.
- Has a passion for statistics, systems theory, psychology (of individuals and teams), and the philosophy of science.
- Understands all process improvement methodologies; for example, PDSA cycle, DMAIC model, Lean, and so on.
- Has excellent communication and writing skills.
- Works well in a team format.
- Can manage meetings.
- Has a pleasant personality and is fun to work with.

- Communicates in the language of the client and does not use technical jargon.
- Is not intimidated by upper management (power).
- Has a customer focus.

The responsibilities of a Black Belt include the following:

- Help to prepare a project charter.
- Communicate with the Champion and Process Owner about progress of the project.
- Lead the project team.
- Select the team members for the project with the cooperation of the Process Owner.
- Schedule meetings and coordinate logistics.
- Help team members design experiments and analyze the data required for the project.
- Provide training in tools and team functions to project team members.
- Help team members prepare for reviews by the Champion and Executive Committee.
- Recommend additional Six Sigma projects.
- Lead and coach Green Belts leading projects limited in scope.

A Black Belt is a full-time quality professional who is mentored by a Master Black Belt but may report to a manager for his tour of duty as a Black Belt. An appropriate time frame for a tour of duty as a full-time Black Belt is two years. Black Belt skills and project work are critical to the development of leaders and high potential people within the organization.

Green Belt

A Green Belt is an individual who works on projects part-time (25%), either as a team member for complex projects or as a project leader for simpler projects. Green Belts are the "work horses" of Six Sigma projects. Most managers in a mature Six Sigma organization are Green Belts. Green Belt certification is a critical prerequisite for advancement into upper management in a Six Sigma organization. Managers act as Green Belts for their entire careers, as their style of management, unless they and their manager decide that he should pursue a Black Belt. Green Belts leading simpler projects have the following responsibilities:

- Refine a project charter.
- Review the project charter with the project's Champion.
- Select the team members for the project with the cooperation of the Process Owner.
- Communicate with the Champion, Master Black Belt, Black Belt, and Process Owner throughout all stages of the project.
- Facilitate the team through all phases of the project.
- Schedule meetings and coordinate logistics.

- Analyze data through all phases of the project.
- Train team members in the basic tools and methods through all phases of the project.

Green Belt Versus Black Belt Projects

Black Belt and Green Belt Six Sigma projects differ. Green Belt projects meet the following five criteria:

1. They tend to be less involved (for example, they have one CTQ and few Xs).
2. They do not deal with political issues.
3. They do not require many organizational resources.
4. They do not require significant capital investment or IT solutions to realize the gains identified during the project.
5. They utilize only basic statistical methods.

On the other hand, Black Belt projects tend to deal with more complex situations that may involve two or more possibly conflicting CTQs and many Xs, involve substantial political issues or are cross-functional in nature, require substantial organizational resources, need substantial capital investment to realize the gains made during the project, and utilize sophisticated statistical methods. One exception is in organizations where executives act as Green Belt team leaders because they control large budgets and are responsible for major systems/issues. In these cases, the executives get assistance from Black Belts or Master Black Belts.

Another exception is where a Process Owner with great area expertise takes on the mantle of team leader. This occurs if area expertise is more critical than Six Sigma expertise in the conduct of a project. In this situation the Black Belt takes more of a nonvoting, facilitator role, while the Process Owner/team leader is a voting member, more involved in the content. Black Belts take more of a formal leadership role early in the project, when the team is forming and storming and needs a lot of direction and support. As the team becomes more self-directed and comfortable with the tools, and as they begin to implement things, the process knowledge of the team leader becomes more important to success, and the Black Belt can become more of an observer, coach, and mentor to the team.

Six Sigma Management Terminology

Six Sigma practitioners use a lot of jargon. You must know the language of Six Sigma management if you want to use it (Gitlow and Levine, 2004). This is unfortunate, but necessary!

CTQ—CTQ is an acronym for Critical-to-Quality characteristic for a product, service or process. A CTQ is a measure of what is important to a stakeholder of a process, for example, waiting time in a physician's office by day, percentage of errors with ATM transactions for a bank's customers per month, or number of car accidents per month on a particular stretch of highway for a Department of Traffic, might all be CTQs. Six

Sigma projects are designed to improve CTQs. CTQs are the problematic metrics on a dashboard; they are also referred to as Ys.

Unit—A unit is the item (e.g., product or component, service or service step) or area of opportunity (e.g., an item, a service, a time period, a geographical area) to be studied with a Six Sigma project.

Defective—A nonconforming unit that is out of specification limits, which prevents the unit from performing its purpose. The unit must be reworked, thrown away, or sold at a lower price.

Defect—A defect is a nonconformance on one, of many possible, quality characteristics of a unit that causes customer dissatisfaction. For a given unit, each quality characteristic is defined by translating customer desires into specifications. It is important to operationally define each defect for a unit. A defect can make a unit defective, but not necessarily; that is, the unit may still serve its intended purpose. For example, if a word in a document is misspelled, that word may be considered a defect. A defect does not necessarily make a unit (in this case, the document) defective. For example, a water bottle can have a scratch on the outside (defect) and still be used to hold water (not defective). However, if a customer wants a scratch-free water bottle, that scratched bottle could be considered defective. A defect is a nonconforming aspect of a unit that does not cause the unit to fail to perform its function.

Defect opportunity—A defect opportunity is each circumstance in which a CTQ can fail to be met. There may be many opportunities for defects within a defined unit. For instance, a service may have four component parts. If each component part contains three opportunities for a defect, the service has 12 defect opportunities in which a CTQ can fail to be met. The number of defect opportunities generally is related to the complexity of the unit under study. Complex units experience greater opportunities for defects to occur than simple units. Hence, one change concept important to Six Sigma projects is the reduction of complexity. Reduction of complexity creates fewer defect opportunities.

Defects per unit (DPU)—Defects per unit refers to the average of all the defects for a given number of units; that is, the total number of defects for n units divided by n. If you are producing a 50-page document, the unit is a page. If there are 150 spelling errors, DPU is 150/50 or 3.0. If you are producing ten 50-page documents, then the unit is a 50-page document. If there are 75 spelling errors in all ten documents, DPU is 75/10 or 7.5.

Defects per opportunity (DPO)—Defects per opportunity refers to the number of defects divided by the number of defect opportunities. If there are 20 errors in 100 services with 1 defect opportunity per service, the DPU is 0.20 (20/100). However, if there are 12 defect opportunities per service, there would be 1,200 opportunities in 100 services. In this case, DPO would be 0.0167 (or, 20/1,200). (DPO may also be calculated by dividing DPU by the total number of opportunities.)

Defects per million opportunities (DPMO)—DPMO equals DPO multiplied by one million. Hence, for the previous the DPMO is (0.0167) x (1,000,000), or 16,700 defects per million opportunities.

Yield—Yield is the proportion of units within specification divided by the total number of units; that is, if 25 units are produced and 20 are good, the yield is 0.80 (20/25).

Rolled throughput yield (RTY)—Rolled throughput yield is the product of the yields from each step in a process. It is the probability of a unit passing through all "k" steps of a process and incurring no defects. RTY = $Y_1 * Y_2 ... Y_K$, where k = number of steps in a process, or the number of component parts or steps in a product or service. Each yield Y for each step or component must be calculated to compute the RTY. For those steps in which the number of opportunities is equal to the number of units then Y = 1-DPU. For those steps in which a large number of defects are possible, but only a few are observed (e.g., number of typographical, grammatical, spelling, errors in a document), the yield Y (the probability of no defects in a unit) can be found by $Y = e^{-DPU}$, where DPU is the defects per unit for the step. The values for $Y=e^{-DPU}$ for values of DPU from 1 through 10 are shown in Table 9.1.

Table 9.1 Values of $Y=e^{-DPU}$

Row	DPU	$Y=e^{-DPU}$
1	1	0.367879
2	2	0.135335
3	3	0.049787
4	4	0.018316
5	5	0.006738
6	6	0.002479
7	7	0.000912
8	8	0.000335
9	9	0.000123
10	10	0.000045

For example, if a process has three steps and the yield from the first step (Y_1) is 99.7%, the yield from the second step (Y_2) is 99.5%, and the yield from the third step (Y_3) is 89.7%, then the rolled throughput yield (RTY) is 88.98% (0.997 X 0.995 X 0.897).

Process sigma—Process sigma is a measure of the process performance determined by using DPMO and a stable normal distribution. Process sigma is a metric that allows for process performance comparisons across processes, departments, divisions, companies, and countries, assuming all comparisons are made from stable processes whose output follows the normal distribution. In Six Sigma terminology, the sigma value of a process is a metric used to indicate the number of defects per million opportunities, or how well the process is performing in respect to customer needs and wants.

The left side of Table 9.2 is used to translate DPMO statistics for a stable, normally distributed process with no shift in its mean (0.0 shift in mean) over time into a process sigma metric, assuming that defects occur at only one of the specifications if there are lower and upper specifications. The right side of Table 9.2 is used to translate DPMO statistics for a stable, normally distributed process that has experienced a 1.5 sigma shift in its mean over time into a process sigma metric.

Table 9.2 Process Sigma—DPMO Table

Assume 0.0 Sigma Shift in Mean				Assume 1.5 Sigma Shift in Mean			
Process σ Level	Process DPMO	Process σ Level	Process DPMO	Process σ Level	Process DPMO	Process σ Level	Process DPMO
0.10	460,172.1	3.30	483.5	0.10	919,243.3	3.10	54,799.3
0.20	420,740.3	3.40	337.0	0.20	903,199.5	3.20	44,565.4
0.30	382,088.6	3.50	232.7	0.30	884,930.3	3.30	35,930.3
0.40	344,578.3	3.60	159.1	0.40	864,333.9	3.40	28,716.5
0.50	308,537.5	3.70	107.8	0.50	841,344.7	3.50	22,750.1
0.60	274,253.1	3.80	72.4	0.60	815,939.9	3.60	17,864.4
0.70	241,963.6	3.90	48.1	0.70	788,144.7	3.70	13,903.4
0.80	211,855.3	4.00	31.7	0.80	758,036.4	3.80	10,724.1
0.90	184,060.1	4.10	20.7	0.90	725,746.9	3.90	8,197.5
1.00	158,655.3	4.20	13.4	1.00	691,462.5	4.00	6,209.7
1.10	135,666.1	4.30	8.5	1.10	655,421.7	4.10	4,661.2
1.20	115,069.7	4.40	5.4	1.20	617,911.4	4.20	3,467.0
1.30	96,800.5	4.50	3.4	1.30	579,259.7	4.30	2,555.2
1.40	80,756.7	4.60	2.1	1.40	539,827.9	4.40	1,865.9
1.50	66,807.2	4.70	1.3	1.50	500,000.0	4.50	1,350.0
1.60	54,799.3			1.60	460,172.1	4.60	967.7
1.70	44,565.4	Process σ Level	Defect per billion opportunities	1.70	420,740.3	4.70	687.2
1.80	35,930.3			1.80	382,088.6	4.80	483.5
1.90	28,716.5			1.90	344,578.3	4.90	337.0
2.00	22,750.1	4.80	794.4	2.00	308,537.5	5.00	232.7
2.10	17,864.4	4.90	479.9	2.10	274,253.1	5.10	159.1
2.20	13,903.4	5.00	287.1	2.20	241,963.6	5.20	107.8
2.30	10,724.1	5.10	170.1	2.30	211,855.3	5.30	72.4
2.40	8,197.5	5.20	99.8	2.40	184,060.1	5.40	48.1
2.50	6,209.7	5.30	58.0	2.50	158,655.3	5.50	31.7

Assume 0.0 Sigma Shift in Mean				Assume 1.5 Sigma Shift in Mean			
Process σ Level	Process DPMO	Process σ Level	Process DPMO	Process σ Level	Process DPMO	Process σ Level	Process DPMO
2.60	4,661.2	5.40	33.4	2.60	135,666.1	5.60	20.7
2.70	3,467.0	5.50	19.0	2.70	115,069.7	5.70	13.4
2.80	2,555.2	5.60	10.7	2.80	96,800.5	5.80	8.5
2.90	1,865.9	5.70	6.0	2.90	80,756.7	5.90	5.4
3.00	1,350.0	5.80	3.3	3.00	66,807.2	6.00	3.4
3.10	967.7	5.90	1.8				
3.20	687.2	6.00	1.0				

The 1.5 sigma shift in the mean is a standard developed by the Motorola Corporation but may not be appropriate for your process. Consequently, the authors do not recommend using the process sigma as a useful metric of process performance. However, it is valuable in explaining what happens to the DPMO as the process mean shifts away from nominal in a generalized sense.

For example, suppose a process has three independent steps, each with a 95% yield. The RTY for the process is 85.74% ($0.95 \times 0.95 \times 0.95$) and the DPO is 0.1426 (DPO = 1.0 − RTY = 1.0 − 0.8574), assuming each step has only one opportunity so that DPU and DPO are the same. The DPMO for the process is 142,600 (DPMO = DPO × 1,000,000). The process sigma metric is obtained, assuming a 1.5 sigma shift in the process mean over time, by looking down the DPMO column to the two numbers bracketing to 142,600. The actual process sigma metric lies between the corresponding two bracketing process sigma metrics. In this example, 142,600 is bracketed by a DPMO of 135,661 and a DPMO of 158,655. The corresponding bracketing process sigma metrics are 2.60 and 2.50. Hence, the actual process sigma metric is approximately 2.55.

Next Steps: Understanding the DMAIC Model

Now that you understand the fundamentals of Six Sigma, the next step is to take you through each phase of the DMAIC model. Then you will be able to execute Six Sigma projects to get your processes as close to Six Sigma quality as possible.

Takeaways from This Chapter

- Six Sigma management is the relentless and rigorous pursuit of the reduction of variation in all critical processes.
- In a Six Sigma process with a 0.0 shift in the mean there are six standard deviations between the process mean and the nearest specification limit. This will result in one defect per billion opportunities beyond each specification limit.

- In a Six Sigma process with a 1.5 sigma shift in the mean there will be only 3.4 defects per million opportunities beyond the nearest specification limit.
- The benefits of Six Sigma are numerous and include financial impact and increased customer and employee satisfaction.
- The number one key success factor to implementation of Six Sigma is commitment of top management. The number two key success factor for implementation of Six Sigma is a disciplined workforce that can follow and improve the process in which they work.
- There are numerous roles and associated responsibilities that are key to successful Six Sigma implementation.
- Six Sigma is full of jargon that you must understand.

References

Gitlow, H., A. Oppenheim, R. Oppenheim, and D. Levine (2015), *Quality Management: Tools and Methods for Improvement*, 4th ed. (Naperville, IL: Hercher Publishing Company). This book is free online at hercherpublishing.com.

Gitlow, H. and D. Levine (2004), *Six Sigma for Green Belts and Champions: Foundations, DMAIC, Tools and Methods, Cases and Certification* (Upper Saddle River, NJ: Prentice-Hall).

Additional Readings

Hahn, G. J., N. Dogannaksoy, and R. Hoerl, "The Evolution of Six Sigma," *Quality Engineering*, 2000, 12, 317–326.

Harry, Mikel, and Rich Schroeder (2000), *Six Sigma: The Breakthrough Management Strategy Revolutionizing the World's Top Corporations* (New York: Doubleday) ISBN 0385494378.

Kubiak, T. M. and Donald W. Benbow (2009), *The Certified Six Sigma Black Belt Handbook*, 2nd ed. (Milwaukee, WI: ASQ Quality Press).

Lowenthal, Jeffrey N. (2002), *Six Sigma Project Management: A Pocket Guide* (Milwaukee, WI: ASQ Quality Press).

Naumann, Earl and Steven Hoisington (2001), *Customer Centered Six Sigma: Linking Customers, Process Improvement, and Financial Results* (Milwaukee, WI: ASQ Quality Press).

Pyzdek, Thomas (2003), The *Six Sigma Handbook*, 2nd ed. (New York: McGraw-Hill).

Snee, R. D., "The Impact of 'Six Sigma' on Quality," *Quality Engineering*, 2000, 12, ix–xiv.

Appendix 9.1 Technical Definition of Six Sigma Management

The Normal Distribution. The term "Six Sigma" is derived from a stable and normal distribution in statistics. Many observable phenomena can be graphically represented as a bell-shaped curve or a normal distribution as illustrated in Figure 9.1.

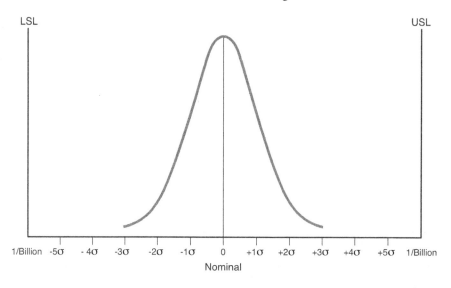

Figure 9.1 Normal distribution with mean ($\mu=0$) and standard deviation ($\sigma=1$)

When measuring any process, it can be shown that its outputs (services or products) vary in size, shape, look, feel, or any other measurable characteristic. The typical value of the output of a process is measured by a statistic called the *mean* or *average*. The variability of the output of a process is measured by a statistic called the *standard deviation*. In a normal distribution, the interval created by the mean plus or minus 2 standard deviations contains 95.44% of the data points, or 45,600 data points per million (or sometime called parts per million denoted ppm) are outside the area created by the mean plus or minus 2 standard deviations [(1.00 - .9544 = .0456) x 1,000,000 = 45,600]. In a normal distribution the interval created by the mean plus or minus 3 standard deviations contains 99.73% of the data, or 2700 ppm are outside the area created by the mean plus or minus 3 standard deviations [(1.00 - .9973 = .0027) x 1,000,000 = 2,700]. In a normal distribution the interval created by the mean plus or minus 6 standard deviations contains 99.9999998% of the data (refer to Figure 9.1), or 2 data points per billion data points (ppb) outside the area created by the mean plus or minus 6 standard deviations.

Relationship Between VoP and VoC. Six Sigma management promotes the idea that the distribution of output for a process (the Voice of the Process) should take up no more than half of the tolerance allowed by the specification limits (the Voice of the Customer), assuming measurement data from a stable and normal distribution of output whose mean can shift by as much as 1.5 standard deviations over time. A shift of 1.5 standard deviations in a process's mean over time has been suggested, by organizations such as Motorola, General Electric, and Allied-Signal, as a common phenomenon early in an organization's Six Sigma effort. Statisticians may argue about the correctness of the assumption of a 1.5 sigma shift in mean, but for the purposes of this text, we accept this common industrial assumption.

Figure 9.2 shows the Voice of the Customer as spoken in the language of a nominal value, m, and lower and upper specification limits, LSL and USL, for a quality characteristic. It also shows the Voice of the Process as spoken in the language of a distribution of process output for a quality characteristic. Figure 9.2 assumes that the distribution of process output (the Voice of the Process) is measurement data, is stable, is normally distributed, that the average process output is equal to nominal, and that the distance between nominal and either specification limit is three times the standard deviation of the process's output. If these five conditions are met, then we can call the process presented in Figure 9.2 a 3-sigma process with no shift in its mean. A 3-sigma process with no shift in its mean is a process that will generate 2,700 defects per million opportunities; that is, 99.73% of its output will be between the lower specification limit (LSL) and the upper specification limit (USL), at least in the near future; in other words, 1,350 DPMO will be above the upper specification limit and 1,350 DPMO will be below the lower specification limit. Recall that 99.73% of the output from a stable and normally distributed process will be between the mean plus three standard deviations ($\mu + 3\sigma$) and the mean minus three standard deviations ($\mu - 3\sigma$). To state this statistically, $P(\mu - 3\sigma < X < \mu + 3\sigma) = .9973$.

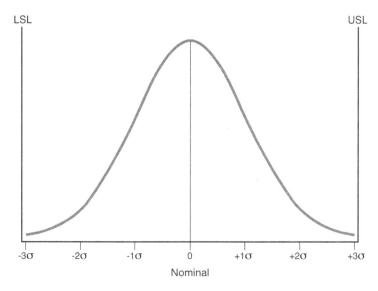

Figure 9.2 Comparison of Voice of the Customer and Voice of the Process for a 3-sigma process with no shift in its mean

Figure 9.3 shows the distribution of the number of days to complete a monthly accounting report. The distribution is stable and normally distributed, with an average of 7 days and a standard deviation of 1 day. Figure 9.3 also shows a nominal value of 7 days, a lower specification limit of 4 days, and an upper specification limit of 10 days. The accounting reporting process is referred to as a 3-sigma process without a shift in mean because the process mean plus or minus three standard deviations is equal to the specification limits: in other words, USL= $(\mu + 3\sigma)$ and LSL = $(\mu - 3\sigma)$. As stated earlier, this scenario will yield 2,700 defects per million opportunities, or one early or late monthly report in 30.86 years [(1 / .0027) / 12].

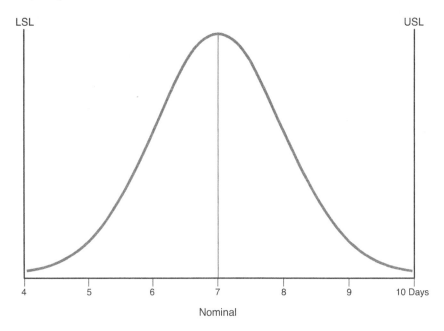

Figure 9.3 3-sigma accounting reporting process with no shift in its mean

Figure 9.4 shows the Voice of the Customer and the Voice of the Process for a quality characteristic that is represented by measurement data, is stable, is normally distributed, has an average process output which can shift by as much as 1.5 standard deviations on either side of nominal, and has a distance between nominal and either specification limit of three standard deviations of process output. When these five conditions are met, we call the process a 3-sigma process with a 1.5-sigma shift in the mean. Such a process will generate 66,807 defects per million opportunities outside of the nearest specification limit, at least in the near future.

Figure 9.5 shows the accounting scenario of Figure 9.3, but the process average shifts by 1.5 standard deviations (the process average is shifted down or up by 1.5 standard deviations, or 1.5 days, from 7.0 days to either 5.5 days or 8.5 days) over time. The 1.5 standard deviation shift in the mean results in 66,807 defects per million opportunities, or one early or late monthly report in 1.25 years [(1 / .066807) / 12].

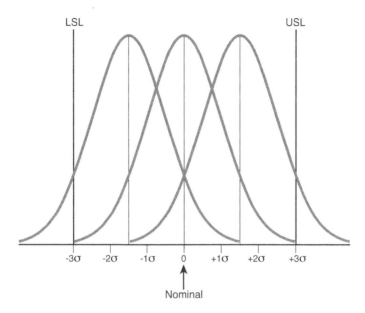

Figure 9.4 Comparison of Voice of the Customer and Voice of the Process for a 3-sigma process with a 1.5 sigma shift in its mean

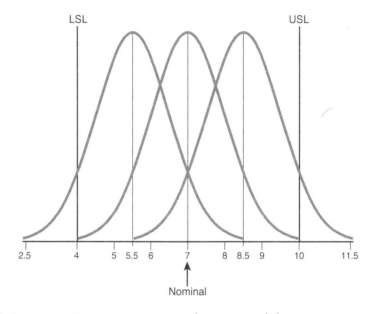

Figure 9.5 3-sigma accounting reporting process with a 1.5 sigma shift in its mean

In Figure 9.6 we have the scenario of Figure 9.2, except the Voice of the Process only takes up half the distance between the specification limits. The process mean remains the same as in Figure 9.2, but the process standard deviation has been reduced to 1/2 day through application of process improvement tools and methods. In this case, the resulting output will exhibit two defects per billion opportunities, or 99.9999998% of its output will be between the lower and upper specification limits, at least in the near future; in other words, one defect per billion opportunities will be outside the upper and lower specification limits. This is called a 6-sigma process with no shift in its mean.

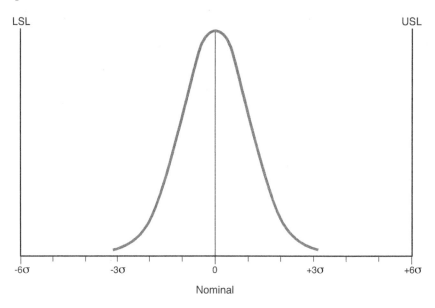

Figure 9.6 6-sigma process with no shift in its mean

Figure 9.7 shows the output of the monthly accounting process in Figure 9.3, except its standard deviation has been reduced to 1/2 day through process improvement activities. This accounting process is now a 6-sigma process with no shift in mean. Its output will exhibit two defects per billion opportunities, or one early or late monthly report in 41,666.667 years [(1 / .000000002) / 12].

Figure 9.8 shows the Voice of the Customer and the Voice of the Process for a quality characteristic that is represented by measurement data, is stable, is normally distributed, where the average process output can shift by as much as 1.5 standard deviations on either side of nominal, and where the distance between nominal and either specification limit is six standard deviations of process output. If these five conditions are met, then we can call the process presented in Figure 9.8 a 6-sigma process with a 1.5-sigma shift in the mean. This is a process that will generate 3.4 defects per million opportunities outside of the nearest specification limit, at least in the near future.

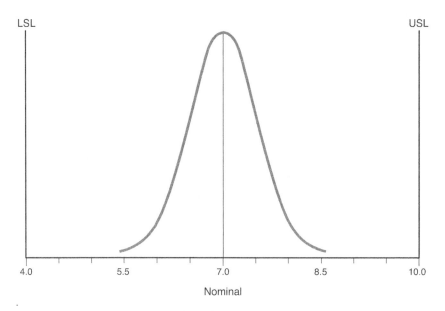

Figure 9.7 6-sigma accounting process with no shift in its mean

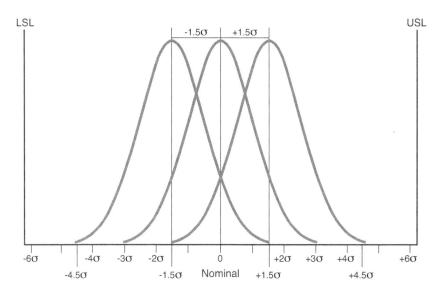

Figure 9.8 6-sigma process with 1.5 sigma shift in its mean

Figure 9.9 shows the same scenario for the accounting process: The process average shifts by 1.5 standard deviations (the process average is shifted down or up by 1.5 standard deviations, or 0.75 days = 1.5 × 0.5 days, from 7.0 days to either 6.25 days or 7.75 days) over time. The 1.5 standard deviation shift in the mean results in 3.4 defects per million opportunities outside of the nearest specification limit, or one early or late monthly report in 24,510 years [(1 / .0000034) / 12]. This is the definition of Six Sigma quality.

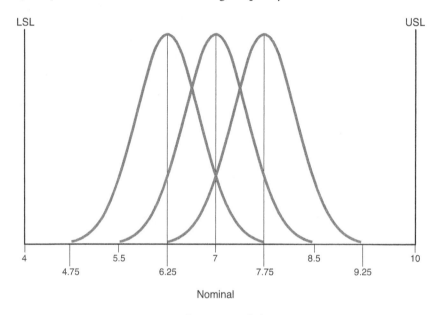

Figure 9.9 Six Sigma accounting process with 1.5 sigma shift in its mean

Does "Six Sigma" Matter? The difference between a three sigma process (66,807 defects per million opportunities) and a six sigma process (3.4 defects per million opportunities) can be seen in a service with 20 component steps. If each of the 20 component steps has a quality level of 66,807 defects per million opportunities, assuming each step does not allow rework, the likelihood of a defect at each step is 0.066807 (66,807/1,000,000). By subtraction, the likelihood of a defect free step is 0.933193 (1.0 − 0.066807). Consequently, the likelihood of delivering a defect free final service is 25.08 percent. This is computed by multiplying 0.933193 by itself 20 times ([1.0 − 0.066807]^{20} = 0.2508 = 25.08%). However, if each of the 20 component parts has a quality level of 3.4 defects per million opportunities (0.0000034), the likelihood of delivering a defect free final service is 99.99932% ([1.0 − 0.0000034]^{20} = 0.99999966^{20} = 0.9999932 = 99.99932%). A three sigma process generates 25.08% defect-free services, while a six sigma process generates 99.99932% defect-free services. The difference between the three sigma process and the Six Sigma process is dramatic enough to certainly believe that Six Sigma level of performance matters, especially with more complex processes with a greater number of steps or activities.

10
DMAIC Model: "D" Is for Define

What Is the Objective of This Chapter?
The objective of this chapter is to take you through the various steps of the Define phase of the Six Sigma DMAIC model so that you can apply them on projects at your organization. We use a case study to demonstrate how the steps of each phase are executed in real world projects.

Purpose of the Define Phase
Remember, the Six Sigma DMAIC model is based on the scientific method and it is used to improve a current process, product, or service. The easiest way to explain the DMAIC model is to think of it in terms of (CTQ or Y) (the problematic metric you are trying to improve) is a function of one or more Xs (the steps in the process that are contributing to the CTQ being problematic).

For example, suppose that it is believed that patient waiting time is affected by insurance type, physician, and availability of an examination room. This could be stated as follows:

> Patient waiting time is a function of insurance type, physician, and availability of an examination room—each of them could be responsible for a long wait time. If we call patient waiting time the CTQ, insurance type (X_1), physician (X_2), and availability of an examination room (X_3), we can write the relationship between the (CTQ) and the Xs as follows:

- CTQ = $f(X_1, X_2, X_3)$ where:
  - CTQ is a measure of patient waiting time.
  - f is the symbol of relationship.
  - X_1 is insurance type.
  - X_2 is physician.
  - X_3 is availability of an examination room.

The five phases—Define, Measure, Analyze, Improve, and Control—take a series of steps aimed at either moving your CTQ in a specified direction or reducing variation in the output of CTQ.

This chapter focuses on the first phase, the Define phase, whose purpose is to define the problem, the rationale for doing the project, and the metrics to measure it—CTQ (Critical to Quality characteristic).

The four main deliverables for the Define phase are

- Project charter
- SIPOC analysis
- Voice of the Customer analysis
- Definition of the CTQ and initial draft of the project objective

At the end of the Define phase, the team conducts a tollgate review with the Project Champion and Process Owner. This is where the team reviews what they have learned in the Define phase and a "go-no go" decision is made by the Project Champion, Process Owner, and Black Belt or Master Black Belt—in other words, is the problem we thought was a problem still a significant enough problem for the team to continue with the project? If yes, and if everyone is satisfied with the team's work, the team proceeds to the Measure phase. If no, the project is abandoned.

TIP

Don't get too emotionally involved with a project because there are times you may need to cut your losses, pull the rip cord, and move on. It's like being on a bad date with someone who at first you thought was Mr. or Miss Wonderful but then realize is more like Mr. or Miss Boring. Save yourself the aggravation later, find the nearest exit ASAP, and ditch them now before the train wreck catches fire. Of course, be polite.

The Steps of the Define Phase

Recall that the major steps to the Define phase of the DMAIC model are the following: activate the Six Sigma team, prepare a project charter, do a SIPOC analysis, do a Voice of the Customer analysis, define the CTQ and create an initial draft of the project objective, and complete a tollgate review. All these steps are necessary to get a Six Sigma project off the ground or stopped.

Activate the Six Sigma Team

Remember from Chapter 8, "Project Identification and Prioritization: Building a Project Pipeline," that we choose the Six Sigma DMAIC methodology if

- The scope is narrow.
- The root cause is not known.
- We want to stabilize (eliminate special causes of variation), decrease common causes of variation, and/or shift the mean of a problematic CTQ.
- The problem is complex.

Our first step is to bring out the high level project charter (see Table 10.1) that we created during the project identification and prioritization phase. The Project Champion, Process Owner, and Master Black Belt review it to make sure it is still accurate and update it if necessary, and they assign a project leader (Black Belt) to lead the project.

Table 10.1 High Level Project Charter

Project Name:	Identify the issue that has been identified as a potential process improvement (PI) opportunity.
Department/Area Name:	What department or area will be impacted by the project?
High Level Problem Description:	Provide a concise description of the issue that needs to be addressed by the project team.
Project Type:	Methodology expected to be used (PDSA, DMAIC, Lean, etc.).
Time to Complete:	Estimated time to complete project using a Gantt chart.
Problematic Objective:	Problematic objective from the dashboard or other source.
Problematic Indicator:	Metric that measures the status of the problematic objective.
Project Scope:	Process steps (flowchart) or functional boundaries of the proposed project. Where does the project start and where does it stop?
Potential Financial Impact:	Projected result of the project and anticipated savings or revenues to be realized.
Current Baseline Performance:	Actual current performance of problematic indicator.
Project Champion:	Senior executive who reviews projects, removes impediments for the team, and can secure adequate resources and support for the project team.
Process Owner:	The manager of the process under study who has the authority to change the process.
IT Champion:	Member of the IT department who can help the team estimate the IT impact of the project.
Finance Champion:	Member of the Finance department who can help the team estimate financial impact of the project.
Project Leader:	Process improvement expert who will be the project leader.
Potential Team Members:	Employees who will work on the project part-time.

Project Champions and Master Black Belts rely on their understanding of the organization and their employees to assign the proper project leader (Black Belt) to lead the project, as well as assign team members to the project to leverage strengths, weaknesses, and time availability to execute the project.

Once the project leader (Black Belt) is assigned she rallies the troops with a kickoff meeting that includes

- Background on the project (review high level project charter)
- Roles and responsibilities of the project, which is part of the project charter created in the following section
- Methodology to be used for the project (e.g., PDSA, DMAIC, etc.)
- Project plan with milestones, which is part of the project charter
- Creation of a communication plan, which is part of the project charter
- Creation of a risk abatement plan, which is part of the project charter
- Agree on weekly meeting times

Now that the team is activated, the project leader (Black Belt) assigns tasks, and the team works through the steps of the Define phase described in the following sections.

Project Charter

The first step in the Define phase is completing the project charter (Gitlow and Levine, 2004). The project charter consists of the following:

- Business case
- Problem statement
- Goal statement
- Scope
- A project plan with milestones
- High level list of benefits
- Roles and responsibilities

Business Case

The purpose of the business case is to get the Six Sigma project team members to answer a set of partially redundant questions on the benefits and costs of a project. The questions are

- **What is the name of the process?** What is the process that this project is centered around?
- **Who is the Process Owner?** Who can change the structure (flowchart) of the process under his or her signature? Hint: the Process Owner.

> **TIP**
>
> If you can't identify the Process Owner or if the Process Owner is not 100% committed to the project, run away from the project because you will spend a ton of time on the project and chances are that your findings will never get implemented due to the lack of a Process Owner or his commitment. Sometimes there are multiple Process Owners; this is called a cross-functional process. In this case, all the Process Owners must buy into and be involved in the project, or your time spent on the project will be lost and your findings will be dead on arrival (DOA).

- **What is the aim of the process?** What is the aim of the process that this project is centered around?
- **Why do the project at all?** Why are we doing the project in the first place?

> **TIP**
>
> Be careful with politics. This is where the politics can come into a project. Sometimes politics shows itself as one person trying to show what an idiot another person is by initiating a project in their area with the secret intention of making them look foolish. Be careful you are not being used as a political pawn to be sacrificed in the name of politics.

- **Why do the project now?** There has to be a reason that the project is being executed now; what is it?
- **What are the consequences of not doing the project?** If we don't execute on this project, what is going to happen?
- **What other projects have higher or equal priority, if any?** When we come to the go-no go decision this helps us ascertain if perhaps another project has a higher priority. Be careful of politics here also!
- **What business objectives are supported by the project?** Are any objectives in our annual or long-term strategic plan supported by this project?

Problem Statement

The purpose of the problem statement is to describe the problem or opportunity or objective in clear, concise, and measurable terms. The problem statement answers the question: What pain is the organization experiencing that this Six Sigma project could reduce or eliminate?

> **TIP**
>
> Remember, pain is felt at all levels in an organization differently. However, the most pain is felt at the lowest levels of the organization because the workers are the ultimate recipients of policies and procedures. For example, policies and procedures for routine janitorial work result in clean and nice smelling restrooms for the C-suite, but dirty and smelly restrooms for the manual labor force. If you don't believe us, just check the next time you are in a factory. On second thought, you may just want to take our word on it!

Goal Statement

The goal statement describes the team's improvement objective. It begins with a verb such as *reduce, eliminate, control,* or *increase*; for example:

- Reduce costs.
- Increase revenues.
- Eliminate waste.
- Decrease variation.
- Control deviations from the budget; this assumes a rational budget.

Scope

The project scope is created by answering the following questions:

- **What are the process boundaries?** What are the starting and stopping points of the project, where will the high level flowchart start and where will it stop?
- **What, if anything, is out of bounds?** What is outside the scope of the project? This may include political issues or resources issues, to name a couple of issues.
- **What resources are available for the project?** What do you need to execute on this project? Ask now or forever hold your peace.

  For the most part, in this book, we will be talking about projects that do not require capital investment to resolve the process problems being addressed by a project.

- **Who can approve expenditures?** Who has the cash if you need it?
- **How much can the team spend beyond its budget without authority?** How much more resources can you get if you need it? This refers to dollars, facilities, personnel, and any other resources needed by a project.
- **What are the obstacles and constraints of the project?** There are always obstacles and constraints; better to start thinking about them sooner than later. Make sure to keep a running list using an issues and risk log so your Black Belt or your Champion

can take care of them before they become big problems. This is where your Project Champion and Process Owner earn their salary as they may need to step in if anything gets escalated up the food chain.

- **What time commitment is expected of team members?** How much time are different team members going to work on the project? Projects need time and effort from the personnel who are on the teams. If their managers will not release them from their regular work to work on the project, then you really need a strong project Champion to kick some supervisory butt to get them the time.
- **What will happen to each team member's regular job while he or she is working on the project?** Are they expected to work more time to participate on the project, with no compensation? Will someone cover their regular job for them?

> **TIP**
>
> When creating the project scope and the business case you probably want to sit down with someone who really knows the process under study such as the Process Owner. It is important to make sure the scope is manageable and narrow enough that the team can execute. Scoping a project is critical. You can't pick a problem like solving world hunger or boiling the ocean because it is way too huge; on the other hand, you don't want to pick a project like moving the soda machine to the optimal position in the factory because it is way too unimportant. On second thought, if you are a big soda drinker, it may be important to you.

Project Plan with Milestones

One of the most important tasks the project (team) leader has is the creation of a project plan to keep the team on track and keep momentum and interest in the project strong.

A Gantt chart is used by the project leader (Black Belt) to construct a schedule for the project and list any milestones. It is a bar chart that plots tasks and subtasks against time. Once a list of tasks and subtasks has been created for each phase of a project, responsibilities can be assigned for each task or subtask. Next, beginning and finishing dates can be scheduled for each task and subtask. Finally, any milestones relevant to a task or subtask are placed on the Gantt chart. Finally the project leader (Black Belt) adds in the status for each task and subtask such as "TBD" for to be done, "WIP" for work in progress, or "Done" for completed tasks or subtasks. It is the responsibility of the project leader (Black Belt) to constantly update the project plan throughout the project and modify it when necessary.

A generic Gantt chart is shown in Table 10.2.

Table 10.2 Generic Six Sigma Project Gantt Chart

Task	Responsible	Status	Jan	Feb	Mar	Apr	May	Jun	Jul	Aug
Define	List person or persons responsible for each task and subtask.	Note status of each task or subtask such as TBD, WIP, Done.								
List Define phase tasks and subtasks.										
Measure										
List Measure phase tasks and subtasks.										
Analyze										
List Analyze phase tasks and subtasks.										
Improve										
List Improve phase tasks and subtasks.										
Control										
List Control phase tasks and subtasks.										

Benefits and Costs

Six Sigma projects have both soft benefits and hard benefits. Examples of soft benefits include improving quality and morale, and decreasing cycle time. Examples of hard (financial) benefits include increasing revenues or decreasing costs. The Master Black Belt or Process Improvement Executive "guesstimates" the dollar impact of the process improvement project so management has an idea of its financial impact during the project prioritization process. This "guesstimate" is refined through iterative learning as the project proceeds and more investigation is done.

There are two taxonomies for classifying the potential cost related benefits that may be realized from a process improvement project.

Taxonomy 1: Cost Reduction Versus Cost Avoidance

Cost reduction includes costs that fall to the bottom of the profit and loss statement. A cost reduction can be used to offset price, increase profit, or can be reinvested elsewhere by management. Cost reductions are calculated by comparing the most recent accounting period's actual costs with the previous accounting period's actual costs.

Cost avoidance includes those costs that can be reduced, if management chooses to do so, but until action is taken no real costs are saved. Examples include reducing labor hours needed to produce some fixed volume of work. Unless the headcount is reduced, or an increase in the volume of work is completed with the same headcount, there are no real savings. It is important to understand that decreasing headcount cannot be associated with Six Sigma. The minute employees feel Six Sigma is being used to get rid of employees, it is the end of Six Sigma. So any headcount reduction means that the employees who can be eliminated must be redeployed elsewhere in the organization at no loss to them. It may mean that they are held on for a period of time with full pay until they can be redeployed. This is critical! The impact of cost avoidance is not visible on the profit and loss statement and is difficult to define but is still important in meeting organizational goals.

Taxonomy 2: Tangible Costs Versus Intangible Costs

Tangible costs are easily identified, for example, the costs of rejects, warranty, inspection, scrap, and rework. Intangible costs are difficult to measure—for example, the costs of long cycle times, many setups, expediting costs, low productivity, engineering change orders, low employee morale, turnover, low customer loyalty, lengthy installations, excess inventory, late delivery, overtime, lost sales, and customer dissatisfaction, to name a few. It is important to realize that some of the most important benefits are unknown and unknowable. Hence, the "guesstimate" of benefits in the Define phase often identifies a minimum estimate of intangible benefits.

The Master Black Belt or Process Improvement Executive develops a formula to "guesstimate" the potential benefits that the organization may realize due to the process improvement project.

For example, here is a possible formula:

Cost reductions   _____
PLUS Cost Avoidance   _____
PLUS Additional Revenue   _____
LESS Implementation Costs   _____
EQUALS Financial Benefits   _____

Roles and Responsibilities

The roles and responsibilities of team members must be clearly specified and agreed to by appropriate managers and supervisors. The roles and responsibilities are shown in Table 10.3.

Table 10.3 Roles and Responsibilities

Role	Responsibility	Stakeholder Signature	Date	Supervisor's Signature
Project Champion				
Process Owner				
Project Leader (Black Belt)				
Finance Rep				
IT Rep				
Team Member 1				
Team Member 2				
Team Member 3				
Team Member 4				

Table 10.3 indicates the supervisor's commitment to the project when she signs the form. Remember, the supervisor must be committed, not just supportive. Supportive just doesn't cut it. The difference between commitment and support is the same as the difference between bacon and eggs. The pig is committed, while the chicken is supportive. This is serious business!

Communication Plan

Recall from Chapter 6, "Non-Quantitative Techniques: Tools and Methods," that a communication plan is created for a project to identify, inform, and appease concerns regarding a process improvement event because many people will typically be affected by it. Implementing new processes typically involves the need for employees to change their behaviors.

Having a communication plan allows the team to be proactive in informing people in the organization as to what is going on so there are no surprises and so no one is caught off guard when it is time to implement solutions. For your convenience, a generic communication plan form is shown in Table 10.4.

Table 10.4 Communication Plan Outline

#	Event/ Communication	Participants/ Audience	Medium	Frequency	When	Lead	Scheduled?	Status	Notes
1									
2									
3									
4									
5									
6									

Risk Abatement Plan

A Failure Modes and Effects Analysis (FMEA) is then created to identify and mitigate risks to the success of the project as a whole as seen in Table 10.5. When using FMEA here the critical parameters are the various risks to the project's success.

Top management's commitment is one of the critical parameters in the Failure Modes and Effect Analysis shown in Table 10.5, regardless of the process being improved. As stated previously, support just doesn't cut it. Two other additions to the FMEA are the commitment of the Process Owner and the Process Champion. If any of these three critical parameters have a high Risk Priority Number, the project is in serious trouble, and the team should consider abandoning the project.

SIPOC Analysis

The second main step of the Define phase requires that team members perform a SIPOC analysis (Gitlow and Levine, 2004). SIPOC analysis is discussed in detail in Chapter 6. Recall, a SIPOC analysis is a simple tool for identifying the Suppliers and their Inputs into a process, the high level steps of a Process, the Outputs of the process, and the Customer segments interested in the outputs.

Table 10.5 Failure Modes and Effects Analysis

Critical Parameter	Potential Failure Mode	Potential Failure Effect	SEV	Potential Causes	OCC	Current Controls	DET	RPN	Recommended Action	Responsibility and Target Date	Date Action Taken	SEV	OCC	DET	RPN

> **TIP**
>
> *Customer segments* is a term of art that really means stakeholder segments; for example, employee segments (top management, supervisors, and hourly workers), investors (individuals, institutions, etc.), regulators, suppliers, and the environment, to name a few. You must also consider the needs and wants of today's stakeholders and future stakeholders. Remember, today's 5-year-old who wants McDonald's for dinner every night may want salmon and broccoli in 30 years. Do you want to be the company doing business with her?

The format of a SIPOC analysis is shown in Table 10.6.

Suppliers: Team members identify relevant suppliers by asking the following questions:

- Where does information and material come from?
- Who are the suppliers?
- Don't forget the Human Resources department as a source of inputs into the process (employees)!

Inputs: Team members identify relevant inputs by asking the following questions:

- What do your suppliers give to the process?
- What effect do the inputs (Xs) have on the process?
- What effect do the inputs (Xs) have on the CTQs (Outputs)?
- Don't forget the Human Resources department as a source of inputs into the process (employees)!

Process: Team members create a high level flowchart of the process taking particular care to identify the beginning and ending points of the process.

Outputs: Team members identify outputs of the process by asking the following questions:

- What products or services does this process make, both intended and unintended?
- What are the outputs critical to the customer's perception of quality? Or, lack of quality? These outputs are called *critical to quality (CTQ) characteristics* of the process.
- Are there any unintended outputs from the process that may cause problems?

Customers: Team members identify relevant customers (market segments) by asking the following questions:

- Who are the customers (stakeholders) or market segments of this process?
- Have we identified the outputs (CTQs) for each market segment?

Table 10.6 SIPOC Analysis

SUPPLIERS (providers of required resources)	INPUTS (resources required by the process)	PROCESS	OUTPUTS (deliverables from the process)	CUSTOMERS (stakeholders who put requirements on the outputs)
Here we list all the suppliers of the inputs to the process.	Here we list all the inputs into the process.	This is a high level flowchart of the process (we create a more detailed one in the Analyze phase).	Here we list all the outputs from the process, both intended and unintended.	Here we list all the stakeholders of the outputs.

Process name:
Process Owner:

Voice of the Customer Analysis

The third main step in the Define phase is the Voice of the Customer (VoC) analysis where we take into consideration the needs and wants of our stakeholders when coming up with our critical to quality metric, or our CTQ (Gitlow and Levine, 2004). Their perspective also helps us create a nominal value and the upper and lower specification limits (USL and LSL) for the CTQ. VoC analysis is discussed in detail in Chapter 6.

There are times you go into a project knowing what your CTQ will be. For example, based on the experience of our Emergency Department staff they can see that patients are not moving through the ED quickly enough, so we need to do a project to decrease our CTQ, which is length of stay in our Emergency Department.

However, there are times when we know we have problems due to patient or physician complaints. But sometimes we need to conduct a VoC analysis and speak to our Emergency Department stakeholders to identify problematic CTQ, which in this example is length of stay in the Emergency Department.

The VoC is also valuable for identifying potential Xs (remember Xs are the factors that cause our CTQ to be problematic). You are probably asking yourself, why are we discussing Xs now; I thought Xs were identified in the Analyze phase? You are correct, but many times stakeholders that you speak to during the VoC analysis in the Define phase give you clues on some of the potential root causes of the problem, so it is always smart to keep a list and start thinking ahead.

There are typically four stages when conducting the VoC analysis, as shown here:

286   A Guide to Six Sigma and Process Improvement for Practitioners and Students

Define/Segment the Market

The first thing we need to do is figure out from whom are we going to get our VoC data; that is, who are the stakeholder segments to be interviewed that can provide input on the CTQs of interest.

The simplest method for segmenting a market is to study the SIPOC analysis and focus attention on the outputs and on the customers (stakeholders) by asking

- What are the outputs of your process?
- Who are the customers of those outputs?
- Are there particular groups of customers whose needs focus on specific outputs?

Plan the VoC

Next we have to plan the VoC. There are two types of VoC data, reactive data and proactive data. Reactive VoC data arrives regardless of whether the organization collects it or not—for example, customer complaints, product returns or credits, contract cancellations, market share changes, customer defections and acquisitions, customer referrals, closure rates of sales calls, web page hits, technical support calls, and sales. Proactive VoC data arrives only if it is collected by personnel in the organization: this is the type of VoC data we need to plan to collect. It is data obtained through positive action—for example, data gathered through interviews, focus groups, surveys, comment cards, sales calls, market research, customer observations, or benchmarking. Some of the steps involved in planning the VoC are

- Choosing stakeholders within each market segment to interview
- Creating interview questions
- Setting up interviews
- Assigning interviews to team members

Collect the Data

This step involves conducting the interviews themselves to collect VoC data from your stakeholders. A few things to keep in mind when conducting the interviews:

- Record their answers in bullet point form; it makes the data easier to organize later.
- Let them talk; this is the time for you to understand their perspective on the process.
- Write down exactly what they say. You don't want anything to get lost in translation. This is important!

Organize and Interpret the Data

The last step involves organizing and interpreting all the raw VoC data (comments) you have collected. First, you create an affinity diagram of all of the raw VoC data by market segment. Second, you name each affinity diagram cluster in each market segment; these cluster names are called *focus points*. Focus points are the underlying themes for one or more data points in the language of the market segment. Third, you determine the cognitive issues underlying each focus point. Cognitive issues are the detailed, unambiguous, qualitative statements of focus points in the language of designers (design engineers). Cognitive issues are turned into one or more operationally definable CTQs or are put on the list of potential Xs to be investigated further in the Analyze phase.

Anecdote on Voice of the Customer Analysis

One bitter cold Sunday, an old farmer trudged for miles through a blizzard to reach the small mountain church he attended. No one else showed up, except the preacher. Looking around the empty pews, the clergyman leaned over the pulpit and suggested to his lone congregant that it hardly seemed worth proceeding with the service with such a low turnout. "Perhaps we'd do better if we returned to our nice, warm homes and had a hot drink," he said in a tone that blatantly encouraged the old farmer to agree.

The old farmer looked at the preacher and said, "I'm just a simple farmer, but when I go to feed my herd, if only one cow shows up, I sure don't let her go hungry." The preacher felt embarrassed and a bit guilty, so he conducted the entire service—hymns, readings, announcements, and a sermon. The whole thing lasted over an hour.

After the service, he said to the farmer, "I hope that met your needs." The farmer said, "I'm just a simple farmer, but when I go to feed my herd, if only one cow turns up, I sure don't force her to eat everything I brought for the lot of them." (Source: http://www.fastcompany.com/999915/preacher-farmer-and-dynamic-communication)

Definition of CTQ(s)

Once you have completed your VoC analysis you should be at a point where you know your CTQ (Gitlow and Levine, 2004). Now you need to officially define it!

Recall, CTQ is an acronym for Critical to Quality characteristic for a product, service, or process. Also recall that a CTQ is a measure of what is important to a customer; for example, waiting time in a physician's office by day, percentage of errors with ATM transactions for a bank's customers per month, or number of car accidents per month on a particular stretch of highway for a Department of Traffic, might all be CTQs. Six Sigma projects are designed to improve CTQs.

Your Six Sigma project has at least one CTQ. You want to state your CTQ in terms of unit, defect or defective, and defect opportunity, so that everyone is clear on this critical project metric.

Unit—A unit is the item (e.g., product or component, service or service step, or time period) to be studied with a Six Sigma project.

Defective—A nonconforming unit that is out of specification limits that prevents the unit from performing it purpose. The unit must be reworked, thrown away, or sold at a lower price.

Defect—A defect is a nonconformance on one, of many possible, quality characteristics of a unit that causes customer dissatisfaction. For a given unit, each quality characteristic is defined by translating customer desires into specifications. It is important to operationally define each defect for a unit. For example, if a word in a document is misspelled, that word may be considered a defect. A defect does not necessarily make a unit defective. For example, a water bottle can have a scratch on the outside (defect) and still be used to hold water (not defective). However, if a customer wants a scratch-free water bottle, that scratched bottle could be considered defective.

Defect opportunity—A defect opportunity is each circumstance in which a CTQ can fail to be met. There may be many opportunities for defects within a defined unit. For instance, a service may have four component parts. If each component part contains three opportunities for a defect, the service has 12 defect opportunities in which a CTQ can fail to be met. The number of defect opportunities generally is related to the complexity of the unit under study. Complex units experience greater opportunities for defects to occur than simple units.

An example of a CTQ being defined is shown in Table 10.7.

Table 10.7 Defining the CTQ

CTQ	Definition of a Unit	Definition of Opportunity for a Defect	Definition of a Defect
Percent of patient no shows in the Psychiatric Clinic by month	A scheduled appointment	A scheduled appointment	When a patient does not show up for an appointment

Create an Initial Draft of the Project Objective

Finally we create an initial draft of the project objective so that the team and its stakeholders are all on the same page. A draft project objective is a statement of the purpose of a Six Sigma project. It contains five key elements: process, CTQ measure, CTQ target, CTQ direction, and a deadline. A project objective is SMART (Specific, Measurable, Attainable, Relevant, and Time Bound). *Specific* refers to the area being targeted for improvement. *Measurable* refers to what you use to measure progress or how you quantify success (CTQ). *Attainable* refers to the degree to which the objective is achievable. *Relevant* refers to the degree to which the objective aligns with the mission of the organization. *Time Bound* refers to the degree to which the objective can be attained in a timely manner.

The Project Champion, project leader (Green or Black Belt), Process Owner, and team members define a draft project objective.

Example: Draft project objective for a project whose aim is to decrease OR turnover time:

To decrease (direction) the time it takes to turn over an operating room between surgeries (measure) in the ORs at ABC Hospital (process) by 30% (goal) by January 1, 2016 (deadline).

Note that the goal of 30% is likely an arbitrary figure. We say this because it ends on a zero. The authors believe that any goal that ends in a zero is suspect of being arbitrary. Our real intention is to decrease the turnover time to as close to zero as possible by January 1, 2015.

Tollgate Review: Go-No Go Decision Point

Once all the tasks and subtasks of the Define phase have been completed the project leader (Black Belt) reviews the project with the Master Black Belt who critiques the Six Sigma project. Then a member of the Finance department critiques the financial impact of the project on the bottom line, a member of the Information Technology department critiques the computer/information related aspects of the project, a member of the Legal department comments on legal issues, if any, the Process Owner critiques the process knowledge aspect of the project, and the Champion critiques the political/resource aspects of the project.

Finally, if all the elements are acceptable, a tollgate review is scheduled where the Six Sigma team presents its project to the Champion and the Process Owner for approval; see Table 10.8 for tollgate review checklist. After each DMAIC phase the team presents the project in a similar way.

Tollgate reviews focus on whether a Six Sigma team properly completed each step of a given phase of the DMAIC model. If the team has completed all steps of each phase, the Project Champion and Process Owner sign off on their work and ask them to continue to the next phase. If a step was missed or not completed to their satisfaction they have the team go back and complete the problematic step.

At this point a go-no go decision on the project must be made. Are the benefits what we thought they were? Do they justify the time and resources we are spending on the project to the level that we want to continue? Do we want to change the focus of the project? The Project Champion and Process Owner typically make this decision keeping in mind the mission of the organization.

Table 10.8 Define Phase Tollgate Review Checklist

Define Phase Component	Champion/Process Owner Sign-Off	Date	Comments
Project Background			
Project Charter			
Project Plan			
SIPOC			
Voice of the Customer analysis			
Definition of CTQ			
Final Project Objective			
Proceed to Measure Phase			

Keys to Success and Pitfalls to Avoid in the Define Phase

- Make sure to get your Project Champion and Process Owner on board and engaged from the very start. The most critical key success factor to any process improvement project is not merely the involvement of top management, but more importantly the commitment from top management. So the Project Champion needs to be your best friend. If the Project Champion or Process Owner is unavailable and constantly cancelling meetings or not being completely supportive, the project should likely be abandoned.
- Roles and responsibilities are super important. Make sure everyone is clear right from the start regarding who is responsible for what, and when it needs to be done. You don't want to have to throw anyone under the bus later—it's not good business practice!
- Pay particular attention to the project scope; do not let scope creep set in. Scope creep refers to continuous growth or uncontrollable changes in the scope of a project.
- Make sure the scope is manageable and that you are always keeping an eye on it. Don't try to boil the ocean!
- Make sure to keep the process map in the SIPOC high level. You can go nuts and create a process map that gets into detail in the Analyze phase.
- When doing VoC interviews always keep things in the exact words of the stakeholder no matter how ridiculous or insane it may be. You don't want to lose any good gold nuggets in translation. Your job at this point is to listen to the interviewee, take notes, and understand their perspective.

- Typically the part of the Define phase that can slow you down the most is the VoC interviews. So do yourself a favor and make sure you identify your critical stakeholders early on so you can schedule a meeting with them to sit down and get the details on the process from their perspective.

Case Study of the Define Phase: Reducing Patient No Shows in an Outpatient Psychiatric Clinic—Define Phase

We have created a fictitious case study to take you through all phases of the DMAIC model. Each DMAIC chapter concludes with the relevant part of the case study for that respective phase.

Patients not showing up for their appointments, also called *no shows*, has always been a big problem for the Outpatient Psychiatric Clinic at XYZ Hospital. Over the last few months it has become so problematic that the hospital administration needed to take action.

Across the nation, no show rates at similar clinics average 5.5%, but at the Outpatient Psychiatric Clinic at XYZ Hospital no show rates over the past several months were nearing 30%! The staff constantly deals with frustrated patients, physicians, and staff. Patients complain frequently that it takes months to schedule an appointment, but the waiting room is empty when they arrive due to other patients not showing up.

Finally administration has had enough and put together a process improvement team to see if they can solve the problem using the Six Sigma DMAIC methodology.

> **NOTE**
>
> No shows in healthcare is a complex problem involving many moving parts. The purpose of this case study is not to get into the intricacies of no shows, but rather to take you through the DMAIC model with a simplified example you can relate to, to help you understand the concepts we are teaching.

Activate the Six Sigma Team

The entire C-suite team at the hospital met with a Master Black Belt to discuss the problem. Then the Master Black Belt met with various stakeholders to scope out the project and create a high level project charter; see Table 10.9.

Table 10.9 High Level Project Charter for No Shows Project

High Level Project Charter	
Project Name:	Reducing patient no shows in the Outpatient Psychiatric Clinic at XYZ Hospital
Department/Area Name:	Department of Psychiatry at XYZ Hospital
High Level Problem Description:	Patient no shows are having a negative impact on the finances of the department and creating enormous frustration for patients, physicians, and employees
Project Type:	Six Sigma DMAIC
Time to Complete:	4 months
Problematic Objective:	To decrease no shows in the Outpatient Psychiatric Clinic at XYZ Hospital
Problematic Indicator:	Percentage of no shows in the clinic by month
Project Scope:	Outpatient Psychiatric Clinic at XYZ Hospital
Potential Financial Impact:	$1,154,150 over the past two years
Current Baseline Performance:	Close to a 25% no show rate over the past 2 years
Project Champion:	CEO of Hospital XYZ
Process Owner:	Assistant Vice President (AVP) of Behavioral Health, XYZ Hospital
Physician Champion:	Dr. X
Nursing Champion:	Nurse Y
Finance Champion:	CFO
Project Leader:	Black Belt
Potential Team Members:	A, B, C, D, E, F, G

Once this was completed, the next step was to activate the Six Sigma Team and get started.

Project Charter

The project charter section of the Define phase of the DMAIC model has nine parts: business case, problem statement, goal statement, scope, project plans with milestones, financial impact, roles and responsibilities, communication plan, and risk abatement plan. Each part is significant in that it helps focus the project on an actionable and doable project.

Business Case

- **What is the name of the process?** Patient scheduling process in the Outpatient Psychiatric Clinic at XYZ Hospital.
- **Who is the Process Owner?** The Process Owner is the Assistant Vice President of Behavioral Health at XYZ Hospital: He has the authority to change the process with his signature.

- **What is the aim of the process?** The aim of the process is to have all available patient appointment slots filled with patients who show up and are seen by their physician.
- **Why do the project at all?** There has been a huge increase in patients not showing up for their appointments, which has put an enormous financial strain on the department and the hospital while causing frustration for patients, physicians, and staff alike.
- **Why do the project now?** The budget for the hospital is extremely tight given the changing healthcare landscape, so every penny counts. Also, the number of complaints from patients and physicians has increased to the point that something needs to be done now.
- **What are the consequences of not doing the project?** The financial stability of the department is being compromised, and if current trends continue it may cause the department to miss budget. Also, the nearest competitor, Hospital ABC, has just opened up a brand new, state of the art clinic, which may lure some of our patients and our best physicians away if we don't get our act together.
- **What other projects have higher or equal priority, if any?** None! For the Department of Behavioral Health, this is a crisis and the most important initiative on our radar.
- **What organizational business objectives are supported by the project?** Meeting budget, giving great customer service, and providing our employees with a great working environment. These objectives come from the hospital dashboard.

Problem Statement

Patient no shows are causing financial problems as well as patient, physician, and employee dissatisfaction in the Outpatient Psychiatric Clinic at XYZ Hospital.

Goal Statement

To reduce patient no shows in the Outpatient Psychiatric Clinic at XYZ Hospital from its current level of close to 30% to an industry standard of 5.5% (or as close to zero as possible); this will in turn increase revenue to the clinic.

Scope

The project scope is created by answering the following questions:

- **What are the process boundaries?** The process starts when a patient calls the clinic to make an appointment, and the process ends at the time of the appointment.
- **What, if anything, is out of bounds?** Only the Outpatient Psychiatric Clinic at XYZ Hospital is part of this project; other clinics are out of bounds, although when we are successful here we will replicate the success at other clinics. Inpatient stays are not part of this project.

- **What resources are available for the project?** Employees including the process improvement team at the hospital, other employees chosen by the Black Belt, Project Champion, and Process Owner.
- **Who can approve expenditures?** The Process Owner is the only one who can approve expenditures.
- **How much can the team spend beyond its budget without authority?** None, but the Process Owner will do whatever it takes to support the project, as long as the team members can justify the needed resources.
- **What are the obstacles and constraints of the project?** The team must work quickly; the Project Champion expects the project to be completed in four months, and there is no budget due to financially tough times. Frequently, this is not a problem since most solutions do not require capital investment. Most improvements are caused by changes to the process such as eliminating unnecessary signatures on a form or finding duplicate work in the process.
- **What time commitment is expected of team members?** The process improvement team will dedicate a Black Belt to spend 20% of his time on the project and a Green Belt to spend 25% of her time on the project. Other team members are authorized to work 5 hours per week on the project.
- **What will happen to each team member's regular job while he or she is working on the project?** Team members will meet with their supervisors at the end of each week to prioritize their workload for the next week realizing they have 5 hours less to do their regular job due to their participation on the project. Management understands that sometimes you need to sacrifice in other areas to improve one particular area.

Project Plan with Milestones

Next, the Black Belt project leader creates a high level project plan; refer to Table 10.9. The real plan created for this project was done by week, not by month. However, for the sake of brevity we are showing you the plan by month. Different team members are assigned various aspects of the project with the Black Belt being responsible and held accountable for all tasks. Notice the team has a pretty aggressive goal of completing this project in four months, so it is critical that the Black Belt keeps the team focused to meet the deadline. Based on his experience, the Black Belt projects the team to finish both Define and Measure in February as he knows stakeholders for the VoC will be easy to schedule and the baseline data in the Measure phase will be easy to collect out of the clinic database. He has given a month for Analyze as he believes there will be many potential Xs to identify and a substantial amount of analysis needed to determine which ones are critical. His plan is to complete a month-long pilot test to optimize the most critical Xs. Interestingly, he has given himself a month for the Control phase, which will not take that long, but he knows nothing ever goes as planned, so he built in a time buffer to meet the deadline. This Gantt chart (timeline) can and will change and be updated by the Black Belt as the project goes on (see Table 10.10).

Table 10.10 Gantt Chart for No Shows Project

No Shows Six Sigma Project Gantt Chart

Task	Responsible	Status	Jan	Feb	Mar	Apr	May	Jun	Jul	Aug
Define										
Activate the team/kickoff meeting	BB	TBD								
Project charter	BB	TBD								
SIPOC analysis	GB	TBD								
Voice of the Customer analysis	GB	TBD								
Definition of CTQ	BB	TBD								
Final project objective	BB	TBD								
Tollgate review	BB, PO*, C**	TBD								
Measure										
Data collection plan for CTQs	BB	TBD								
Collect baseline data	GB	TBD								
Analyze baseline data	BB	TBD								
Estimate process capability for CTQs	BB	TBD								
Tollgate review	BB, PO, C	TBD								
Analyze										
Detailed flowchart of current state process	GB	TBD								
Identification of potential Xs	GB	TBD								
Data collection plan for Xs	BB	TBD								
Test of theories to determine critical Xs	BB	TBD								
Develop hypotheses	BB	TBD								
Tollgate review	BB, PO, C	TBD								

* Process Owner
** Champion

No Shows Six Sigma Project Gantt Chart

Task	Responsible	Status	Jan	Feb	Mar	Apr	May	Jun	Jul	Aug
Define										
Improve										
Generate potential solutions	GB	TBD								
Select solutions	BB	TBD								
Create flowchart for new improved process	GB	TBD								
Identify and mitigate risks for new process	BB	TBD								
Run pilot test of new process	BB	TBD								
Pilot test data collected and analyzed	BB	TBD								
Implement new process full scale	BB	TBD								
Tollgate review	BB, PO, C	TBD								
Control										
Develop control plan	BB									
Document costs and benefits of project	CFO									
Mistake proof optimal configuration	BB									
Document the revised system	GB									
Create standard operating procedures (SOPs)	GB									
Turn over new system to Process Owner	BB									
Put results in database	GB									
Tollgate review/project signoff by Champion/Process Owner	BB, PO, C									
Disband the team and celebrate success!	BB									
Process Owner PDSAs new process forever	PO									

Financial Impact

The financial impact was initially guesstimated in the high level project charter. The Black Belt meets with the CFO to make sure the assumptions remain the same, which they have as seen in Table 10.11. The Black Belt and the CFO finalize the financial impact during the Control phase.

Table 10.11 Financial Impact for No Shows Project

Type of Appointment	Number of No Shows	Net Revenue per Slot	No Shows x Net Revenue/Slot
First appointment	2435	$250	$608,750
Regular patient appointment	3636	$150	$545,400
	Total financial impact of no shows for past two years =		$1,154,150

Roles and Responsibilities

The Black Belt meets with the Process Owner and various supervisors to define the roles and responsibilities of team members, which are agreed upon by all managers and supervisors as seen in Table 10.12. Since this is a healthcare project the team includes a Physician Champion and a Nursing Champion to make sure we are taking the Voice of the Physician and Voice of the Nurse into consideration on the project. Also when it comes time to implement changes to the process, we need their commitment. Finally, the team includes a Finance Representative to make sure that the financial impact of the project is calculated properly, and an IT/Data Representative to facilitate collection of data from any databases.

Table 10.12 Roles and Responsibilities for No Shows Project

Role	Responsibility	Stakeholder Signature	Date	Supervisor's Signature
Project Champion	CEO	CEO	2/1/15	
Process Owner	AVP	AVP	2/1/15	CEO
Project Leader (Black Belt)	BB1	BB1	2/1/15	CEO
Physician Champion	Doctor X	DX	2/1/15	CMO
Nursing Champion	Nurse Y	NY	2/1/15	CNO
Finance Rep	CFO	CFO	2/1/15	CEO
IT/Data Rep	CIO	CIO	2/1/15	CEO
Green Belt Team Member	Team Member 1	GB1	2/1/15	Manager 1
Green Belt Team Member	Team Member 2	GB2	2/1/15	Manager 2
Green Belt Team Member	Team Member 3	GB3	2/1/15	Manager 2
Green Belt Team Member	Team Member 4	GB4	2/1/15	Manager 3

Communication Plan

The team members created a communication plan, shown in Table 10.13, to identify and address stakeholder concerns and ensure that not only is the process aspect of the project covered but the cultural aspect of the project is addressed as well.

> **NOTE**
> Opinion leaders are the most active and respected voices in a specific community who have a substantial amount of influence over their peers and are asked for their advice often.

Risk Abatement Plan (FMEA)

A Failure Modes and Effects Analysis, as seen in Table 10.14, was then created by the team to identify risk elements that may impede the success of the project as a whole, and then they then came up with action items to mitigate those risk elements.

The FMEA in Table 10.14 indicates that buy-in from top management, physicians, and nurses are the critical factors to a successful project.

SIPOC Analysis

The team then conducted a SIPOC analysis, as seen in Table 10.15, to identify the suppliers and their inputs into the process, the high level steps of the process, the outputs from the process, and the stakeholder segments interested in those outputs.

Voice of the Customer Analysis

The next step for the team was to conduct a Voice of the Customer (VoC) analysis so they could get a 360-degree understanding of the process, identify their Critical to Quality metric (CTQ), and begin to identify Xs they could leverage later in the Analyze phase. Remember, the VoC analysis consists of four substeps: defining and segmenting the market, planning the VoC, collecting the data, and organizing and interpreting the data.

Table 10.13 Communication Plan for No Shows Project

#	Event/ Communication	Participants/ Audience	Medium	Frequency	When	Lead	Scheduled?	Status	Notes
1	General announcement memo to staff	All staff in Department of Behavioral Health	Email and fax blast	Once	TBD	AVP, BH	n/a	TBD	Draft communication to inform staff of project, sent by AVP and CEO.
2	Weekly email to stakeholders	All project stakeholders	Email	Weekly	Every Friday	Black Belt	n/a	Ongoing	Black Belt to delegate to team member.
3	Meetings with opinion leaders	Opinion leaders	Face to face	Once	TBD	Black Belt	N	TBD	Identify opinion leaders and work with them to promote the project.
4	Poster placed in clinic lunch room	All staff in Department of Behavioral Health	Printed poster	Once	TBD	Communications Director	N	TBD	Need to create and place poster in clinic lunch room.
5	Presentation at physician staff meeting	All physicians in Department of Behavioral Health	Face to face	Once	TBD	Black Belt	N	TBD	Need to create presentation and schedule time to present.
6	Update hospital intranet site	All staff in Department of Behavioral Health	Intranet website	Once	TBD	Black Belt	N	TBD	Create content and meet with webmaster to post to intranet.

Table 10.14 Failure Modes and Effects for No Shows Project

Critical Parameter	Potential Failure Mode	Potential Failure Effect	SEV	Potential Causes	OCC	Current Controls	DET	RPN	Recommended Action	Responsibility and Target Date	Date Action Taken	SEV	OCC	DET	RPN
Lack of buy-in by top management.	Project doesn't get supported.	Project unexecutable.	10	Lack of commitment.	10	None.	10	1000	Top management must have skin in the game.	CEO	2/16/2015	2	4	10	80
Project exceeds budget.	Black Belt gets fired!	Need to hire new BB.	3	BB lacks financial expertise.	8	BB does his best.	5	120	Finance rep on team responsible for budget.	Finance rep	2/16/2015	2	5	5	50
Takes too long to get data.	Project is delayed.	Heads roll.	8	BB lacks IT expertise.	8	BB does his best.	5	320	IT rep on team responsible for data.	IT rep	2/16/2015	4	4	5	80
Lack of buy in from physicians.	Implementation issues.	Unsuccessful implementation.	7	Entitled physicians.	8	Lots of bitching back and forth.	8	448	Physician Champion part of team.	Physician Champion	2/16/2015	4	6	8	192
Lack of buy in from nurses.	Implementation issues.	Unsuccessful implementation.	7	Entitled physicians.	8	Lots of bitching back and forth.	8	448	Nursing Champion part of team.	Nursing Champion	2/16/2015	4	6	8	192
Lack of awareness.	Lack of cooperation.	Misaligned team.	5	Lack of communication.	5	Rumor mill in overdrive.	2	50	Communication plan created.	Black Belt	2/16/2015	1	5	2	10
Scope too broad.	Project becomes unmanageable.	Project unexecutable.	7	Trying to boil the ocean.	7	Planning by Black Belt.	3	147	Strong planning effort by Black Belt.	Black Belt	2/16/2015	3	3	3	27

Chapter 10 DMAIC Model: "D" Is for Define

Table 10.15 SIPOC Analysis for No Shows Project

Process Name: Patient Scheduling Process in the Outpatient Psychiatric Clinic at XYZ Hospital
Process Owner: Assistant Vice President of Behavorial Health

SUPPLIERS (Providers of required resources)	INPUTS (Resources required by the process)	PROCESS	OUTPUTS (Deliverables from the process)	CUSTOMERS (Stakeholders who put requirements on the outputs)
Referring physicians Patients/families Appointment schedulers Insurance providers	Referral Phone call Patient information Patient and physician availability Insurance plan	Call comes in. ↓ Patient entered into electronic medical record system. ↓ Appointment scheduled. ↓ Insurance verified. ↓ Reminder call if time. ↓ Patient either shows or no shows.	New patient record in electronic medical record system Scheduled appointment Verified insurance Reminder Arrived visit Patient no show	Patient registration reps in clinics Nurses Physicians Insurance providers Finance office Patients/families

Define/Segment the Market

The team went back to the SIPOC analysis to define and segment the suppliers and stakeholders. They decided to focus on the following segments during the VoC to help them better understand the process:

- Patients/families
- Appointment schedulers
- Insurance providers
- Patient registration reps in clinics
- Nurses
- Physicians
- Finance office

Plan the VoC

Next up was planning the VoC. The team decided to focus on collecting proactive VoC data since no reactive VoC data was available for this process. They then planned the VoC by doing the following:

Choosing Stakeholders within Each Market Segment to Interview

The team decided that if they interviewed the following number of stakeholders within each segment, it would give them a great understanding of the process:

- Five patients
- Two appointment schedulers
- Two insurance providers
- Two patient registration reps in clinics
- Five nurses
- Five physicians
- One finance office

The sample size in each segment is a function of the budget available to perform the VoC. It is an arbitrary number that in the opinion of an expert yields reliable information about that segment.

Creating Interview Questions

The team then decided on the questions to ask related to the patient scheduling process in the Outpatient Psychiatric Clinic at XYZ Hospital to try to understand the process as much as possible:

- How do you feel about the patient scheduling process as relates to patient no shows?
- What issues/problems do you see with the patient scheduling process that may lead to patient no shows?
- What solutions/recommendation do you have to decrease the number of patient no shows?
- What emotions or images come to mind when you think about the patient no shows?

Setting Up Interviews

The Black Belt assigned one of the Green Belts the task of setting up the interviews with the various stakeholders making sure to follow the timeline set out by the Black Belt in the project plan.

Assigning Interviews to Team Members

The Black Belt then assigned interviews to different team members based on his understanding of each team member, his personality, and his relationships with various stakeholders.

Collect the Data

The various team members then went out and conducted the interviews to collect the VoC data from their respective stakeholders. The Black Belt gave the team specific instructions:

- Record their answers in bullet point form because it makes the data easier to organize later.
- Let the interviewees talk. This is time for you to understand their perspective on the process. However, be careful that the interview doesn't turn into a therapy session. Gently keep the interviewee on task.
- Write down exactly what the interviewees say because you don't want anything to get lost in translation.

Organize and Interpret the Data

We discussed the VoC Summary Table earlier in this chapter. Here you get to see exactly how it works.

VoC Analysis Summary Table

Team members analyzed the VoC data by the stakeholder segments listed previously. Next, they used all the raw VoC data points (see column 2 of Table 10.16) to create affinity diagram themes (see column 3 in Table 10.16). Next, team members identified the underlying themes of all the raw VoC data in each affinitized group. Team members named each affinitized group, and these names become focus points. Focus points are stated in the language of the process improver, and they are called cognitive issues (see column 4 in Table 10.16). Finally, members converted each cognitive issue into one or more CTQ characteristics (see column 5 in Table 10.16).

The team identified two CTQs via the VoC interviews: patient no show rate (which they kind of knew all along would be a CTQ) and turnaround time to schedule appointments. They decided to focus on the first CTQ, patient no show rate, because they agreed that turnaround time to schedule appointments could be a function of the no show rate. In other words, turnaround time to schedule appointments could be a critical X for the patient no show rate CTQ.

Table 10.16 Voice of the Customer Summary Table for No Shows Project

Selected Market Segment	Raw VoC Data	Affinity Diagram Theme (focus point)	Cognitive Issues	CTQ
Patients	"I can't get an appointment for three months. I get here and the waiting room is empty. No wonder I am crazy!" (1) Note: **The number in parentheses is a number that indicates a cluster of raw VoC data comments; see the next column for the name of each cluster.**	Variation in patients showing up for their appointments (1)	Patients not showing up for their appointments affects everyone.	Patient no show rate by month
	"They don't take no shows seriously, so I don't feel bad if something else comes up." (1)	Variation in scheduling patients for appointments (2)	Time between date patient calls to schedule appointment and date of appointment is too long.	Turnaround to schedule appointments by appointment
	"Everyone else seems to miss appointments and they aren't held accountable, so I don't feel bad if I do." (1)	Miscellaneous comments (X)		
	"Why does it take so long to get an appointment here?" (2)			
Appointment schedulers	"These no shows make our job so hard." (1)			
	"Tough to predict who will no show and who won't." (1)			
	"The variation in who shows and who doesn't makes it tough to schedule." (1)			
	"I know we do reminders sometimes. I think we should do reminders for all; they work!" (X)			
	"We hear complaints from everyone. I am glad they are finally doing something about it!" (1)			
	"We have our usual suspects." (X)			

Chapter 10 DMAIC Model: "D" Is for Define

Selected Market Segment	Raw VoC Data	Affinity Diagram Theme (focus point)	Cognitive Issues	CTQ	
Insurance providers	"I would charge for missed appointments like my dentist does; that will solve the problem." (X)				
	"We have heard that copays can cause no shows." (X)				
	"No shows seem to be a problem that a lot of hospitals struggle with." (1)				
	"They aren't good for anybody." (1)				
Patient registration representatives	"We see a lot of frustration when patients no show—from the docs to the nurses to other patients to staff. We need to decrease them asap!" (1)				
	"Sometimes we are busy; sometimes we aren't" (1)				
	"It makes it hard for our boss to schedule us. I feel bad for her sometimes." (1)				
	"Sometimes we get sent home if there are too many no shows. I get it but it still sucks." (1)				
Nurses	"I have physicians giving me crap for times they have nothing to do, like it's my fault!" (1)				
	"Sometimes I am super busy; sometimes it's slow" (1)				
	"I hear patients complaining that they had to wait a few months to get in. I have no idea why." (2)				
	"I don't know why they schedule younger patients in the morning; they don't show up!" (X)				
Physicians	"I am the most expensive resource and half the time I feel like I am sitting on my butt." (1)				

Selected Market Segment	Raw VoC Data	Affinity Diagram Theme (focus point)	Cognitive Issues	CTQ
	"I would schedule older patients earlier in the day and younger patients later in the day." (X)			
	"New patients are flakier. They make an appointment, find a sooner one somewhere else, and then don't call to cancel." (X)			
Finance office	"It definitely has a financial impact and with the way things have been going we need every penny." (1)			
	"They are directly affecting our bottom line." (1)			
	"We need to figure out how to decrease them and fast!" (1)			
	"I am cool with imposing some type of penalty. Patients need to be held accountable too; this is a business" (X)			

Potential Xs

The team also identified some potential Xs that will be explored later in the Analyze phase:

- Patients who no show frequently
- No penalty fees (could be a solution)
- Lack of reminders
- Age of patient and time of appointment
- Insurance company
- New versus existing patients

Definition of CTQ(s)

Next the team defined the CTQ in terms of a unit, opportunity for a defect, and finally a defect to ensure everyone is on the same page as seen in Table 10.17.

Table 10.17 Definition of CTQ for No Shows Project

CTQ	Definition of a Unit	Definition of Opportunity for a Defect	Definition of a Defect
Percent of patient no shows in the Psychiatric Clinic by month	A scheduled appointment	A no show for a scheduled appointment	When a patient does not show up for their appointment

Initial Draft Project Objective

To decrease (direction) the number of patient no shows (measure) in the scheduling process for the Outpatient Psychiatric Clinic at XYZ Hospital (process) from 25% to as close to 0% as possible (goal) by January 1, 2016 (deadline).

Tollgate Review: Go-No Go Decision Point

The team conducted the Define phase tollgate review (see the tollgate review checklist in Table 10.18) with the Project Champion (the hospital CEO), the Process Owner, and the rest of the team on hand. The CEO is pleased with the progress, gives the thumbs up to move to Measure, and encourages the team to press on!

Table 10.18 Define Phase Tollgate Review Checklist for No Shows Project

Define Phase Component	Champion/Process Owner Sign-Off	Date	Comments
Project Background	Champion and Process Owner	Feb 21/15	
Project Charter	Champion and Process Owner	Feb 21/15	
Project Plan	Champion and Process Owner	Feb 21/15	
SIPOC	Champion and Process Owner	Feb 21/15	
Voice of the Customer analysis	Champion and Process Owner	Feb 21/15	
Definition of CTQ	Champion and Process Owner	Feb 21/15	
Final Project Objective	Champion and Process Owner	Feb 21/15	CEO thinks team should have some type of penalty for patients who no show frequently. Master Black Belt reminds him we aren't coming up with solutions yet!
Proceed to Measure Phase	Champion and Process Owner	Feb 21/15	

Takeaways from This Chapter

- The Define phase is the first of the five phases in the DMAIC model.
- The steps of the Define phase are the following:

  - Activate the Six Sigma team by identifying team members and holding a kickoff meeting.
  - Create a project charter to give structure to the project.
  - Perform a SIPOC Analysis to understand the process at a high level.
  - Perform a VoC analysis to understand the needs and wants of all stakeholders.
  - Define the CTQ(s) to give the team a way to measure success.
  - Create a final project objective, which is a statement on the clear purpose of the project.
  - Hold a tollgate review, which includes a go-no go decision on whether to continue with the project.

References

Gitlow, H. and D. Levine (2004), *Six Sigma for Green Belts and Champions: Foundations, DMAIC, Tools and Methods, Cases and Certification* (Upper Saddle River, NJ: Prentice-Hall).

Additional Readings

Center for Quality Management, "A Special Issue on: Kano's Methods for Understanding Customer-defined Quality," *The Center for Quality Management Journal*, Volume 2, Number 4, Fall 1993 (Cambridge, MA), p. 13.

Gitlow, H., A. Oppenheim, R. Oppenheim, and D. Levine (2015), *Quality Management: Tools and Methods for Improvement*, 4th ed. (Naperville, IL: Hercher Publishing Company). This book is free online at hercherpublishing.com.

Kano, N. and F. Takahashi, "The Motivator Hygiene Factor in Quality," JSQC, 9(th) Annual Presentation Meeting, Abstracts, pp. 21-26, 1979.

Kano, N., N. Seraku, F. Takahashi, and S. Tsuji, "Attractive Quality and Must-Be Quality," *Hinshitsu (Quality)*, 1984a, vol. 14, no. 2, pp. 147-156.

Rasis, D., H. Gitlow, and E. Popovich, "Paper Organizers International: A Fictitious Six Sigma Green Belt Case Study – Part 2," *Quality Engineering*, vol. 15, no. 2, pp. 259-274.

Rasis, D., H. Gitlow, and E. Popovich, "Paper Organizers International: A Fictitious Six Sigma Green Belt Case Study – Part 1," *Quality Engineering*, volume 15, number 1, 2002, pp. 127-145.

11

DMAIC Model: "M" Is for Measure

What Is the Objective of This Chapter?

The objective of this chapter is to take you through the various steps of the Measure phase of the Six Sigma DMAIC model so that you can apply them on projects at your organization. We use a case study to demonstrate how the steps of each phase are executed in real world projects.

Purpose of the Measure Phase

Let's go back to our equation CTQ is a function of one or more Xs or CTQ = $f(X_1, X_2, X_i,.... X_n)$ where:

- CTQ is a measure of your problematic key indicator.
- X_i represents the ith factor that causes your CTQ to be problematic.

We have now successfully completed the Define phase and have identified what our CTQ(s) is/are. This chapter focuses on the second phase, the Measure phase, whose purpose is for team members to concretely understand the definition, measurement system, and current capability of each CTQ. Team members must clearly understand each CTQ and their behaviors if they are going to improve them in the future. How do we know whether we are successful in improving the process if we don't have anything to compare our improved process to?

The five main deliverables for the Measure phase are

- Operational definition of the CTQ(s)
- Data collection plan for the CTQ(s)
- Measurement analysis of the CTQ(s)
- Collect baseline data for each CTQ
- Estimate process capability for each CTQ(s)

At the end of the Measure phase, the team conducts a tollgate review with the Project Champion and Process Owner. This is where the team reviews what they have learned in the Measure phase and a "go-no go" decision is made by the Project Champion. In other words, after looking at the baseline data, is the problem we thought was a problem still a significant enough problem for the team to continue with the project? If yes, and if everyone is satisfied with the team's work, the team proceeds to the Analyze phase. If no, the project is abandoned or sent back to redo the Measure phase.

The Steps of the Measure Phase

The Measure phase of the DMAIC model has five steps: operational definitions of the CTQ(s), data collection plan for the CTQ(s), validate measurement systems for the CTQ(s), collect baseline data for the CTQ(s), and estimate process capability for the CTQ(s). All of these are used to determine if there is a significant problem with the CTQ(s).

Operational Definitions of the CTQ(s)

Operational definitions give meaning to a word or term such that all stakeholders of the word or term agree on its definition (Gitlow et al., 2015; Gitlow and Levine, 2004). This prevents endless bickering and unpleasantness later on in the project. For example, John is told he will receive 10% of his yearly sales as a commission. He sells one million dollars and expects $100,000 in commission, but he only gets $90,000. He asks why and is told he had $10,000 of returns. John says he was not informed of this rule and is unhappy with management. Operational definitions prevent this type of event from occurring.

As you learned in Chapter 6, "Non-Quantitative Techniques: Tools and Methods," an operational definition contains three parts: a criterion to be applied to an object or group, a test of the object or group, and a decision as to whether the object or group meets the criterion.

- **Criteria**—Operational definitions establish Voice of the Process language for each CTQ and Voice of the Customer specifications for each CTQ.
- **Test**—A test involves comparing Voice of the Process data with Voice of the Customer specifications for each CTQ, for a given unit of output.
- **Decision**—A decision involves making a determination whether a given unit of output meets Voice of the Customer specifications.

Data Collection Plan for CTQ(s)

The next step of the Measure phase is to create a data collection plan to lay out how you will collect the baseline data on the CTQ in terms of defining what you are going to measure, how you are going to measure it, who will collect it, and the sampling plan.

The collection of baseline data requires a data collection plan; see Table 11.1 with an example of a CTQ. The elements of a data collection plan include a data collection form, a sampling plan (sample size, sample frequency), and sampling instructions (who, where, when, and how).

Defining What to Measure
- **Metric**—What is the name of the CTQ you are collecting data on?
- **Type of metric**—Is the data for this CTQ attribute (classification or count) or measurement data?

Defining How to Measure
- **Measurement method**—Will the data be collected visually or via automated collection (i.e., extracted from a database)?
- **Data tags needed to stratify the data**—Data tags are the various fields you are going to collect data on to help define the measure, such as item, patient, time, date, location, tester, line, customer, buyer, operator, and so on.
- **Data collection method**—Will the data be collected manually, or on a spreadsheet via a computer?

Defining Who Will Collect It
- **Person(s) assigned**—Who will be assigned and held accountable for the data collection?

Sampling Plan
- **Where**—What is the location where the data will be collected?
- **When**—When and how often will the data be collected? How long will the data be collected for?
- **How many**—How many data points will be collected for each sample?

Validate Measurement System for CTQ(s)

Measurement systems analysis studies are used to calculate the capability of a measurement system for a CTQ to determine whether it can accurately deliver data to team members. For our purposes we examine a Gage run chart to understand the ability of a measurement system to deliver useful data (Gitlow and Levine, 2004).

A Gage run chart is used to determine whether inspectors (employees) are consistent in their measurement of a particular item and whether they agree on the measurements of a particular item. This definition is grossly simplified, but it serves our purpose for this book; see Gitlow and Levine (2004) for details.

Table 11.1 Data Collection Plan for CTQs

Define What to Measure			Define How to Measure		Define Who Will Collect It		Sampling Plan	
Metric	Type of Metric	Measurement Method	Data Tags Needed to Stratify the Data	Data Collection Method	Person(s) Assigned	Where?	When?	How Many?
CTQ #1: Turnaround time for lab test	Measurement data	Automated from database	Date, time, specimen number, patient identifier, patient last name, patient first name	Spreadsheet	Bri	Lab	Every day for 3 months	One data point per lab test

The data required by a Gage run chart should be collected so that it represents the full range of conditions experienced by the measurement system. For example, the most senior employee doing a job and the most junior employee doing a job should repeatedly measure each item selected in the study. The data should be collected in random order to prevent inspectors from influencing each other.

An example of a Gage run chart follows. Two Inspectors independently use a gauge to measure each of five units four separate times; this results in 40 measurements. The data to be collected by the two inspectors is shown in Table 11.2. Table 11.2 presents the standard order, or the logical pattern, for the data to be collected by the team members.

Table 11.2 Standard Order for Collecting Data for the Gage Repeatability and Reproducibility (R&R) Study

Row	Unit	Inspector	Measurement	Row	Unit	Inspector	Measurement
1	1	Mike	To be collected	10	2	Mike	To be collected
2	1	Mike	To be collected	11	2	Mike	To be collected
3	1	Mike	To be collected	12	2	Mike	To be collected
4	1	Mike	To be collected	13	2	Dan	To be collected
5	1	Dan	To be collected	14	2	Dan	To be collected
6	1	Dan	To be collected	15	2	Dan	To be collected
7	1	Dan	To be collected	16	2	Dan	To be collected
8	1	Dan	To be collected	17	3	Mike	To be collected
9	2	Mike	To be collected	18	3	Mike	To be collected

Row	Unit	Inspector	Measurement	Row	Unit	Inspector	Measurement
19	3	Mike	To be collected	30	4	Dan	To be collected
20	3	Mike	To be collected	31	4	Dan	To be collected
21	3	Dan	To be collected	32	4	Dan	To be collected
22	3	Dan	To be collected	33	5	Mike	To be collected
23	3	Dan	To be collected	34	5	Mike	To be collected
24	3	Dan	To be collected	35	5	Mike	To be collected
25	4	Mike	To be collected	36	5	Mike	To be collected
26	4	Mike	To be collected	37	5	Dan	To be collected
27	4	Mike	To be collected	38	5	Dan	To be collected
28	4	Mike	To be collected	39	5	Dan	To be collected
29	4	Dan	To be collected	40	5	Dan	To be collected

Table 11.3 shows the random order used by team members to collect the measurement data required for the measurement study. Random order is important because it removes any problems induced by the structure of the standard order. Table 11.3 is an instruction sheet to the team members actually collecting the data.

Table 11.3 Random Order for Collecting Data for the Gage Run Chart

Random Order	Standard Order	Unit	Inspector	Random Order	Standard Order	Unit	Inspector
1	36	5	Mike	16	24	3	Dan
2	5	1	Dan	17	38	5	Dan
3	30	4	Dan	18	17	3	Mike
4	29	4	Dan	19	32	4	Dan
5	26	4	Mike	20	11	2	Mike
6	28	4	Mike	21	27	4	Mike
7	6	1	Dan	22	19	3	Mike
8	8	1	Dan	23	10	2	Mike
9	4	1	Mike	24	33	5	Mike
10	3	1	Mike	25	37	5	Dan
11	18	3	Mike	26	2	1	Mike
12	20	3	Mike	27	35	5	Mike
13	40	5	Dan	28	23	3	Dan
14	9	2	Mike	29	13	2	Dan
15	31	4	Dan	30	7	1	Dan

Random Order	Standard Order	Unit	Inspector
31	15	2	Dan
32	22	3	Dan
33	14	2	Dan
34	34	5	Mike
35	1	1	Mike

Random Order	Standard Order	Unit	Inspector
36	12	2	Mike
37	16	2	Dan
38	39	5	Dan
39	21	3	Dan
40	25	4	Mike

Table 11.4 shows the data collected in the Gage R&R study in random order.

Table 11.4 Data for Gage R&R Study

Random Order	Order	Unit	Inspector	Measure
1	36	5	Mike	21.85
2	5	1	Dan	21.19
3	30	4	Dan	23.14
4	29	4	Dan	23.09
5	26	4	Mike	23.28
6	28	4	Mike	23.23
7	6	1	Dan	21.29
8	8	1	Dan	21.24
9	4	1	Mike	21.24
10	3	1	Mike	1.33
11	18	3	Mike	22.28
12	20	3	Mike	22.34
13	40	5	Dan	21.78
14	9	2	Mike	21.65
15	31	4	Dan	23.02
16	24	3	Dan	22.17
17	38	5	Dan	21.84
18	17	3	Mike	22.31
19	32	4	Dan	23.19
20	11	2	Mike	21.67
21	27	4	Mike	23.24
22	19	3	Mike	22.31
23	10	2	Mike	21.60
24	33	5	Mike	21.84

Random Order	Order	Unit	Inspector	Measure
25	37	5	Dan	21.76
26	2	1	Mike	21.29
27	35	5	Mike	21.93
28	23	3	Dan	22.14
29	13	2	Dan	21.50
30	7	1	Dan	21.21
31	15	2	Dan	21.51
32	22	3	Dan	22.23
33	14	2	Dan	21.55
34	34	5	Mike	21.89
35	1	1	Mike	21.34
36	12	2	Mike	21.56
37	16	2	Dan	21.55
38	39	5	Dan	21.81
39	21	3	Dan	22.18
40	25	4	Mike	23.27

A visual analysis of the data in Table 11.4 using a Gage R&R run chart from Minitab reveals the results shown in Figure 11.1. In Figure 11.1, each dot represents Mike's measurements and each square represents Dan's measurements. Multiple measurements by each inspector are connected with lines. Good repeatability for an inspector is demonstrated by the low variation in the squares connected by lines and the dots connected by lines for each unit. Figure 11.1 indicates that repeatability is good. Good reproducibility is demonstrated by the similarity of the squares connected by lines and the dots connected by lines for each unit. Reproducibility is good. The Gage run chart shows that most of the observed total variation in gage readings is due to differences between units.

Collect and Analyze Baseline Data for the CTQ(s)

Next the team needs to put the data collection plan to work by collecting baseline data to determine the stability and capability of each CTQ (Gitlow et al., 2015; Gitlow and Levine, 2004). If a dashboard or balanced scorecard is being used in an organization, baseline data already exists for each CTQ (see the key indicators on the dashboard or balanced scorecard). If a dashboard or balanced scorecard is not being used in an organization, baseline data may have to be collected for each CTQ.

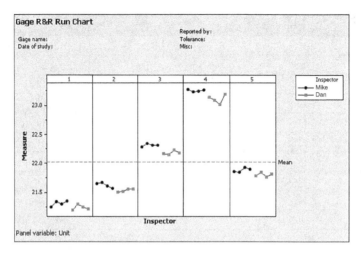

Figure 11.1 Gage R&R run chart obtained from Minitab

Team members collect *passive baseline data* for each CTQ. Passive baseline data is collected as a direct consequence of doing business. There is no intervention into the process to control the conditions that generate outputs. For example, waiting times to have a prescription filled in a hospital pharmacy are measured by a clock that is accurate to the second. A wait time longer than 5 minutes (300 seconds) is deemed unacceptable by patients (VoC). A sample of the last five patient wait times every hour (the clinic is open 8 hours per day) is collected for two weeks (Monday through Friday). A list of the 400 wait times is listed in sequential order (across the rows) in Table 11.5.

Table 11.5 Sample of Patient Wait Times

81	185	274	232	239	112	148	155	174	166	149	261	172
121	181	114	169	160	183	124	89	23	119	132	172	217
240	190	294	171	200	231	162	218	198	170	207	168	183
132	307	138	246	77	161	221	181	102	154	258	193	23
262	191	223	308	279	198	199	108	238	222	235	229	300
148	185	162	150	192	181	220	256	167	253	153	178	127
95	260	168	166	238	236	121	236	334	90	189	116	258
224	74	302	162	151	224	153	204	67	188	214	251	203
210	120	110	186	108	140	166	175	170	184	117	140	225
140	104	250	176	146	172	112	217	243	226	228	246	124
65	176	118	142	177	188	132	248	162	262	90	155	228
80	180	195	246	75	310	144	125	85	168	264	237	106
226	191	128	83	206	52	217	140	148	20	190	179	105
133	157	226	186	201	211	298	144	133	269	128	157	136

149	123	119	120	283	186	319	155	105	160	151	215	127
111	58	243	52	196	159	160	211	226	214	169	81	188
159	179	136	110	191	230	141	146	187	206	109	142	149
164	255	100	122	32	212	174	163	120	201	184	205	233
249	188	139	129	146	230	127	123	159	352	172	277	254
109	149	139	207	153	250	250	257	124	263	200	147	221
155	167	5	174	231	223	174	196	192	142	215	201	178
129	256	118	129	42	187	67	265	176	240	171	240	187
112	175	255	173	111	225	122	113	238	134	188	147	138
241	173	124	203	155	208	215	164	209	197	320	174	174
267	163	136	146	253	89	188	113	219	100	192	131	133
154	110	201	225	116	69	136	269	178	209	119	217	183
232	166	215	221	259	73	206	199	121	201	163	233	97
78	177	257	201	283	225	193	233	193	190	224	108	157
120	228	180	104	208	143	186	167	203	98	44	162	190
309	162	175	246	218	223	245	140	206	79	118	201	129
84	102	110	77	223	122	171	241	113	112			

After collecting the data, team members answer the following questions regarding the baseline data using specific statistical tools.

Team members analyze the baseline data to answer the following questions:

1. Does the baseline data for the CTQ exhibit any patterns over time? A run chart (see Chapter 4, "Understanding Data: Tools and Methods") is used to study raw baseline data over time.

2. Is the baseline data for the CTQ stable? Does it exhibit any special causes of variation? Control charts (see Chapter 5, "Understanding Variation: Tools and Methods") are used to determine the stability of a process.

3. If the CTQ is not stable (exhibits special causes of variation), where are the special causes of variation so appropriate corrective actions can be taken by team members to stabilize the process? Again, a control chart is used to identify where and when special causes of variation occur. However, they are not used to identify the causes of special variations. Tools such as log sheets, brainstorming, and cause and effect diagrams are used to identify the causes of special variations (see Chapter 6, "Non-Quantitative Techniques: Tools and Methods").

4. If the baseline data for the CTQ is stable, what are the characteristics of its distribution? In other words, what is its spread (variation), shape (distribution), and center (mean, median, and mode)? Basic descriptive statistics are discussed in Chapter 4.

The answers to the preceding questions for the waiting times in a pharmacy baseline data are as follows.

1. Does the baseline data for the waiting times in a pharmacy exhibit any patterns over time?

   Figure 11.2 shows a run chart of the wait times for each subgroup.

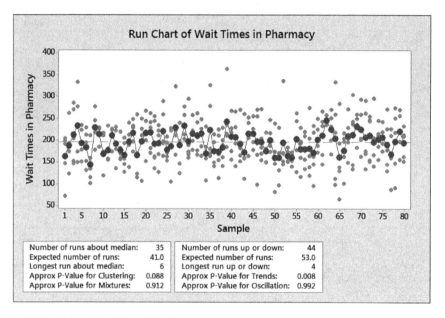

Figure 11.2 Run chart of wait times in pharmacy

Figure 11.2 reveals individual wait times randomly scattered around a mean wait time of about 175 seconds. The variability of wait times seems to be relatively constant over all of the subgroups and fluctuates roughly between 60 seconds and 350 seconds. Additionally, Figure 11.2 shows a connected plot of points that are the subgroup averages computed from the actual data.

2. Is the baseline data for the waiting times in a pharmacy stable? Does it exhibit any special causes of variation? If waiting times in a pharmacy are stable, what are their characteristics?

   An X Bar and R chart of the subgroups of five customer wait times is shown in Figure 11.3. Figure 11.3 shows that both the R chart and the X Bar chart are stable; that is, neither chart exhibits any special causes of variation in the wait time in a pharmacy. This control chart analysis verifies the findings from the run chart in Figure 11.2.

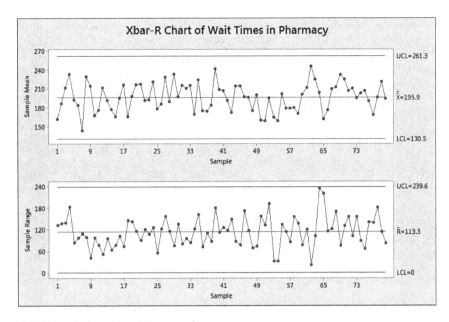

Figure 11.3 X Bar R chart of wait times in pharmacy

A basic statistical analysis of the wait times in a pharmacy data from Minitab are given in Table 11.6.

Table 11.6 Basic Statistical Analysis from Minitab of Wait Times in a Pharmacy Data

Variable	Count	Mean	St. Dev.	Minimum	Median	Maximum	Range
Wait times in pharmacy	400	195.90	49.04	64.00	194.00	364.00	300.00

The median and the mean are approximately the same indicating the distribution of wait times in a pharmacy is symmetric around the mean of 195 seconds. Further, the standard deviation is approximately 50 seconds. Figure 11.4 shows a histogram of the wait times.

The histogram shows an approximate normal distribution with a mean of 195 seconds with a standard deviation of 50 seconds.

Estimate Process Capability for CTQ(s)

Process capability compares the distribution of the output of a process (called Voice of the Process) with the customer's specification limits for the outputs (called Voice of the Customer). A process must be stable (as shown in a control chart) to determine its capability, but it still may be out of specification limit(s) (Gitlow and Levine, 2004; Gitlow et al., 2015).

To understand the capability of the process we first need to properly define the unit, the opportunity for a defect, and the defect with respect to the CTQ. Recall from Chapter 9, "Overview of Six Sigma Management," the following definitions:

Unit—A unit is the item (e.g., product or component, service or service step) or area of opportunity (e.g., a time period, a geographical area, the size of a unit, etc.) to be studied with a Six Sigma project.

Defect opportunity—A defect opportunity is each circumstance in which a CTQ can fail to be met. There may be many opportunities for defects within a defined unit.

Defect—A defect is a nonconformance on one, of many possible, quality characteristics of a unit that causes customer dissatisfaction.

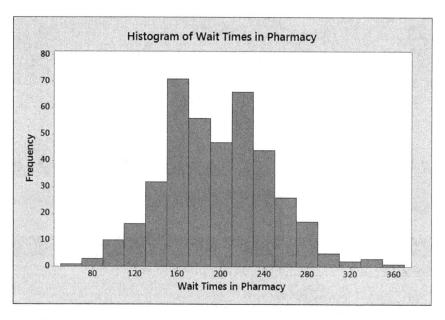

Figure 11.4 Histogram of wait times in a pharmacy

To further explain process capability for a CTQ let's go back to our example of waiting time in a pharmacy; see Table 11.7.

- Our CTQ would be Wait Time in the Pharmacy measured by wait time in seconds after the patient arrives at the pharmacy to after the patient hands her prescription to the pharmacist by patient.
- The definition of a unit would be a prescription order.
- The definition of an opportunity for a defect would be a prescription order.
- A defect in this case would be any time greater than 300 seconds.

Table 11.7 Defining the CTQ for Wait Times in Pharmacy

CTQ	Definition of a Unit	Definition of Opportunity for a Defect	Definition of a Defect
Wait time in seconds after the patient arrives at the pharmacy to after the patient hands her prescription to the pharmacist by patient	A prescription order	A prescription order	When a patient must wait longer than 300 seconds to receive a filled prescription

To calculate process capability for wait time in the pharmacy we would do the following:

- Defects (wait times over 300 seconds) = 7
- Units (number of prescription orders) = 400
- Process capability = 393/400 = 98.25%

Therefore 98.25% of the wait times in the pharmacy are less than or equal to 300 seconds. This is what the waiting time in the pharmacy process is capable of generating; it is the identity of the process.

Tollgate Review: Go-No Go Decision Point

Once all the tasks and subtasks of the Measure phase have been completed, the project leader (Black Belt) reviews the project with the Master Black Belt who critiques the Six Sigma theory and method aspects of the project. Then a member of the Finance department critiques the financial impact of the project on the bottom line, a member of the Information Technology department critiques the computer/information related aspects of the project, a member of the Legal department comments on legal issues, if any, the Process Owner critiques the process knowledge aspect of the project, and the Champion critiques the political/resource aspects of the project.

Finally, if all the elements are acceptable a tollgate review is scheduled where the Six Sigma team presents its project to the Champion and the Process Owner for approval. After each phase the team presents the project in a similar way.

Tollgate reviews focus on whether a Six Sigma team properly completed each step of a given phase of the DMAIC model. If the team has completed all steps of each phase, see the Measure phase checklist in Table 11.8, the Project Champion, Process Owner, and Black Belt sign off on their work and ask them to continue to the next phase. If there is a step that was missed or not completed to their satisfaction they have the team go back and complete the problematic step.

At this point a go-no go decision on the project must be made. Are the benefits what we thought they were? Do they justify the time and resources we are spending on the project to the level that we want to continue? Do we want to change the focus of the project? The Project Champion and Process Owner typically make this decision keeping in mind the mission of the organization while making this decision. If the answer is a go, we continue to the Analyze phase of the DMAIC model. If the answer is a no-go, the Measure phase is redone or the project is terminated.

Table 11.8 Measure Phase Tollgate Review Checklist

Measure Phase Component	Champion/Process Owner/ Black Belt Sign-Off	Date	Comments
Data Collection Plan for CTQ(s)			
Collect and Analyze Baseline data			
Estimate Process Capability for CTQ(s)			
Proceed to Analyze Phase			

Keys to Success and Pitfalls to Avoid

- Data collection can take time, so as soon as you know what you need, ask for it or start collecting it!

- Don't rush to collect data. Spending time up front on creating a sound data collection plan saves you time later!

- Make sure all key stakeholders agree on the operational definition of the CTQ to avoid any misunderstandings later!

- When analyzing baseline data certain statistical tools are used depending on whether you have attribute or measurement data; make sure to use the correct tools.

- Make sure you have read Chapters 4 and 5 to ensure you understand which tools to use to understand data and variation.

- Become familiar with a statistical software package such as Minitab. It is crucial to be able to analyze data, so make sure to practice and go through the how-to sections in Chapters 4 and 5.

- Many times data collection is viewed by employees as a burden. So when creating your communication plan it may be wise to have an initiative to educate and create awareness among stakeholders to facilitate their cooperation and buy-in.

Case Study: Reducing Patient No Shows in an Outpatient Psychiatric Clinic—Measure Phase

This section presents the Measure phase of the DMAIC model for the psychiatric clinic example concerning patient no show rates. This example utilizes all the steps of the Measure phase discussed previously.

Operational Definition of the CTQ(s) and Data Collection Plan for the CTQ(s)

The first step of the Measure phase is to create an operational definition of the CTQ. Then you create a data collection plan to lay out how you will collect the baseline data on the CTQ, defining how you are going to measure it, who will collect it, and the sampling plan.

The operational definition of the CTQ (patient no shows by month) is as follows:

Criteria: If a patient shows up for a scheduled appointment.

Test: Did the patient show up for the scheduled appointment?

Decision: The patient showed up or didn't show up.

Each month the percentage of no shows is calculated to provide data for the CTQ; see Table 11.9.

The collection of baseline data requires a data collection plan. The elements of a data collection plan include a data collection form, a sampling plan (sample size, sample frequency), and sampling instructions (who, where, when, and how); see Table 11.10.

Table 11.9 Percentage of No Shows by Month

Month	No Shows	Total Scheduled Patients	% No Shows
Feb 2013	263	1050	0.2505
Mar 2013	233	1003	0.2323
Apr 2013	295	1130	0.2611
May 2013	248	1070	0.2318
Jun 2013	232	900	0.2578
Jul 2013	247	980	0.2520
Aug 2013	251	1030	0.2437
Sep 2013	275	1090	0.2523
Oct 2013	255	1010	0.2525
Nov 2013	273	1090	0.2505
Dec 2013	231	950	0.2432
Jan 2014	221	880	0.2511
Feb 2014	286	1240	0.2306
Mar 2014	278	1110	0.2505
Apr 2014	157	660	0.2379
May 2014	267	1130	0.2363
Jun 2014	241	970	0.2485
Jul 2014	255	1090	0.2339

Month	No Shows	Total Scheduled Patients	% No Shows
Aug 2014	243	990	0.2455
Sep 2014	262	1000	0.2620
Oct 2014	248	960	0.2583
Nov 2014	263	1020	0.2578
Dec 2014	270	1120	0.2411
Jan 2015	277	1110	0.2495

Table 11.10 Data Collection Plan for CTQs

Define What to Measure		Define How to Measure			Define Who Will Collect It	Sampling Plan		
Metric	Type of Metric	Measurement Method	Data Tags Needed to Stratify the Data	Data Collection Method	Person(s) Assigned	Where?	When?	How Many?
CTQ_1: Patient no show % by month	Attribute classification data	Automated	Date, patient scheduled, patient arrived	Electronic medical record into spreadsheet	Black Belt	Outpatient Psychiatric Clinic at XYZ Hospital	Two years' worth of data from February 2013 to January 2015	All scheduled patients

Validate Measurement System for CTQ(s)

Four patients were sampled from the baseline data every fourth day resulting in six days of data. Two coders (called inspectors) counted the number of shows (not no shows) twice for each of the four patients. The data in Table 11.11 is in a format that is easy to understand what is going on. However, Table 11.11 was randomized so the coders did not know what patient they were checking on.

Table 11.11 Random Order for Collecting Data for the Gage Run Chart

Row	Proper Show Codes	Inspector	Record Number	Row	Proper Show Codes	Inspector	Record Number
1	10	1	1	13	10	1	4
2	10	2	1	14	10	2	4
3	10	1	1	15	10	1	4
4	10	2	1	16	10	2	4
5	10	1	2	17	10	1	5
6	10	2	2	18	10	2	5
7	10	1	2	19	10	1	5
8	10	2	2	20	10	2	5
9	10	1	3	21	10	1	6
10	10	2	3	22	10	2	6
11	10	1	3	23	10	1	6
12	10	2	3	24	10	2	6

The Gage run chart in Figure 11.5 shows complete agreement as to the number of shows for each coder (each coder checked each patient twice in random order, hence the two circles and two squares next to each other in each panel) for each patient and between both coders (inspectors) for the six selected days. Consequently, the measurement system is considered good.

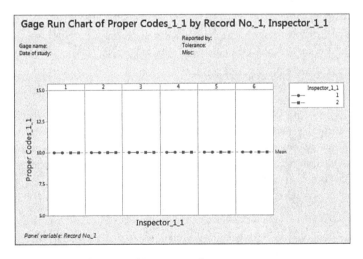

Figure 11.5 Gage R&R run chart obtained from Minitab

Collect and Analyze Baseline Data

Next, the team put the data collection plan to work by collecting baseline data to determine the stability and capability of each CTQ or Y. Team members collected passive baseline data for each CTQ. Remember, passive baseline data is collected as a direct consequence of doing business. After collecting the data, team members answered the following questions regarding the baseline data using specific statistical tools:

1. **Does the baseline data for the CTQ exhibit any patterns over time?** Using the line graph in Figure 11.6 seems like there are no trends up or down over the past two years.

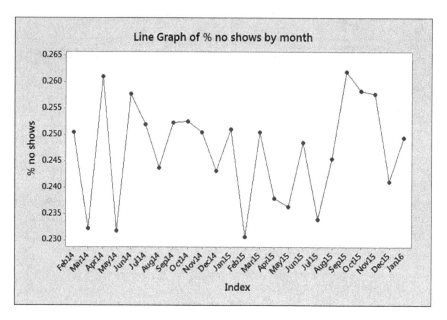

Figure 11.6 Line graph of % patient no shows

2. **Is the baseline data for the CTQ stable? Does it exhibit any special causes of variation?** Yes, the baseline data is stable as evidenced by the control chart in Figure 11.7, which shows no special causes of variation present over the past 24 months. This means all variation in the process is common variation or variation due to the process itself. The team can focus on improving the process to decrease the % no shows. Sometimes it is difficult to do nothing when the data seems to scream something is happening.

3. **If the CTQ is not stable (exhibits special causes of variation), where are the special causes of variation so appropriate corrective actions can be taken by team members to stabilize the process?** There are no special causes of variation present in this process over the past 24 months.

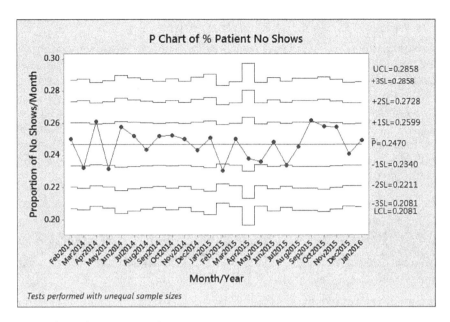

Figure 11.7 P chart of % patient no shows

4. **What is the distribution of the data?** Since the data on our CTQ is classification attribute data (proportion by month), it does not make sense to make histograms or dot plots; these are used to examine the distribution of measurement data.

5. **If the baseline data for the CTQ is stable, what are the characteristics of its distribution? In other words, what is its spread (variation), shape (distribution), and center (mean, median, and mode)? Is the baseline data what we expected when undertaking the project?** Descriptive statistics on the proportion of no shows per month

   Mean = .2471 (or 24.71%)

   Median = .2500 (or 25.00%)

   Standard deviation = .0095 (or 0.95%)

NOTE

Since the mean and median are similar we can assume that there are no months that are outliers (either low or high). The standard deviation is small compared to the mean and median, meaning that there is low variation in the process.

Tollgate Review: Go-No Go Decision Point

The team conducts the Measure phase tollgate review using the Measure phase checklist seen in Table 11.12 with the Project Champion (the hospital CEO), the Black Belt, the Process Owner, and the rest of the team on hand. Again, the CEO is very pleased with the progress; he is surprised at how high the no show rate is at almost 25%. Being a Six Sigma Black Belt himself, he looks at the control chart and realizes the process is in control and exhibits no special causes of variation, meaning that all the variation in the process is common cause. This leads to his final comment, "Well, folks, since all the variation in the process is common, it means that it is management's responsibility to fix it. So let's get going so we can give our front-line employees a better process to work with. They deserve it!"

Table 11.12 Measure Phase Tollgate Review Checklist

Measure Phase Component	Champion/Process Owner Sign-Off	Date	Comments
Data Collection Plan for CTQ(s)	CEO of Hospital XYZ	Feb 28/15	
Collect and Analyze Baseline data	CEO of Hospital XYZ	Feb 28/15	The CEO wonders if some doctors have higher no show rates than others, and the Master Black Belt reminds him that we aren't in the Analyze phase quite yet!
Estimate Process Capability for CTQ(s)	CEO of Hospital XYZ	Feb 28/15	
Proceed to Analyze Phase	CEO of Hospital XYZ	Feb 28/15	

Takeaways from This Chapter

- The Measure phase is the second of the five phases in the DMAIC model.
- The steps of the Measure phase are
    - Operationally define the CTQ(s) to ensure everyone is on the same page with regard to the definition of the CTQ
    - Create a sound data collection plan that ensures you are collecting the right data, the right amount of data, and the right type of data
    - Make sure the data is accurate using a Gage run chart
    - Collect and analyze baseline data and estimating process capability will help you understand the stability and capability of the current process so you have something to compare to once you improve the process later in the Improve phase.

References

Gitlow, H., A. Oppenheim, R. Oppenheim, and D. Levine (2015), *Quality Management: Tools and Methods for Improvement*, 4th ed. (Naperville, IL: Hercher Publishing Company). This book is free online at hercherpublishing.com.

Gitlow, H. and D. Levine (2004), *Six Sigma for Green Belts and Champions: Foundations, DMAIC, Tools and Methods, Cases and Certification* (Upper Saddle River, NJ: Prentice-Hall).

Additional Readings

Berenson, M. L., D. M. Levine, and T. C. Krehbiel (2004), *Basic Business Statistics: Concepts and Applications*, 9th ed. (Upper Saddle River, NJ: Prentice-Hall).

Gitlow, H., and R. Oppenheim (1986), *STATCITY: Understanding Statistics Through Realistic Applications*, 2nd ed. (Homewood, IL: Richard D. Irwin).

Rasis, D., H. Gitlow, and E. Popovich, "Paper Organizers International: A Fictitious Six Sigma Green Belt Case Study – Part 1," *Quality Engineering*, volume 15, number 1, 2002, pp. 127-145.

Rasis, D., H. Gitlow, and E. Popovich, "Paper Organizers International: A Fictitious Six Sigma Green Belt Case Study – Part 2," *Quality Engineering*, vol. 15, no. 2, pp. 259-274.

12
DMAIC Model: "A" Is for Analyze

What Is the Objective of This Chapter?
The objective of this chapter is to take you through the various steps of the Analyze phase of the Six Sigma DMAIC model so that you can apply them on projects at your organization. We use a case study to demonstrate how the steps of each phase are executed in real world projects.

Purpose of the Analyze Phase
Let's go back to our equation CTQ is a function of one or more Xs or CTQ = $f(X_1, X_2, X_i, \ldots X_n)$ where:

- CTQ is a measure of your problematic key indicator.
- X_i represents the ith factor that causes your CTQ to be problematic.

We completed the Define phase and idevntified our CTQ(s). Next, we completed the Measure phase by operationally defining, conducting measurement systems analysis, and collecting baseline data for our CTQ(s).

This chapter focuses on the third phase, the Analyze phase, whose purpose is for team members to identify the factors or Xs that cause our CTQ to be problematic.

The eight main deliverables for the Analyze phase are

- Detailed flowchart of the process
- Identification of potential Xs for the CTQs
- Failure Modes and Effects Analysis (FMEA) to reduce the number of Xs
- Operational definitions of Xs
- Data collection plan for Xs
- Validate the measurement system for Xs

- Test of theories to determine critical Xs
- Develop hypotheses/takeaways about the relationships between the critical Xs and CTQ(s)

At the end of the Analyze phase, the team conducts a tollgate review with the Project Champion, Black Belt, and Process Owner. This is where the team reviews what they have learned in the Analyze phase and makes a go-no go decision on the project. If everyone is satisfied with the team's work the team proceeds to the Improve phase.

The Steps of the Analyze Phase

The purpose of the Analyze phase is to determine the critical Xs that will change the Voice of the Process for the CTQ(s) that are relevant to the project. There are eight steps to this phase: develop a detailed flowchart of the current state process, identify the potential Xs for each CTQ, perform an FMEA to identify the critical Xs to improve the distribution of the CTQ(s), operationally define the critical Xs, develop a data collection plan for each critical X, validate the measurement system for each critical X, test theories about the relationship between each critical X and each CTQ, develop takeaway theories about the relationships between the Xs and the CTQ9s), and do a tollgate review.

Detailed Flowchart of Current State Process

The first step in the Analyze phase is for the team to complete a detailed flowchart of the process under study. Remember from Chapter 3, "Defining and Documenting a Process," a flowchart is a pictorial summary of the steps, flows, and decisions that comprise a process. There are two types of flowcharts you can use to create a detailed flowchart of your process:

- **Process flowchart**—A flowchart that lays out process steps and decision points in a downward direction from the start to the stop of the process.

  An example of a process flowchart for an adult health assessment is shown in Figure 12.1 with starbursts representing opportunities for improvement in the process.

- **Deployment flowchart**—A flowchart organized into "lanes" that show processes that involve various departments, people, stages, or other categories.

  Figure 12.2 shows an example of a deployment flowchart for a medication reconciliation process with starbursts representing opportunities for improvement in the process.

We recommend using a deployment flowchart if you have a process with multiple departments or employees responsible for different parts of the process, as well as tracking the number and location of handoffs within the process; otherwise a process flowchart will do the trick.

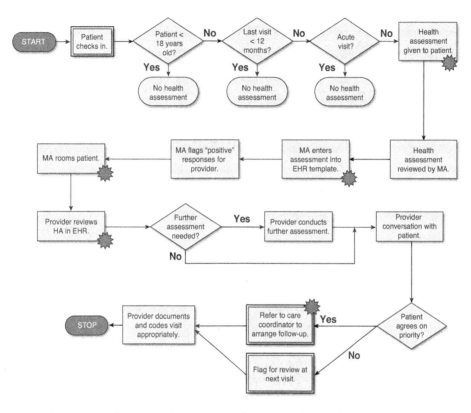

Figure 12.1 Process flowchart example (Source: Modified from ahrq.gov)

Identification of Potential Xs for CTQ(s)

The next step in the Analyze phase is to identify potential Xs for your CTQ(s). There are various ways to identify potential Xs including from a flowchart, from the Internet, from talking to experts, from benchmarking, from cause and effects diagrams, from data analysis, from the list of 70 change concepts, and from other sources.

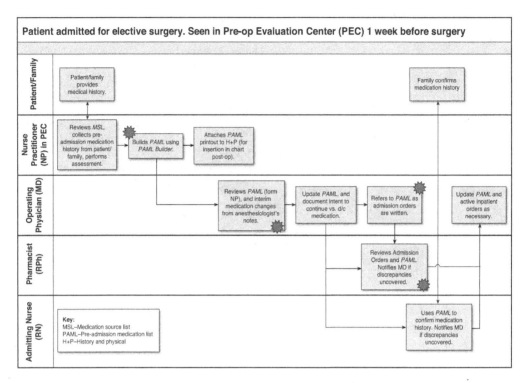

Figure 12.2 Deployment flowchart example (Source: Modified from ahrq.gov)

From the Flowchart

A common method used by team members to identify the Xs in a process is from a detailed process flowchart created in the first step of the Analyze phase (Gitlow and Levine, 2004; Gitlow et al., 2015). The purpose of a flowchart in the Analyze phase is to provide a "detailed" picture of the process under study. The definition of "detailed picture" is that the flowchart should provide the necessary specificity required to identify all potential Xs that might affect each CTQ. It is important to understand the current process to be able to effectively standardize and improve it. Understanding the current process requires a flowchart with enough detail so that team members can identify the points in the process impacted by the Xs. Team members manipulate the Xs to create the improved process.

Team members should go through the detailed flowchart step by step asking the following questions to identify potential Xs:

- Why are steps done? How else could they be done?
- Is each step necessary?
  - Are they value added and necessary? Are they repetitive?
  - Would a customer pay for this step specifically? Would they notice if it's gone?
  - Is it necessary for regulatory compliance?

- Does the step cause waste or delay?
- Does the step create rework?
- Could the step be done in a more efficient and less costly way?
- Is the step happening at the right time (in sequence or parallel)?
- Are the right people doing the right thing?
- Could steps be automated?
- Does the process contain proper feedback loops?
- Are roles and responsibilities for the process clear and well documented?
- Are there obvious bottlenecks, delays, waste, or capacity issues that can be identified and removed?
- What is the impact of the process on all stakeholders; this includes other processes?

From Experts

Another way to identify potential Xs is to ask people who are experts in the process under study. Many times you will be able to go back to your Voice of the Customer (VoC) interviews from the Define phase and go through them to see whether there are any potential Xs there. Now that you have a better understanding of the process and have analyzed some data you may want to go back and re-interview certain stakeholders that are experts in the process.

From the Internet

If you are trying to solve a specific problem, odds are you aren't the first person in the world trying to solve it! So many times there is no need to reinvent the wheel, just go to your favorite search engine on the Internet and see what you can find.

From Brainstorming

Brainstorming is a way to elicit a large number of ideas from a group of people in a short period of time (Gitlow and Levine, 2004; Gitlow et al., 2015). Members of the group use their collective thinking power to generate ideas and unrestrained thoughts.

Brainstorming is discussed in detail in Chapter 6, "Non-Quantitative Techniques: Tools and Methods."

From Cause and Effect Diagram

A cause and effect diagram is used if there is only one CTQ in a Six Sigma project, while a cause and effect matrix is used if there are two or more CTQs in a Six Sigma project. A cause and effect (C&E) diagram, also known as an *Ishikawa* or *fishbone diagram,* is a tool used to organize the possible sources of variation (Xs) in a CTQ and assist team members in the identification of the most probable causes (Xs) of the variation. Cause and effect diagrams are discussed in detail in Chapter 6. Recall, the data for a cause and effect diagram can come from a flowchart. Frequently, the data for a cause and effect diagram comes from a brainstorming session, but for our purposes, we will think of a flowchart as a tool useful in identifying the Xs related to a CTQ (Gitlow and Levine, 2004; Gitlow et al., 2015).

Failure Modes and Effects Analysis (FMEA) to Reduce the Number of Xs

Many times you will identify an enormous amount of Xs in the first part of the Analyze phase (Gitlow and Levine, 2004; Gitlow et al., 2015). However, operationally defining and collecting data on all of those Xs is not feasible due to the amount of time it takes. One way to reduce the number of Xs is to create a Failure Modes and Effects Analysis, which you learned about in Chapter 6, for all the identified Xs.

An internal committee of experts in the process under study is typically assembled to assign values for severity (how severe is the X), occurrence (how often does it happen), and detection (how easy is it to detect) for each of the Xs you identified. You can use the scales for severity, occurrence, and detection shown in Chapter 6, or you may want to create your own that make more sense to the context of your business.

Multiplying the values for severity, occurrence, and detection gives you a Risk Priority Number (RPN), which helps you prioritize and select Xs that you want to explore further to see if they are critical. One way you can compare the RPNs is to take the RPNs from the FMEA and put them into a Pareto diagram to help prioritize them.

Operational Definitions for the Xs

Once you have reduced your list of Xs to a manageable number, the next step is to operationally define each X (Gitlow and Levine, 2004; Gitlow et al., 2015).

Recall, an operational definition contains three parts: a criterion to be applied to an object or group, a test of the object or group, and a decision as to whether the object or group meets the criterion.

- **Criteria**—Operational definitions establish Voice of the Process language for each X and "Voice of the Customer" specifications for each X.
- **Test**—A test involves comparing Voice of the Process data with Voice of the Customer specifications for each X, for a given unit of output.
- **Decision**—A decision involves making a determination whether a given unit of output meets Voice of the Customer specifications.

Problems, such as endless bickering and ill-will, can arise from the lack of an operational definition. A definition is operational if all relevant users of the definition agree on the definition.

Operational definitions are discussed in detail in Chapter 6.

Data Collection Plan for Xs

Once you have operationally defined the Xs, the next step is to create a data collection plan to lay out how you will collect the data on the Xs in terms of defining what you are going to measure, how you are going to measure it, who will collect it, and the sampling plan. See the example in Table 12.1 on creating a data collection plan for the number of courier pickups per day of pathology slides from various labs in a hospital.

Defining What to Measure

- **Measure**—What is the name of the X you are collecting data on?
- **Type of metric**—Is the data for this X attribute or measurement data?

Defining How to Measure

- **Measurement method**—Will the data be collected visually or via automated collection (i.e., extracted from a database)?
- **Data tags needed to stratify the data**—Data tags are defined for the measure, such as time, date, location, tester, line, customer, buyer, operator, and so on.
- **Data collection method**—Will the data be collected manually, on a spreadsheet, via a computer?

Defining Who Will Collect It

- **Person(s) assigned**—Who will be assigned and held accountable for the data collection?

Sampling Plan

- **Where?**—What is the location where the data will be collected?
- **When?**—When and how often will the data be collected? How long will data be collected for?
- **How many?**—How many data points will be collected for each sample?

Table 12.1 Data Collection Plan for Xs

Define What to Measure		Define How to Measure			Define Who Will Collect It	Sampling Plan		
Metric	Type of Metric	Measurement Method	Data Tags Needed to Stratify the Data	Data Collection Method	Person(s) Assigned	Where?	When?	How Many?
X_1: Number of courier pickups per day	Attribute data	Clinic log	Date, time, number of samples picked up, clinic identifier	Spreadsheet	Merce	Labs	Every day for three months	One data point per day
X_2: ...								

Validate Measurement System for X(s)

The next step in the Measure phase is to validate the measurement system we are using to measure the baseline data for our Xs. Measurement systems analysis is discussed in detail in Chapter 6 and in the Measure phase in Chapter 11, "DMAIC Model: 'M' Is for Measure" (Gitlow and Levine, 2004).

Test of Theories to Determine Critical Xs

Once you have collected data for each X, the next step is to then develop hypotheses that determine whether the selected Xs truly impact the stability, shape, variation, and mean of the CTQ(s) so that you can come up with solutions in the Improve phase.

The way to do this is by testing of the theories that we have for each critical X identified earlier in the Analyze phase. When you initially identified Xs you had a theory on how each X affected the CTQ. Now that you have collected data on each X, it is time to put that theory to the test to see which Xs affect the CTQ(s).

Test of theories consists of three elements:

1. **Theory**—This is where you state the theory for each X in terms of how you believe it impacts the CTQ.
2. **Analysis**—This is where you determine whether each X is critical to the CTQ via statistical methods, process knowledge, or a review of the literature.

3. **Conclusion**—Based on your analysis this is where you state whether the X is critical to the stability, shape, variation, and mean of the CTQ.

Statistical Methods

Team members can use statistical methods to test a theory between an X and a CTQ. For example, after the team has collected baseline data in the Analyze phase, they analyze that data to determine whether that X affects the center, spread, shape, and stability of the CTQ. The following statistical methods can be used to help determine whether an X is critical or not:

- **Does the data for the X exhibit any patterns over time? Tools:** A line graph or a run chart is used to study raw baseline data over time.
- **Is the data for the X stable? Does it exhibit any special causes of variation? Tools:** Control charts are used to determine the stability of a process.
- **If the X is not stable (exhibits special causes of variation), where are the special causes of variation so that appropriate corrective actions can be taken by team members to stabilize the process? Tools:** Again, a control chart is used to identify where and when special causes of variation occur. However, they are not used to identify the causes of special variations. Tools such as log sheets, brainstorming, and cause and effect diagrams are used to identify the causes of special variations.
- **What is the distribution of the data? Tools:** A histogram or a dot plot helps us understand the distribution of the data.
- **If the baseline data for the CTQ is stable, what are the characteristics of its distribution? In other words, what is its spread (variation), shape (distribution), and center (mean, median, and mode)? Is the baseline data what we expected when undertaking the project? Tools:** Basic descriptive statistics such as mean, median, standard deviation.
  - **Mean**—The process average; is the process average what we expected it to be?
  - **Median**—The middle number.
  - **Standard deviation**—Tells us about the spread or variation of our data about the mean.

Process Knowledge

Often, teams do not have access to good data and due to their familiarity with the process are able to use process knowledge to determine whether an X is critical. Process knowledge is the result of studying the theory of a process, or the result of experience with a process that reinforces the theory of the process. For example, *lean manufacturing* theory explains the direct relationship between batch size and cycle time; if batch size is decreased, cycle time is decreased. Sometimes the solution is obvious or seems obvious, so you just make your best guess; try it and see what happens.

Review of the Literature

A review of the literature can be used to develop a hypothesis. For example, in the trade journal of the Linen Supply Association of America (LSAA), *The Linen Supply News*, an article reported a study that suggests a statistically significant negative relationship between dryness of sheets after processing and thread count of sheets.

Develop Hypotheses/Takeaways about the Relationships between the Critical Xs and CTQ(s)

Team members develop hypotheses that explain the relationships between specific critical Xs and each CTQ. A hypothesis states a premise about a CTQ, for example, the mean value of CTQ > 25 units, or about a relationship between variables. For example, CTQ = a $- b_1X_1 + b_2X_2$. CTQ = a $- b_1X_1 + b_2X_2$ is a hypothetical statement of "If X_1 is increased by 1 unit, then the CTQ will decrease by b_1 units. Further, if X_2 is increased by 1 unit, then the CTQ will increase by b_2 units." Both statements assume that there is no interaction between X_1 and X_2.

Say for example, you have 10 Xs, X_1 through X_{10}, and you believe only that X_3, X_5, and X_8 are critical Xs with X_5 being a main driver. Your hypothesis would be CTQ = f (X_3, X_5, X_8) with X_5 being the primary driver.

Go-No Go Decision Point

Once all the tasks and subtasks of the Analyze phase have been completed the project leader (Black Belt) reviews the project with the Master Black Belt who critiques the Six Sigma theory and method aspects of the project. Then, a member of the Finance department critiques the financial impact of the project on the bottom line, a member of the Information Technology department critiques the computer/information related aspects of the project, a member of the Legal department comments on legal issues, if any, the Process Owner critiques the process knowledge aspect of the project, and the Champion critiques the political/resource aspects of the project.

Finally, if all the elements are acceptable a tollgate review is scheduled where the Six Sigma team presents its project to the Champion and the Process Owner for approval; see Table 12.2. After each phase the team presents the project in a similar way.

At this point a go-no go decision on the project must be made. Are the benefits what we thought they were? Do they justify the time and resources we are spending on the project to the level that we want to continue? Do we want to change the focus of the project? The Project Champion and Process Owner typically make this decision keeping in mind the mission of the organization.

Table 12.2 Analyze Phase Tollgate Review Checklist

Analyze Phase Component	Champion/Process Owner/ Black Belt Sign-Off	Date	Comments
Detailed flowchart of current state process			
Identification of Xs for CTQ(s)			
FMEA to reduce number of Xs			
Operational definitions for Xs			
Data collection plan for Xs			
Validate measurement system for Xs			
Test of theories for Xs			
Develop hypotheses on Xs			
Proceed to Improve Phase			

Keys to Success and Pitfalls to Avoid

- Flowchart the process in as much detail as possible to really understand where the pain points and opportunities for improvement lie. Also make sure to verify and validate with the process experts to make sure it is 100% correct.

- Use your stakeholders to your advantage! Much of what they say during the VoC analysis in the Define phase can be used to identify Xs during the Analyze phase.

- To help give structure to the identification of your Xs take advantage of tools and methods such as brainstorming, affinity diagrams, cause and effect diagrams.

- If you are having a problem, odds are someone else has had the same problem, so use the Internet or other resources to help identify Xs.

- Sometimes gathering data on an X is not possible or is too time consuming or an X and its solution is just plain obvious.

- As in the Measure phase, data collection can take time, so as soon as you know what you need ask for it or start collecting it!

- Don't rush to collect data. Spending time up front creating a sound data collection plan saves you time later. Only collect data you need. Using an FMEA to reduce the number of Xs you identify to only ones with potential of being critical Xs will save you a lot of time collecting data.

- Make sure all key stakeholders agree on the operational definition of the Xs to avoid any misunderstandings later.

- Many times data collection is viewed by employees as a burden. So when creating your communication plan it may be wise to have an initiative to educate and create awareness among stakeholders to facilitate their cooperation and buy-in.
- Not all tools are used when testing theories. A good Black Belt knows which ones to use in different situations.
- You should end up with only a few critical Xs that really impact the CTQ.

Case Study: Reducing Patient No Shows in an Outpatient Psychiatric Clinic—Analyze Phase

This section presents the Analyze phase of the DMAIC model for the psychiatric clinic example concerning patient no show rates. This example utilizes all the steps of the Analyze phase discussed previously.

Detailed Flowchart of Current State Process

In the first step of the Analyze phase the team completed a detailed process map of the scheduling process for the Outpatient Scheduling Clinic at XYZ Hospital. The team created a deployment flowchart to illustrate the process in detail; see Figure 12.3. Notice the lanes, which depict who is responsible for different parts of the process, as well as the starbursts, which represent opportunities for improvement.

Identification of Xs for CTQ(s)

After creating a detailed flowchart, the team used the following methods to identify Xs.

Cause and Effect Diagram

The team sat down with team members and created the cause and effect diagram shown in Figure 12.4 to identify Xs. After some discussion the team agreed that they would investigate a few of the Xs that came out of the cause and effect diagram that made the most sense, namely:

- X_1: Reminder calls made to patients?
- X_2: How far out appointment scheduled?
- X_3: Physician to be seen
- X_4: New versus established patients
- X_5: Age of patient and time of appointment

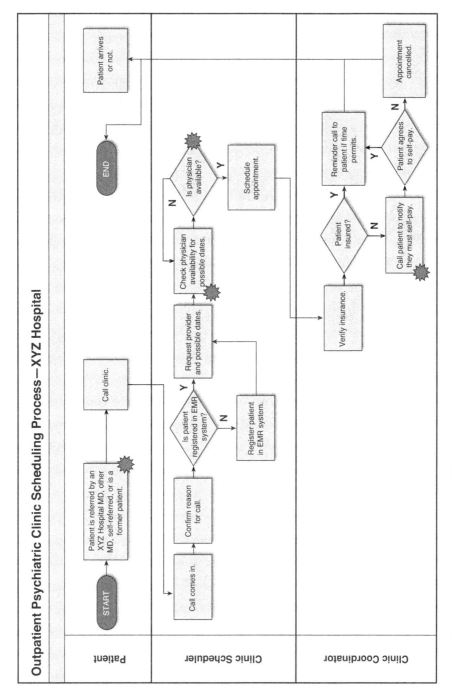

Figure 12.3 Detailed flowchart (Note: EMR = Electronic Medical Record)

Chapter 12 DMAIC Model: "A" Is for Analyze 345

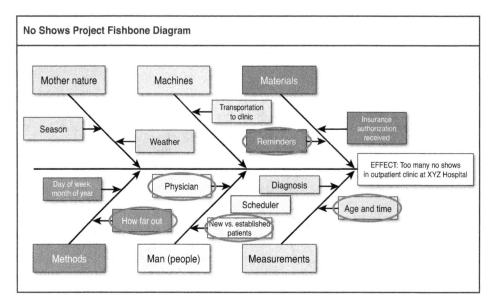

Figure 12.4 Cause and effect (fishbone) diagram for no shows

Brainstorming

After doing some further brainstorming the team added another X:

- X_6: Insurance company (payer)

They thought perhaps insurance company (X_6) could be a potential X as different providers may have different copays, which may de-incentivize patients from showing up.

Failure Modes and Effects Analysis (FMEA) to Reduce the Number of Xs

Due to the small number of Xs identified, there was no need for the team to reduce the number of Xs using FMEA.

Operational Definitions of the Xs

The team then created operational definitions of the Xs.

X1: Reminder Calls Made to Patients

- **Criteria**—Has a reminder phone call been made to each patient prior to their visit, yes or no?
- **Test**—Is the Reminder Call Made field in the database populated with a Y for each patient?

- **Decision**—If Reminder Call Made field is populated with a Y, patient received a reminder call. If Reminder Call Made field is not populated with a Y, patient did not receive a reminder call. Note: Y = yes.

X2: How Far Out Is Appointment Scheduled

- **Criteria**—The number of days between the date the appointment is scheduled and the appointment date itself.
- **Test**—Select a patient and subtract the Appointment Date field in the database from the Date the Appointment Is Scheduled field in the database.
- **Decision**—The result of the subtraction in the preceding test is how far out the appointment is scheduled.

X3: Physician to Be Seen

- **Criteria**—The name of the physician the patient is scheduled to see.
- **Test**—Open the patient record and locate the Physician Name field in the database.
- **Decision**—If the name in the field is the physician the patient is scheduled to see, the physician is correct. If not, the physician is incorrect.

X4: New Versus Established Patients

- **Criteria**—Has patient made a previous visit to this clinic, yes or no?
- **Test**—Select a patient and determine whether the patient had a Previous Visit to Clinic noted in the database.
- **Decision**—If Previous Visit to Clinic field is populated with a Y, the patient is an established patient. If Previous Visit to Clinic field is not populated with a Y, the patient is a new patient. Note: Y = yes.

X5: Age of Patient and Time of Appointment

- **Criteria**—The age of the patient at the time of the visit and the time of the appointment.
- **Test**—Open the patient record and locate the Patient Age field and the Time of Appointment field in the database.
- **Decision**—The value in the Patient Age field is the age of the patient, and the time in the Time of Appointment field is the time of appointment.

X6: Insurance Company (Payer)

- **Criteria**—The name of the insurance company the patient has.
- **Test**—Open the patient record and locate the Insurance Company field in the database.
- **Decision**—The name in that field is the patient's insurance company.

Data Collection Plan for Xs

The team was confident that at least some of these Xs were responsible for increasing the no show rate in the Outpatient Psychiatric Clinic at XYZ Hospital. The next step was to create the data collection plan shown in Table 12.3 for each X so that the team could test their theories to determine which of these Xs were in fact critical.

Validate Measurement System for Xs

Since the baseline data for the Xs are coming right out of the database from the hospital's electronic medical record system the team felt that it was unnecessary to complete a measurement systems analysis. Measurement systems analysis is discussed in detail in Chapter 6 and in Chapter 11.

Test of Theories to Determine Critical Xs

After collecting data for each X, the next step for the team was to then determine which Xs affected the stability, shape, variability, and mean of the CTQ in the Improve phase.

They did this is by testing the theories they had for each critical X identified in the Analyze phase.

X_1: Reminder Calls Made to Patients

Theory: The office makes reminder calls to patients regarding their appointments when the staff has time, and the team thinks there may be a correlation between reminders and no shows. They believe that many patients forget they have an appointment and if given a reminder call will show up.

Analysis: The team analyzed data in Table 12.4 to see whether reminder calls decreased the amount of no shows in the clinic.

Of those 8,423 patients who were given a reminder call by the staff, 912 or 11% ended up no showing. Of the 16,160 who were not given a reminder call by the staff, 5,159 or 32% ended up no showing. No shows are almost three times lower when a reminder call is made absent any interaction effects.

Conclusion: Based on the preceding analysis it appears that X_1: Reminders is likely a critical X that affects no show rate.

X_2: How Far Out Appointment Scheduled

Theory: Team members hypothesized that the farther out an appointment is made, the greater the chance that a patient would no show. The reason is that if an appointment is made too far out the patient would shop around at other hospitals to see if they could get an earlier appointment, and if they could, they would take it and not bother to cancel or show up.

Table 12.3 Data Collection Plan for Xs

Define What to Measure			Define How to Measure		Define Who Will Collect It	Sampling Plan		
Metric	Type of Metric	Measurement Method	Data Tags Needed to Stratify the Data	Data Collection Method	Person(s) Assigned	Where?	When?	How Many?
X_1: Reminder calls made to patients?	Attribute	Automated	Reminder call made? (Y/N)?	Electronic medical record into spreadsheet	Green Belt 1	Outpatient Psychiatric Clinic at XYZ Hospital	Two years' worth of data from February 2013 to January 2015	All scheduled patients
X_2: How far out is appointment scheduled	Measurement	Automated	Appointment date, date appointment made	Electronic medical record into spreadsheet	Green Belt 1	Outpatient Psychiatric Clinic at XYZ Hospital	Two years' worth of data from February 2013 to January 2015	All scheduled patients
X_3: Physician to be seen	Attribute	Automated	Physician name	Electronic medical record into spreadsheet	Green Belt 1	Outpatient Psychiatric Clinic at XYZ Hospital	Two years' worth of data from February 2013 to January 2015	All scheduled patients
X_4: New vs. established patients	Attribute	Automated	Previous visit to clinic? (Y/N)	Electronic medical record into spreadsheet	Green Belt 1	Outpatient Psychiatric Clinic at XYZ Hospital	Two years' worth of data from February 2013 to January 2015	All scheduled patientsv
X_5: Age of patient and time of appointment	Measurement	Automated	Patient age	Electronic medical record into spreadsheet	Green Belt 1	Outpatient Psychiatric Clinic at XYZ Hospital	Two years' worth of data from February 2013 to January 2015	All scheduled patients
X_6: Insurance company (payer)	Attribute	Automated	Insurance company	Electronic medical record into spreadsheet	Green Belt 1	Outpatient Psychiatric Clinic at XYZ Hospital	Two years' worth of data from February 2013 to January 2015	All scheduled patients

Table 12.4 Data on Reminder Calls

Reminder?	Total Patients Scheduled	No Shows	% No Show?
Yes	8,423	912	0.11
No	16,160	5159	0.32

Analysis: The team collected data on no show rate by how far out the appointment was scheduled. Rarely are appointments scheduled more than eight months in advance, so they collected data on no show rates on how far out appointments were scheduled from one to eight months out. The results are listed in Table 12.5 and graphically displayed in the line graph in Figure 12.5. It seems that contrary to what the team thought, the no show rate is not affected by how far out the appointment is scheduled.

Table 12.5 Data on How Far Out Appointment Scheduled

How Far Out Appointment Scheduled	No Show Rate
1 month	0.24
2 months	0.23
3 months	0.22
4 months	0.25
5 months	0.23
6 months	0.26
7 months	0.22
8 months	0.24

Conclusion: Based on the preceding analysis, it appears that X_2: How far out appointment scheduled is not a critical X, as the farther out the appointment is made does not affect the no show rate.

X_3: Physician to Be Seen by Patient

Theory: The team believes that different physicians may have different no show rates due to the fact that some see more new patients who they believe may have a higher no show rate.

Analysis: The team analyzed no show % by physician, shown in Table 12.6, to see whether there was a difference in no show rates among the five physicians in the clinic.

Conclusion: As you can see from the data in Table 12.6 and bar chart in Figure 12.6, it seems likely that X_3: Physician seen by patient is not a critical X for no shows as there is no real difference in no show rates among different physicians.

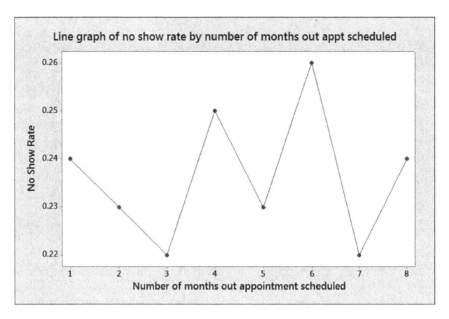

Figure 12.5 Line graph of no show rate by number of months out appointment scheduled

Table 12.6 No Shows by Physician To Be Seen

Physician	Total Patients Scheduled	No Shows	% No Show
Dr. A	5,423	1,290	0.24
Dr. B	4,217	1,099	0.26
Dr. C	3,904	900	0.23
Dr. D	6,438	1,617	0.25
Dr. E	4,601	1,165	0.25

X_4: New Versus Established Patients

Theory: The team had a theory that the new patient population has a much higher no show rate than the established patient population. Part of their reasoning was due to established patient loyalty and patients being comfortable with their physician. Team members further theorized that if new patients couldn't get an appointment soon enough they would "shop around for a physician," and if they found something sooner at another hospital they would go there and not cancel their appointment at XYZ Hospital.

Analysis: The team analyzed data on no shows rates by new versus established patients as seen in Table 12.7.

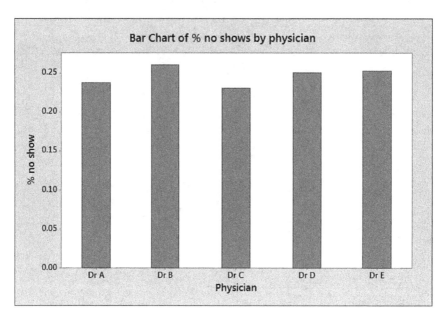

Figure 12.6 Bar chart of % no shows by physician

Table 12.7 No Shows by New Versus Established Patients

Patient Type	Total Patients Scheduled	No Shows	% No Show?
New	6,306	1,501	0.24
Established	18,277	4,570	0.25

They found that new patients no showed at almost the same rate as established patients; 24% for new patients versus 25% for established patients.

Conclusion: Based on the preceding analysis it appears that X_4: New versus established patients is not a critical X.

X_5: Age of Patient and Time of Appointment

Theory: The team thinks that perhaps there is a correlation between age of patient and time of appointment and no show rate. The older the patient, the less likely he is to no show in the morning because an older patient is more responsible and less likely to sleep in. Conversely the younger the patient the more likely he is to no show in the morning because a younger patient has more to do with his time and may have had a late night the previous night.

Analysis: The team analyzed the data in Table 12.8 regarding no show rates by age of the patient and time of the appointment to see if both variables together were a critical X that affected no shows.

Table 12.8 No Shows by Age and Time of Appointment

Age	AM vs. PM appointments	No Show Rate
Older (>=30)	AM	0.24
Older (>=30)	PM	0.21
Younger (< 30)	AM	0.64
Younger (< 30)	PM	0.15

Upon analysis of the data in Table 12.8 it is obvious that an interaction exists between age of patient and time of appointment. That is, younger patients have a much higher no show rate in the morning at 64% than older patients at 24%, while in the afternoon the no show rates are similar at 21% for older patients versus 15% for younger patients.

Conclusion: Based on the preceding analysis it appears that X_5: Age of patient combined with time of appointment is indeed a critical X that affects no show rate.

X_6: Insurance Company of the Patient (Payer)

Theory: The team wondered if the volume of no shows varied by the insurance company the patient had perhaps due to the fact that some had higher co-pays than others.

Analysis: The team collected data on the actual number of no shows by insurance company; see Table 12.9. Next, team members used a Pareto diagram to graphically display the results, as shown in Figure 12.7.

Table 12.9 Data on No Shows by Insurance Company

Insurance Company	Number of No Shows
1	1,061
2	957
3	1,031
4	1,012
5	935
6	931
Other	144

As is evidenced by the data in Table 12.9 and the Pareto diagram in Figure 12.7, the insurance company that the patient has does not seem to affect no show volume or rate.

Conclusion: Based on the preceding analysis, insurance company (payer) of the patient does not seem to be a critical X.

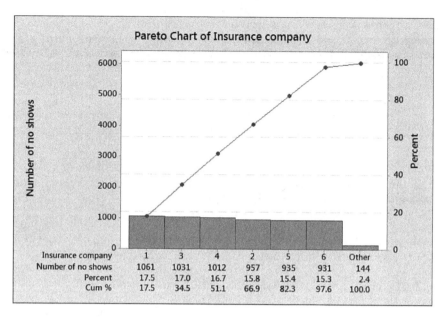

Figure 12.7 Pareto diagram of no shows by insurance company

Develop Hypotheses/Takeaways about the Relationships between the Critical Xs and CTQ(s)

Based on the testing of the preceding theories, the team hypothesized that:

> No show rate is a function of X_1 (reminder calls made to patients) and X_5 (age of patient combined with the time of their appointment).

The team's hypothesis is that no show rate is CTQ = $f(X_1, X_5)$.

Tollgate Review—Go-No Go Decision Point

The team conducts the Analyze phase tollgate review using the Analyze phase checklist seen in Table 12.10 with the Project Champion (the hospital CEO), the Black Belt, the Process Owner, and the rest of the team on hand. The Process Owner, the Assistant Vice President of Behavioral Health, commented that the previous hospital he had worked at had used reminder calls successfully to decrease no shows and was looking forward to the team implementing them in the Improve phase.

Table 12.10 Analyze Phase Tollgate Review Checklist

Analyze Phase Component	Champion/Process Owner/ Black Belt Sign-Off	Date	Comments
Detailed flowchart of current state process	CEO of Hospital XYZ	Mar 27/15	
Identification of Xs for CTQ(s)	CEO of Hospital XYZ	Mar 27/15	
FMEA to reduce number of Xs	N/A		
Operational definitions for Xs	CEO of Hospital XYZ	Mar 27/15	
Data collection plan for Xs	CEO of Hospital XYZ	Mar 27/15	
Validate measurement system for Xs	N/A		
Test of theories for Xs	CEO of Hospital XYZ	Mar 27/15	
Develop hypotheses on Xs	CEO of Hospital XYZ	Mar 27/15	
Proceed to Improve Phase	CEO of Hospital XYZ	Mar 27/15	

Takeaways from This Chapter

- The Analyze phase is the third of the five phases in the DMAIC model.
- The steps of the Analyze phase are
    - Create a detailed flowchart of the process in nauseating detail to really understand current state.
    - Identify Xs or factors that cause your CTQ to be problematic. Xs can be identified by many various methods.
    - If you have a large number of Xs, one way you can quickly eliminate them is by using FMEA.
    - Once you have a list of Xs, you must operationally define them, create a data collection plan for them, validate their measurement system, and then collect data on them.
    - Next you want to test the theories you have on each of them to determine which Xs are critical Xs.
    - Finally you develop hypotheses about the relationships between the critical Xs and the CTQ.

References

Gitlow, H., A. Oppenheim, R. Oppenheim, and D. Levine (2015), *Quality Management: Tools and Methods for Improvement*, 4th ed. (Naperville, IL: Hercher Publishing Company). This book is free online at hercherpublishing.com.

Gitlow, H. and D. Levine (2004), *Six Sigma for Green Belts and Champions: Foundations, DMAIC, Tools and Methods, Cases and Certification* (Upper Saddle River, NJ: Prentice-Hall).

Additional Readings

Rasis, D., H. Gitlow, and E. Popovich, "Paper Organizers International: A Fictitious Six Sigma Green Belt Case Study – Part 1," *Quality Engineering*, vol. 15, no. 1, 2002, pp. 127-145.

Rasis, D., H. Gitlow, and E. Popovich, "Paper Organizers International: A Fictitious Six Sigma Green Belt Case Study – Part 2," *Quality Engineering*, vol. 15, no. 2, pp. 259-274.

13

DMAIC Model: "I" Is for Improve

What Is the Objective of This Chapter?

The objective of this chapter is to take you through the various steps of the Improve phase of the Six Sigma DMAIC model so that you can apply them on projects at your organization. We use a case study to demonstrate how the steps of each phase are executed in real world projects.

Purpose of the Improve Phase

Let's go back to our equation CTQ is a function of one or more Xs, or CTQ = $f(X_1, X_2, X_i, \ldots X_n)$ where:

- CTQ is a measure of your problematic key indicator.
- X_i represents the ith factor that causes your CTQ to be problematic.

First, we completed the Define phase and identified our CTQ(s). Second, we completed the Measure phase by operationally defining, conducting measurement systems analysis, collecting baseline data, and determining the capability of our CTQ(s). Third, we completed the Analyze phase by identifying potential Xs, and then we tested our theories on the Xs to determine which ones are most critical to improve the distribution of the CTQ(s).

This chapter focuses on the fourth phase of the DMAIC model, called the Improve phase. The purpose of the Improve phase is to identify alternative methods for performing each problematic step in a process as measured by an X; the alternative methods are called change concepts. A change concept is a new method for performing a step in a process to which that X is a metric, or a revised method for performing a step in a process for which that X is a metric. Change concepts are aimed at improving the stability, shape, variation, and mean of each CTQ.

The seven main deliverables for the Improve phase are the following:

- Generate alternative methods for performing each step in the process represented by one or more Xs to improve the stability, shape, variation, and mean of all the CTQs.

- Select the best alternative method (change concept) for all the CTQs.
- Create a flowchart of the new improved process.
- Identify and mitigate risk elements for the new process.
- Run a pilot test of the new process.
- Collect and analyze the pilot test data.
- Implement the revised process on a trial basis.

At the end of the Improve phase, the team conducts a tollgate review with the Project Champion, Process Owner, and Black Belt. This is where the team reviews what they have learned in the pilot test phase and a go-no go decision is made by the Project Champion, Process Owner, and Black Belt. In other words, after comparing pilot test data with the baseline data are we happy enough to move to the Control phase and lock in our new process with documentation and training? If yes, and if everyone is satisfied with the team's work, the team proceeds to the Control phase. If not, we return to the Analyze phase and try to find another set of Xs that can improve the distribution(s) of our CTQ(s).

The Steps of the Improve Phase

The Improve phase has seven steps: generate alternative methods for performing each step in the process (other configurations for the current Xs), select the best configuration of the Xs for improving the distribution of the CTQ(s), create a flowchart of the future state process using the revised configurations of the Xs, identify and mitigate potential problems with the new configurations of the Xs, run a pilot study of the process with the new configurations of the Xs, collect and analyze the pilot test data, and make a go versus no go decision on the revised future state flowchart. The purpose of these steps is to determine the optimal configuration of the process to optimize the distribution of the CTQ(s).

Generate Alternative Methods for Performing Each Step in the Process

The first step in the Improve phase is to identify alternative methods for performing each step in the process (called change concepts) represented by one or more critical Xs to improve the stability, shape, variation, and mean of each CTQ. The most important part of this step is to make sure you involve stakeholders who perform the process on a daily basis both for their process knowledge, as well as to get their buy-in when you go to pilot test, and then implement the revised process on a trial basis. There are various ways to identify potential solutions that we cover in the following sections, including brainstorming, list of 70 change concepts, simulation, and experimental design.

Brainstorming

We use brainstorming in the Improve phase to elicit alternative methods for performing each step in the process represented by one or more critical Xs to improve the stability, shape, variation, and mean of each CTQ. Recall that brainstorming is discussed in detail in Chapter 6, "Non-Quantitative Techniques: Tools and Methods."

List of 70 Change Concepts

While all changes do not lead to improvement, all improvement requires change. The ability to develop, test, and implement changes is essential for any individual, group, or organization that wants to continuously improve. But what kinds of changes lead to improvement? Usually, a unique, specific change concept in one or more critical Xs is required to obtain improvement in one or more CTQ(s). Recall that the 70 change concepts are discussed in Chapter 6.

Simulation

Simulation is a tool used to build a model (usually a computerized flowchart with arrival times and services times, etc.) of a process and to perform experiments on the process to understand the behavior of the change concepts (critical Xs) on the stability, shape, variation, and mean of the CTQ(s). A number of off-the-shelf simulation software products (for example, Arena) have significantly reduced the programming effort of building a simulation model. Now, users can allocate more time studying "what-if" scenarios. For more about simulation, see Gitlow, H., Levine, D., and Popovich, E. (2006), *Design for Six Sigma for Green Belts and Champions: Foundations, DMADV, Tools and Methods, Cases and Certification,* Prentice-Hall Publishers (Upper Saddle River, NJ).

Experimental Design

Design of Experiments (DoE) is a collection of statistical methods for studying the relationships between the critical Xs, and their interactions with each other, on the distribution(s) of the CTQ(s). Experimental designs are used to study processes, products, or services. The purpose of an experimental design depends on a level of knowledge available concerning the process, product, or service being studied. For more information about experimental design, see Gitlow and Levine (2004).

Benchmarking

Benchmarking is a technique that can be used to discover alternative settings of the critical Xs (change concepts) to improve the distribution of the CTQ(s). Benchmarking is discussed in detail in Chapter 6.

Lean Tools and Methods

Recall from Chapter 7, "Overview of Process Improvement Methodologies," that Lean tools and methods include the 5Ss, Total Productive Maintenance (TPM), quick changeovers, poka-yoke, and value stream mapping. Although we only discuss these briefly in Chapter 7, it is important to realize that any or all of the lean tools and methods could be used as potential change concepts for one or more Xs that could be used to improve the distribution of the project's CTQ(s). Please see the references in Chapter 7 to learn more about Lean Thinking.

Select the Best Alternative Method (Change Concepts) for All of the CTQs

Now that you have generated change concepts for each critical X that may optimize the stability, shape, variability, and mean of all the CTQs, it is time to select the change concepts that you will implement in your pilot test. Team members should involve the people who work closely on the process to leverage their process knowledge and to facilitate buy-in to the process changes chosen. Often the best solution is obvious, so you just go with it. Other times there may be debate among team members as to which is the best set of change concepts. If this happens, team members should use a more objective method of selecting the optimal configuration of the process. Two alternatives for selecting potential configurations of the process are impact/effort matrices or decision matrices.

Impact/Effort Matrix

An impact/effort matrix, see Figure 13.1, is a tool used to assess the impact of potential configurations of the process based on the effort needed to execute them. It is set up as a 2x2 grid with effort needed to implement the solution on the horizontal axis and impact of the solution on the vertical axis.

After you place each configuration of the process in an appropriate cell, the next step is to determine the best one following impact-effort combinations:

- **Cell 1**—High impact, low effort: These are the solutions you probably want to go with.
- **Cell 2**—High impact, high effort: You need to further study the solutions, perhaps using a decision matrix.
- **Cell 3**—Low impact, low effort: You need to further study the solutions, perhaps using a decision matrix.
- **Cell 4**—Low impact, high effort: These are the solutions you probably want to avoid.

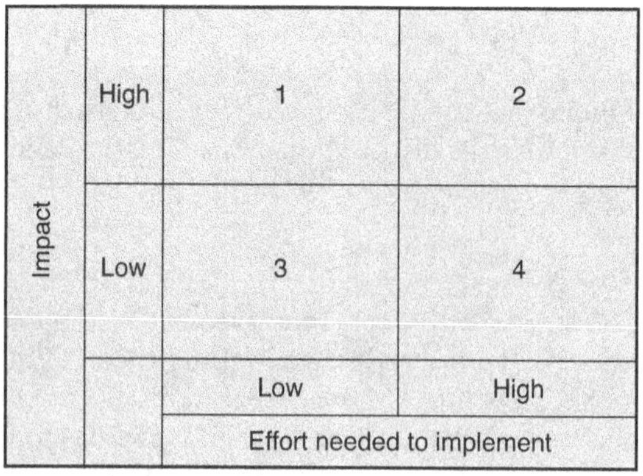

	Low	High
High (Impact)	1	2
Low (Impact)	3	4

Effort needed to implement

Figure 13.1 Impact/effort matrix

Decision Matrix

A decision matrix, see Table 13.1, is a tool used to choose among potential configurations of the process. The potential change concepts for a critical X are selected using a matrix that weighs the importance based upon certain criteria. The criteria and weights are selected by the Process Owner and the Black Belt, with possible assistance from a person from the Finance department and a person from the IT department. You are free to use whatever criterion makes sense to the team members, the Black Belt, the Process Owner, and the Champion.

The cell values are assigned by process experts and are defined as follows: 0 = no relationship, 1 = weak relationship, 3 = moderate relationship, and 9 = strong relationship. The weights of the criteria are multiplied by the values in the cells in each column to get a weighted average for each alternative configuration (flowchart). Team members use this information to determine which is the best configuration for all the CTQs. The example in Table 13.1 clearly indicates that configuration 1 (weighted average = 8.4) is the best configuration for the revised process.

Table 13.1 Decision Matrix

		Alternative Configurations of the Process		
Criteria	Weight	Configuration 1	Configuration 2	Configuration 3
Root cause addressed	0.4	9	3	3
Time needed for implementation	0.2	9	3	3
Cost of implementation	0.1	3	9	9
Ease of implementation	0.2	9	3	1
Buy-in of solution	0.1	9	9	3
TOTAL	1	8.4	4.2	3.2

Create a Flowchart for the Future State Process

The next step in the Improve phase is for the team to complete a detailed future state flowchart of the process under study that takes into effect the change concepts proposed by team members. Remember, flowcharting is discussed in detail in Chapter 6.

The following two questions help you modify the current state flowchart into the future state flowchart:

- What is the revised method (change concept) for each critical X?
- Who is responsible for the revised method for each critical X?

Identify and Mitigate the Risk Elements for New Process

Now that you have selected the best set of change concepts for each X (the best practice flowchart) that optimizes the stability, shape, variability, and mean of all the CTQs, it is smart to use risk management to identify risk elements. Risk elements address potential issues that may arise from the proposed best practice flowchart of the process. One tool that you can use to identify potential risk elements and address potential solutions is a Failure Modes and Effects Analysis (FMEA), discussed in detail in Chapter 6.

Run a Pilot Test of the New Process

For many DMAIC projects it is beneficial to conduct a pilot test of the newly improved process before rolling it out on a full scale (Gitlow and Levine, 2004). The effect of the optimized and risk-proofed revised best practice process is tested using a pilot study. A pilot study is a small scale test of the new best practice process. It can be in a limited number of locations for a trial period of time, or in all locations for a limited period of time. Whichever type of pilot study is used, its purpose is to validate the results from the newly identified best practice process on the distribution of all the CTQs. Additionally, a pilot study is important to facilitate buy-in by stakeholders of the revised best practice process.

The steps in conducting a pilot test of a revised best practice method are as follows:

Step 1: Pilot Test Charter

Team members prepare a charter for the pilot study; see Table 13.2. The charter answers the "5W1H" questions about the pilot study: Who, What, Where, When, Why, and How.

Step #2: Pilot Test Communication Plan

Team members inform all stakeholders of the revised best practice process of the impending pilot study. This communication answers any questions the stakeholders may have about the pilot study. Recall that communication plans are discussed in Chapter 6.

Step 3: Train Relevant Employees

Team members train all relevant employees in the revised best practice process.

Collect and Analyze the Pilot Test Data

Next you are going to assess the extent of the improvement by collecting and analyzing data from a pilot test of the revised process. The steps for the pilot test are shown in the following sections.

Step 1: Conduct Pilot Test

Team members and appropriate employees conduct the pilot study on a small scale and/or for a limited time frame.

Table 13.2 Improve Phase Pilot Test Charter

Pilot Test Charter WHAT?					
Team Name:		Charter Date:			
Pilot Name:		Project Champion:			
Pilot Start Date:		Process Owner:			
Pilot Test Description:					
Previous Pilot?		Where:			
Is there data available?		Data Source:			
Pilot Study Objectives:					
WHY?					
CTQs:			Current	Target (if rational v. directional)	% Change
Cost of Pilot Study:		Financial Impact:			
WHERE AND WHEN?					
Location of Pilot:					
Duration of Pilot:					
WHO?					
Resources:		Name/ Resource	Role/Responsibility		
HOW?					
Roles - skills gap analysis complete?		Due Date:	Date Completed:		
Training scheduled developed?		Due Date:	Date Completed:		
Approvals:					
	Pilot Team Leader		Champion	Black Belt	
	Date		Date	Date	

Step 2: Study Pilot Test Results

Team members evaluate the results of the pilot study on the distribution(s) of the CTQ(s). Several tips that may make a pilot study more effective are the following:

- First, it is important that team members are present when the pilot study is performed to uncover and observe problems.
- Second, team members should make sure that the test conditions are as similar as possible to actual conditions.
- Third, team members and appropriate employees should be sure to record the configuration for the critical Xs to highlight any unanticipated relationships between the critical Xs and CTQ(s).
- Fourth, all involved employees should collect control chart data and log sheet (diary) data from the pilot study.
- Fifth, team members should expect problems in the revised best practice method that did not surface during the pilot study.
- Sixth, team members should compare the before and after control charts to determine the effect of the revised best practice process on the distributions of all the CTQs.

Go-No Go Decision Point

If the results from your pilot test are not to your satisfaction, you need to understand what did not go as planned, tweak the process, and conduct another pilot test. However, if the results of the pilot test meet or exceed your expectations, you should implement the newly improved process full scale.

At this point, once all the tasks and subtasks of the Improve phase have been completed the project leader (Black Belt) reviews the project with the Master Black Belt who critiques the Six Sigma theory and method aspects of the project. Then, a member of the Finance department may want to critique the financial impact of the project on the bottom line, a member of the Information Technology department critiques the computer/information related aspects of the project, a member of the Legal department comments on legal issues, if any, the Process Owner critiques the process knowledge aspect of the project, and the Champion critiques the political/resource aspects of the project. Finally, if all the elements are acceptable a tollgate review is scheduled where the Six Sigma team presents its project to the Champion and the Process Owner for approval; see Table 13.3.

Table 13.3   Improve Phase Tollgate Review Checklist

Improve Phase Component	Champion/Process Owner/ Black Belt Sign-Off	Date	Comments
Generate alternative methods for performing each step in the process represented by one or more critical Xs to improve the stability, shape, variation, and mean for all the CTQs.			
Select solutions.			
Create flowchart for new improved process.			
Identify and mitigate risk elements.			
Run pilot test.			
Pilot test data collected and analyzed.			
Go-no go decision point.			
Proceed to Control Phase.			

At this point a go-no go decision on the project must be made. Are the benefits what we thought they were? Do they justify the time and resources we are spending on the project to the level that we want to continue? Do we want to change the focus of the project? The Project Champion and Process Owner typically make this decision keeping in mind the mission of the organization.

Keys to Success and Pitfalls to Avoid

- Tools like the impact/effect matrix and decision matrix are useful when you have multiple best practice processes.
- Including stakeholders and opinion leaders when deciding on the best practice process helps facilitate critical buy-in to the new best practice process.
- Create a future state map for the new best practice process to help people understand how it works, and what expectations will be placed on them. This facilitates buy-in by all stakeholders.
- Document the revised best practice process so you can train new and future employees.
- You may have to run multiple pilot tests to refine the process. Don't get discouraged; sometimes you have to almost run a mini-PDSA cycle during the Improve phase.
- Make sure the team is present during the pilot test(s) so they understand what is working and what is not working in the revised best practice process.

- It is important to ensure that the conditions experienced during the pilot reflect the conditions of the full process to ensure that all aspects of process variation are tested.
- Implementation of the process full scale is an exercise in project management; the more detailed your action plan can be in terms of tasks, who is responsible, status and timeline, the more successful it will be.

Case Study: Reducing Patient No Shows in an Outpatient Psychiatric Clinic—Improve Phase

In this section we continue the case study of the no show rate in the psychiatric clinic using all the steps of the Improve phase of the DMAIC model.

Generate Alternative Methods for Performing Each Step in the Process

The first step of the Improve phase for the team was to generate potential solutions for each of the critical Xs discovered in the Analyze phase. The following Xs were determined to be critical: X_1: Reminder calls made to patients and X_5: Age of patient and time of appointment. The team used brainstorming and benchmarking to come up with revised change concepts (methods) for the critical Xs.

Brainstorming

X_5: Age of Patient and Time of Appointment

Well, this was a no-brainer! The team again looked at the Analyze phase data for X_5 in Table 13.4 and decided it was obvious that patients younger than 30 years old should be scheduled in the afternoons. Their solution was to modify the decision tree that schedulers use to ensure, if possible, patients younger than 30 years old are no longer scheduled in the morning.

Table 13.4 No Shows by Age and Time of Appointment

Age	AM vs. PM Appointments	No Show Rate
Older (>=30)	AM	0.24
Older (>=30)	PM	0.21
Younger (< 30)	AM	0.64
Younger (< 30)	PM	0.15

Benchmarking
X_1: Reminder Calls Made to Patients

Again, the team looked at the data from the Analyze phase; see Table 13.5. However, the data wasn't as helpful as it was with X_5.

Table 13.5 Data on Reminder Calls

Reminder?	Total Patients Scheduled	No Shows	% No Show?
Yes	8423	912	0.11
No	16160	5159	0.32

In this case the team knew that reminder calls are successful in reducing no shows, but that they are costly in terms of time required by staff to make them. They wondered if there were other ways to remind patients without using so much of their scarce employee resources to make them. The team performed some benchmarking on other options available to them, which included the following:

- Reminder calls by staff to all patients
- Automated reminders
- Email reminders
- Text reminders

The next step was to select the best possible solution that could be implemented by the team.

Select the Best Alternative Method (Change Concept) for All the CTQs

The team chose to use a decision matrix, see Table 13.6 below, as there were several factors they wanted to take into consideration, including cost.

Table 13.6 Decision Matrix for No Shows

		Alternative Solutions			
Criteria	Weight	Reminder Calls by Staff to All Patients	Automated Reminders	Email Reminders	Text Reminders
Root cause addressed	0.3	9	9	1	3
Staff resources required	0.3	1	9	9	9
Cost of implementation	0.3	1	9	9	9
Ease of implementation	0.1	3	3	3	3
TOTAL	1	3.6	8.4	6	6.6

The criteria the team used to select the best solution was:

- **Root cause addressed**—The degree to which the solution addresses the problematic issue.
- **Staff resources needed**—The team created this criteria as they wanted to understand the amount of staff resources needed to execute each solution.
- **Cost of implementation**—The financial cost of implementation for each solution.
- **Ease of implementation**—The ability to quickly and successfully implement each solution.

Reminder calls by staff would definitely address the root cause; however, calling all patients by staff members would take an inordinate amount of time, the labor cost would be substantial, and the ease of implementation would be moderate as staff would either need to be reallocated or hired.

Automated reminders would address the root cause as well as calls by staff but would require no staff. The cost of the service would be an ongoing monthly expense but a minimal one with a return on investment supported by the Finance department, and the ease of implementation would be seamless.

The email reminder and text reminders were seen as alternatives that while fairly cheap and not utilizing staff resources would not be as effective as reminder calls. This was due to the fact that some patients may not check their email frequently enough and some older patients may not be into text messaging.

After assigning weights and completing the matrix, the team found that the automated reminders solution was their best alternative (average weight = 8.4).

Create a Flowchart of the New Improved Process

The next step for the team was to create a deployment flowchart of the future state process (see Figure 13.2). The flowchart reflects the process changes needed by the solutions selected in the previous section. The following two questions helped the team modify the current state flowchart into the future state flowchart:

What is the revised method (change concept) for each critical X?

For X_5: Age of patient and time of appointment. The new method for doing this step in the process is to have an alert in the electronic medical record system that alerts the scheduler if the patient is less than 30 years of age, and if so the scheduler should try to schedule the patient for an appointment in the afternoon, if possible.

For X_1: Reminder calls made to patients. The new method for doing this step in the process is to set up an automated reminder system that calls patients 48 hours prior to their appointment.

Who is responsible for the revised method for each critical X?

For X_5: The schedulers are responsible for X_5 (age of patient and time of appointment).

For X_1: The clinic coordinator is responsible for X_1 (reminder calls made to patients).

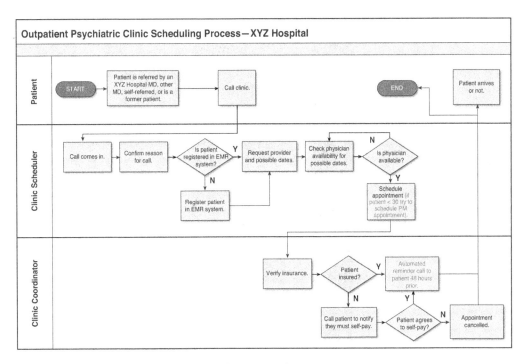

Figure 13.2 Future state deployment flowchart for no shows

Identify and Mitigate the Risk Elements for the New Process

The team then completed a Failure Modes and Effects Analysis, see Table 13.7, to identify risk elements and address potential issues that may arise from proposed best practice method.

Run a Pilot Test of the New Process

Step 1: Pilot Test Charter

Team members prepare a charter for the pilot study; see Table 13.8. The charter answers the "5W1H" questions about the pilot study: Who, What, Where, When, Why, and How.

Step 2: Pilot Test Communication Plan

Team members created a communication plan to inform all stakeholders of the revised best practice process of the impending pilot study.

Table 13.7 FMEA for No Shows Solutions

No Shows Six Sigma Improve Phase Failure Modes and Effects Analysis											
Risk Elements	Failure Mode	SEV	OCC	DET	RPN	Action	SEV	OCC	DET	RPN
Scheduler forgets to take patient's age into consideration when scheduling appointments.	Younger patients receive AM appointments and many of them no show.	8	8	10	640	Put alert in electronic medical record system that alerts scheduler if patient is less than 30 years of age.	8	1	10	80
Hospital may not have most current patient phone number on file, so automated call is ineffective.	Patient does not receive automated reminder call and no shows.	8	7	8	448	Confirm patient phone numbers when scheduling appointments.	6	2	8	96

Step 3: Train Relevant Employees

Team members train all relevant employees in the revised best practice process.

Table 13.8 No Shows Pilot Test Charter

No Shows Pilot Test Charter				
WHAT?				
Team Name: Psychiatric Clinic No Shows		Charter Date: April 10/15		
Pilot Name: No Show Reductions		Project Champion: CEO of XYZ Hospital		
Pilot Start Date: April 15/15		Process Owner: AVP, Behavioral Health		
Pilot Test Description: To test revised process including automated reminder calls and scheduling patients less than 30 years of age in the afternoons.				
Previous Pilot? No		Where: N/A		
Is there data available? N/A		Data Source: N/A		
Pilot Study Objectives:				
WHY?				
CTQs: Percent of patient no shows in the Psychiatric Clinic by month		Current	Target	% Change
		.2471	.05	
Cost of Pilot Study: No cost		Financial impact: TBD		
WHERE AND WHEN?				
Location of Pilot: Psychiatric Clinic				
Duration of Pilot: 3 weeks				
WHO?				
Resources:	Name/Resource	Role/Responsibility		
Schedulers	Scheduler team	Scheduling younger patients in afternoon.		
Registration	Registration staff	Ensure patient phone numbers are current.		
HOW?				
Roles - skills gap analysis complete? Yes		Due Date: Apr 10/15	Date Completed: Apr 8/15	
Training scheduled developed? Yes		Due Date: Apr 10/15	Date Completed: Apr 8/15	
Approvals:				
	PTL#1	CEO	BB#1	
	Pilot Team Leader	Executive Sponsor	Black Belt	

	Apr 10/15		Apr 10/15	Apr 10/15	
	Date		Date	Date	

Collect and Analyze the Pilot Test Data

Step 1: Conduct Pilot Test

Team members and appropriate employees conducted the pilot study for a period of time deemed appropriate by the Black Belt and Master Black Belt.

Step 2: Study Pilot Test Results

Team members evaluated the results of the pilot test in the CTQ(s) by creating the control chart in Figure 13.3. As you can see the pilot test was highly successful, as the no show rate dropped from 24.71% in the Measure phase to 8.44% during the pilot test.

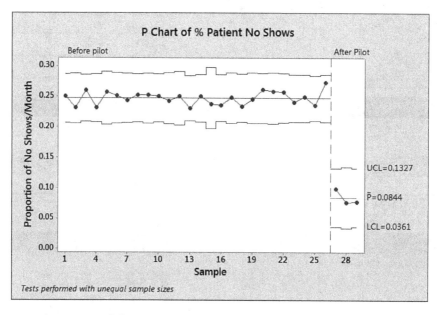

Figure 13.3 Pilot test control chart

Tollgate Review—Go-No Go Decision Point

The team conducts the Improve phase tollgate review using the Improve phase checklist seen in Table 13.9 with the Project Champion (the hospital CEO), the Black Belt, the Process Owner, and the rest of the team on hand. Based on the immense success of the pilot test both the Project Champion and the Process Owner urged the team to move to the Control phase to implement the process full scale and lock it in with documentation and training. Already seeing the potential financial impact, the CFO also reminded the team to quantify the hard benefits of the project in the Control phase.

Table 13.9 Improve Phase Tollgate Review Checklist

Improve Phase Component	Champion/Process Owner/ Black Belt Sign-Off	Date	Comments
Generate alternative methods for performing each step in the process represented by one or more critical Xs to improve the stability, shape, variation, and mean for all the CTQs.	CEO of Hospital XYZ	Apr 30/15	
Select solutions.	CEO of Hospital XYZ	Apr 30/15	
Create flowchart for new improved process.	CEO of Hospital XYZ	Apr 30/15	
Identify and mitigate risk elements.	CEO of Hospital XYZ	Apr 30/15	
Run pilot test.	CEO of Hospital XYZ	Apr 30/15	
Pilot test data collected and analyzed.	CEO of Hospital XYZ	Apr 30/15	
Go-no go decision point.	CEO of Hospital XYZ	Apr 30/15	
Proceed to Control phase.	CEO of Hospital XYZ	Apr 30/15	

Takeaways from This Chapter

- The Improve phase is the fourth of the five phases of the DMAIC model.
- The steps of the Improve phase are
  - Generate potential solutions for each of your critical Xs to stabilize and improve the shape, variation, and mean of each CTQ via various methods such as brainstorming, change concepts, simulation, experimental design, and benchmarking to name just a few.
  - Select the optimal solution for each critical X. Some solutions are obvious while some require more rigorous methods to help select your solution, such as an impact/effort matrix or a decision matrix.

- Create a flowchart of the future state process to help stakeholders understand how the new solutions will impact the process and to help educate and train employees on the new process.
- Identify and mitigate risk elements for the new process using Failure Modes and Effects Analysis (FMEA).
- Run a pilot test of the new process (if necessary).
- Collect and analyze pilot test data.
- Make a go-no go decision on whether to implement the new process full scale.

References

Gitlow, H. and D. Levine (2004), *Six Sigma for Green Belts and Champions: Foundations, DMAIC, Tools and Methods, Cases and Certification,* Prentice-Hall Publishers (Upper Saddle River, NJ).

Additional Readings

Rasis, D., H. Gitlow, and E. Popovich, "Paper Organizers International: A Fictitious Six Sigma Green Belt Case Study – Part 1," *Quality Engineering*, vol. 15, no. 1, 2002, pp. 127-145.

Rasis, D., H. Gitlow, and E. Popovich, "Paper Organizers International: A Fictitious Six Sigma Green Belt Case Study – Part 2," *Quality Engineering*, vol. 15, no. 2, pp. 259-274.

14
DMAIC Model: "C" Is for Control

What Is the Objective of This Chapter?
The objective of this chapter is to take you through the various steps of the Control phase of the Six Sigma DMAIC model so that you can apply them on projects at your organization. We use a case study to demonstrate how the steps of each phase are executed in real world projects.

Purpose of the Control Phase
First, we completed the Define phase by identifying the problematic CTQ(s). Second, we completed the Measure phase by operationally defining, conducting measurement systems analysis, collecting baseline data, and determining the capability of the CTQ(s). Third, we completed the Analyze phase by identifying potential Xs and then testing our theories about the impact of those Xs on the distributions of the CTQ(s). Fourth, we completed the Improve phase to determine the best practice method (which is the optimal configuration of each critical X) that optimizes the stability, shape, variability, and mean of each CTQ.

This chapter focuses on the fifth and final phase of the DMAIC model, the Control phase. The Control phase ensures that the optimized settings of the critical Xs are locked into place and are insensitive to human and environmental noise. Additionally, team members reduce the risks of collateral damage to related processes caused by the newly optimized settings of the critical Xs in the process under study. Team members spread the process improvements throughout the organization and hand the process over to the Process Owner for continual turning of the PDSA cycle.

The seven main deliverables for the Control phase are

- Reduce the effects of collateral damage to related processes.
- Standardize improvements (International Standards Organization [ISO]).
- Develop a control plan for the Process Owner.
- Identify and document benefits and costs of the project.

- Input the project into the Six Sigma database.
- Diffuse the improvements throughout the organization,
- Champion, Process Owner, and Black Belt review the project.

At the end of the Control phase, the team conducts a tollgate review with the Project Champion, Process Owner, and Black Belt. This is where the team reviews what has occurred in the project and is ready to close out the project and celebrate its success.

The Steps of the Control Phase

This section explains the seven steps of the Control phase of the DMAIC model: reduce the effect of collateral damage to related processes, standardize improvements to the process, develop a control plan for process owners, identify and document the benefits and costs of the project, put the project into the organizational project data base, diffuse the project's results to other relevant areas of the organization, and conduct a tollgate review of the project. Once the Control phase is complete, the team turns the revised process over to the Process Owner for continual turning of the PDSA cycle for the process and disbands the process team, and the team celebrates their success. Maintaining process improvements is critical to changing the culture of an organization to a process-oriented culture.

Reduce the Effects of Collateral Damage to Related Processes

The Control phase can be used to explore the risks created by the optimized settings of the critical Xs to other (related) processes that were not the focus of the original project (Gitlow and Levine, 2004). For example, the distribution of cycle times for step 1 in a process is stable, normally distributed with a mean of 100 minutes and a standard deviation of 10 minutes, while the distribution of cycle times for step 2 in the process is stable, normally distributed with a mean of 75 minutes and a standard deviation of 10 minutes. This process is experiencing a bottleneck in step 1 because the cycle times in step 1 are usually greater than the cycle times in step 2. If step 1 is improved through the application of the Six Sigma DMAIC model and the distribution of cycle times becomes stable, normally distributed with a mean of 40 minutes with a standard deviation of 10 minutes, then the bottleneck in the process shifts from step 1 to step 2. This is an example of the collateral damage to related processes that can be caused by Six Sigma projects. This type of collateral damage is not necessarily a bad thing because the new process's capability may be needed by the organization. However, the process improvement in step 1 uncovered a weakness in step 2. The Control phase seeks to mitigate the collateral damage (risks) to processes that are not part of the focus of the current Six Sigma project.

Team members use risk management to identify the *risk elements* of a proposed change to a process, product, or service. Before we discuss risk management, you can determine your attitude toward risk by using the following test:

Ten million dollars is yours (tax free), but there is a 1 in a _____ chance you will experience a quick and painless death and your family will not get to keep the money. What risk would you accept for the opportunity to get the ten million dollars?

    1 in a 1,000,000

    1 in 100,000

    1 in 10,000

    1 in 1,000

    1 in 100

    1 in 10

Some people will not accept the 1 in a million chance, while a few actually will accept 1 in ten. It is important that you take into account all the affected process owners' attitudes toward risk. This will help you avoid land mines that can derail the project. You should know your boss's attitude toward risk before offering advice on solutions to process problems.

Team members identify the risk elements (collateral damage) to related processes:

1. Team members identify all related processes that might experience collateral damage from the process improvements identified in the project under study.
2. Team members identify risk elements for each related process.
3. Team members identify targets for each critical parameter with significant risk elements for each related process.
4. Team members assign and prioritize risk ratings to the risk elements for each of the related processes; see Table 14.1.

Table 14.1   Risk Element Scores

		Severity		
	Risk Element Score	**Low (1)**	**Medium (3)**	**High (5)**
Likelihood	High (5)	5	15	25
	Medium (3)	3	9	15
	Low (1)	1	3	5

5. Team members prioritize the risk elements for each related process; see Table 14.2. The name of each related process is stated in column 1, the risk element(s) (for example, failure modes) for each related process are stated in column 2, the potential source(s) of harm (hazards) for each risk element are listed in column 3, the source(s) or actual injury or damage (harm) are listed in column 4, the likelihood score is shown in column 5, the severity score is shown in column 6, and the risk element score is shown in column 7.

Table 14.2 Prioritization of Risk Elements

1	2	3	4	5	6	7
Related Process 1	Risk Element for the related process 1	Hazard (potential source of harm) for the related process 1	Harm (physical injury to person and/or damage to property) from the related process 1	Likelihood 1 = low 5 = high	Severity 1 = low 5 = high	Risk Element Score 1 to 8 = low 9 to 15 = medium 16 to 25 = high

6. Team members construct risk abatement plans for risk elements with high and medium risk elements; that is, a risk element score of 9 to 25.

7. Team members identify changes to the process under study to reduce the risk for each high and medium risk element in the related processes identified in Table 14.2.

8. Team members estimate the risk element score after the risk abatement plan is set into motion.

9. Team members identify the risk element owner and set a completion date for operationalization of the risk abatement plan.

10. Team members document the risk abatement plans. A format for a risk abatement plan is shown in Table 14.3.

Table 14.3 Format for a Risk Abatement Plan

Potential Risk Elements for Process i	Potential Harm for the Risk Elements from Process i	Risk Element Score		Modification of the Change Concept for the Critical X That Will Resolve the Problem(s) for Some Other Process	Risk Owner	Completion Date
		Before	After			

11. Team members carry out the risk abatement plan.

Standardize Improvements (International Standards Organization [ISO])

ISO is a system of documenting what you do, and doing what you document. It is a critical first step to improvement of a process; it is standardization of a process (Gitlow et al., 2015). Team members standardize process improvements by answering the following questions.

1. Who is involved at the revised step of the process?
2. What should they be doing after standardization of the revised process procedure?
3. Why should they follow the revised process procedure?
4. Where should they be doing the revised process procedure?
5. When should they be doing the revised process procedure?
6. How should they be doing the revised process procedure? Training.
7. How much will it cost to do the revised process procedure?
8. Is additional training needed to perform the revised process procedure?
9. How often should the revised process procedure be monitored?
10. Who will monitor the revised process procedure?
11. Who will make decisions on the future outputs of the revised process procedure?

The answers to the preceding questions are formalized in training manuals (including flowcharts of the best practice method), training programs for existing and new employees, and, if appropriate, International Standards Organization (ISO) documentation.

ISO 9000 and ISO 14000

The ISO 9000 and ISO 14000 families of standards are among ISO's most widely known and successful standards. ISO 9000 has become an international reference for quality requirements in business to business dealings, and ISO 14000 looks set to achieve at least as much, if not more, in helping organizations to meet their environmental challenges.

Generic Table of Contents of an ISO Standard

A generic table of contents for an ISO product specification that may or may not be relevant to your project is shown in Table 14.4.

Table 14.4 Generic Table of Contents for an ISO Type Standard

- Applicable documents
  - Internal documents
    - Drawing of product
    - Drawing of package
    - Specifications for component parts
  - External documents
    - Regulations
    - Accepted standards
- Product description
  - Features of product
  - Variations of product features
- Product provisions
  - Functioning of product (general, operating characteristics, acceptable noise level, acceptable pollution, etc.)
  - Materials specifications
  - Workmanship specifications
  - Safety requirements
  - Dimensions (specifications)
  - Finish appearance
  - Marking
- Manufacture
  - Fabrication
  - Painting
  - Assembly
- Shipping
  - Packaging
    - Requirements
    - Tests
  - Marking and Labeling
- Inspection

Develop a Control Plan for the Process Owner

Team members develop a control plan for the best practice method by teaching the Process Owner and workers how to monitor the critical Xs and CTQs (Gitlow and Levine, 2004). A control plan takes the form of Table 14.5. The purpose of creating a control plan is to create a document with steps needed to follow to control a process to maintain the improvements.

The control plan is utilized in the following manner:

1. Original best practice flowchart is replaced with the revised best practice flowchart in ISO documentation.
2. Appropriate personnel are trained in the working of the revised process using training materials such as training manuals or instructional video.
3. Employees learn operational definitions for the CTQs and critical Xs, including nominal values and specification limits.
4. Employees learn how to monitor the CTQs and critical Xs using approved sampling plans.
5. Employees learn what to do with defective output.
6. Employees use the revised process. Remember, this requires personal discipline, which is one of the key foundation stones for improvement of a process!
7. Employees collect data on the CTQs and critical Xs, and analyze the data to understand the stability and capability of the CTQs and critical Xs.
8. If a CTQ or critical X is stable and capable, then repeat the PDSA cycle for further improvements in the process.
9. If a CTQ or critical X is not stable or capable, then go back to the Plan phase of the PDSA cycle and revise the best practice flowchart.

See the case study at the end of the chapter for an example of a control plan.

The columns for Table 14.5 are as follows:

- **Process step**—The process step that needs to be controlled.
- **CTQ or critical X**—Whether the variable to be controlled is a CTQ or critical X.
- **Metric characteristic**—Description of the CTQ or critical X.
- **Metric specification/requirement**—Upper specification limit, target (nominal level), and lower specification limit.
- **Measurement method**—How the metric will be measured.
- **Where the data points for the metric will be collected**—What is the location where the data on the metric will be collected?
- **Sample size**—How many data points will be collected on the metric?
- **Frequency**—How often data points on the metric will be collected.
- **Who measures**—Who will measure the metric?
- **Where recorded**—Where the measurements of the metric will be recorded.
- **Decision rule/corrective action**—What will be the decision rule to take corrective action and what will be the corrective action taken?

Table 14.5 Control Plan

Process Name:						Project Champion:				
Prepared By:						Process Owner:				
Approved By:						Date Initiated:				
Process Step	CTQ or Critical X?	Metric Characteristic	Metric Specification/ Requirement USL/Target/LSL (nominal)	Measurement Method	Where the Data Points for the Metric Will Be Collected	Sample Size	Frequency	Who Measures	Where Recorded	Decision Rule/ Corrective Action

Identify and Document the Benefits and Costs of the Project

Team members document the actual benefits (realized benefits to date) and potential benefits (future benefits), as well as the hard costs and soft costs, of the Six Sigma project. Benefits include, but are not limited to the following:

1. Improved financial performance
2. Improved safety (fewer accidents and unsafe behaviors)
3. Decreased cycle time, wait time, service time, and so on
4. Identification of additional improvement opportunities (other potential Six Sigma projects)
5. Improved work environment (increase joy in work for employees, as well as all stakeholders of the process).

Input the Project into the Six Sigma Database

The Master Black Belt enters the project into the organization's database. The database is used to spread the newly discovered improvements and/or innovations from Six Sigma projects throughout the entire organization.

Diffuse the Improvements throughout the Organization

The diffusion portion of the Control phase explains how to spread improvements among the different areas within an organization and from one organization to another organization (for example, suppliers, subcontractors, and regulators, to name a few). How to diffuse improvements is not obvious. For example, creating a newsletter or having a meeting for all interested persons is not the way to reliably spread improvements. Other methods are needed. This section discusses such methods for both *inter* (between) and *intra* (within) firm diffusion (Cool et al., 1997; Rogers, 2003).

All potential adopters of process improvements fall into one of five adopter categories: innovator, early adopter, early majority, late majority, and laggard (Rogers, 2003). Innovators are frequently the gatekeepers of new ideas into their organization. They are venturesome, cosmopolite, friendly with a clique of innovators, possess substantial financial resources, and understand complex technical knowledge. However, they may not be respected by the members of their organization. They are considered to be unreliable by their near peers due to their attraction to new things. Early adopters are the embodiment of successful, discrete use of ideas. They are the key to spreading process improvements. Early adopters are well respected by their peers, localite, opinion leaders, and role models for other members of their organization. Early majority deliberate for some time before adopting new ideas and interact frequently with their peers. They are not opinion leaders. Late majority require peer pressure to adopt an improvement. They have limited economic resources that require the removal of uncertainty surrounding an innovation. Laggards are suspicious of change and their reference point is in the past. They are very localite and are near isolates in their organization.

Successful diffusion of an improvement to a process must consider several factors. First, it must involve opinion leaders. Team members identify opinion leaders by asking themselves: "Who would we go to for advice about the process under study within our organization?" They prepare a motivational plan to induce opinion leaders to try the process improvement. The motivational plan must have the commitment of the Champion and Process Owner and should consider a balance of extrinsic (punishments and rewards) and intrinsic (the joy of doing an act) motivators. Second, it must provide a process improvement that is adequately developed and not too costly for potential adopters within the organization. Third, it must not exceed the learning capacity of potential adopters. Fourth, it must not exceed the process owner's ability to communicate the improvement to his/her direct reports. Fifth, it must utilize the intimacy between potential adopters and the team members. If the preceding five factors do not exist, or do not exist effectively, the window of opportunity for the spread of the improvement begins to close. Additionally, team members must develop user friendly training programs (e.g., courseware, locations, time of day, to name a few issues) for both the process owner and his employees concerning the process improvement(s).

Conduct a Tollgate Review of the Project

The Champion, Process Owner, and Black Belt review the Six Sigma project with a final tollgate review; see Table 14.6. If it is acceptable, team members turn the revised and improved process, with a control plan, over to the Process Owner, and the process owner accepts the process by signing the chart shown in Table 14.7. The Process Owner and her staff adhere to the process and endeavor to improve it by turning the PDSA cycle. If it is not acceptable, the Champion, Process Owner, and Black Belt provide constructive feedback to team members who then return to the appropriate phase of the DMAIC model.

Once, the Control phase is complete, the team disbands and celebrates their success. The Process Owner keeps turning the PDSA cycle for the revised process—forever!

Table 14.6 Control Phase Tollgate Review Checklist

Control Phase Component	Champion/Process Owner/ Black Belt Sign-Off	Date	Comments
Reduce the effects of collateral damage to related processes.			
Standardize improvements (ISO).			
Develop a control plan for the Process Owner.			
Identify and document the benefits and costs of the project.			
Input the project into the Six Sigma database.			

Control Phase Component	Champion/Process Owner/ Black Belt Sign-Off	Date	Comments
Diffuse the improvements throughout the organization.			
Champion, Process Owner, and Black Belt review the project.			

Table 14.7 Sign-Off Sheet for a Six Sigma Project

	Signature	Date
Champion		
Process Owner		
Black Belt		

Keys to Success and Pitfalls to Avoid

- Do not let complacency set in on this phase. It is easy to "take your eye off the ball" and revert to the old process, which is why you need to devote the same energy to the Control phase that you did to the previous phases. It is one thing to improve a process, but it is an entirely different thing to maintain the improvement to the process.
- Rigorous follow up will ensure that the gains are sustained. This is *very* important!
- Have a final debrief as a team to review lessons learned so you can be even better on the next project.
- Look for opportunities to replicate what you have done in this project to other areas of your organization; this prevents others from having to reinvent the wheel.
- In the course of your work on the current project you may have noticed opportunities for other projects. Now is a good time to bring them to the attention of the appropriate people to see whether they merit further investigation.
- Make sure you schedule a team meeting for six months from the end of the project to make sure the revised best practice is being maintained and improved.
- Celebrate success and take an opportunity to reward and thank everyone who participated on the project!

Case Study: Reducing Patient No Shows in an Outpatient Psychiatric Clinic—Control Phase

In this section, we continue the case study of reducing the no show rate for the psychiatric clinic using all the steps of the Control phase of the DMAIC model.

Reduce the Effects of Collateral Damage to Related Processes

Team members used risk management and identified one main risk element; see Table 14.8. It is failing to train new schedulers in the revised best practice process to schedule younger patients in the afternoon if possible.

Table 14.8   Risk Elements for Scheduling Process

Risk Elements	Risk Category	Likelihood of Occurrence	Impact of Occurrence	Risk Element Score	
Failing to train new schedulers	Performance	5	5	25	High
Scale: 1–5, with 5 being the highest.					

The preceding risk element must be dealt with in a risk abatement plan. The risk abatement plan for failing to train new schedulers is to document the revised scheduling process in training manuals.

Standardize Improvements (International Standards Organization [ISO])

Team members standardized process improvements by answering the following questions.

1. **Who is involved at the revised step of the process?** Schedulers are involved for the revised scheduling process for younger patients, and registration staffs are involved in confirming patients' phone numbers to ensure the automated reminders reach their intended targets.

2. **What should they be doing after standardization of the revised standard operating procedures?** The schedulers should be trying to schedule younger patients in the afternoon if possible. The registration staff needs to confirm patients' phone numbers to ensure the automated reminders reach their intended targets.

3. **Why should they follow the revised standard operating procedures?** To ensure that the solutions the team came up with in the Improve phase are executed and to sustain improvements made with the new process.

4. **When should they be doing the revised process?** Both when the patients are scheduled and when patients are registered.

5. **For whom should they be doing the revised process?** For each patient scheduled and for each patient who arrives.
6. **How should they be doing the revised process?** As part of the scheduling process and as part of the registration process.
7. **How much will it cost to do the revised process?** There are no costs to do the revised process.
8. **Is additional training needed to perform the revised process?** Yes, for current employees and for new employees.
9. **How often should the revised process be monitored?** Monthly.
10. **Who will monitor the revised process?** The Process Owner.
11. **Who will make decisions on the future outputs of the revised process?** The Process Owner.

Develop a Control Plan for the Process Owner

Team members developed a control plan for no shows. Table 14.9 requires a monthly sampling of

- Reminder calls to see how many have been answered (the vendor has data on this metric)
- Patient appointment times for patients less than 30 years of age to ensure we are scheduling them in the afternoon
- Overall no shows rates to see if the process changes have been successful

Financial Impact

The financial impact in terms of missed net revenue due to no shows was initially guesstimated in the high level project charter and the Define phase; see Table 14.10, assuming a 24.71% no show rate. The Black Belt meets with the CFO to recalculate the financial impact based on the new no show rate. They looked at the original baseline data and recalculated what the financial impact in terms of missed net revenue would have been with the new no show rate of 80%; see Table 14.11. Assuming the number of patients will be the same or greater, the financial impact of reducing no shows will be a decrease of $780,650 in missed net revenue for the next year.

Table 14.9 Control Plan for No Shows Process

Process Name: Outpatient Scheduling Process at XYX Hospital						Project Champion: CEO of XYZ Hospital			
Prepared by: Black Belt						Process Owner: AVP for Behavioral Health at XYZ Hospital			
Approved by: AVP for Behavioral Health at XYZ Hospital						Date Initiated: May 25/15			
Process	CTQ or X?	Metric Characteristic	Metric Specification/ Requirement	Measurement Method USL/Target/LSL	Where the Data Points for the Metric Will Be Collected	Sample Size	Frequency	Who Measures	Where Recorded	Decision Rule/ Corrective Action
Scheduling patients < 30 years of age	X	Time of appointment	LSL = 12pm USL = 5pm	Extract from database	Electronic Medical Record System	Each patient	Each patient	Scheduling staff	Spreadsheet	If less than <75% of younger patients not scheduled between 12 noon and 5 pm must run PDSA cycle to improve to >75%.

Process Name: Outpatient Scheduling Process at XYX Hospital										Project Champion: CEO of XYZ Hospital	
Prepared by: Black Belt										Process Owner: AVP for Behavioral Health at XYZ Hospital	
Approved by: AVP for Behavioral Health at XYZ Hospital										Date Initiated: May 25/15	
Process	CTQ or X?	Metric Characteristic	Metric Specification/ Requirement	Measurement Method		Where the Data Points for the Metric Will Be Collected	Sample Size	Frequency	Who Measures	Where Recorded	Decision Rule/ Corrective Action
				USL/Target/ LSL							
Registration	X	Phone number correct? (Y/N)	LSL = 90% USL = 100%	Sample		Electronic Medical Record System	First 5 of every hour	Once per month	Registration lead	Spreadsheet	If 90% of patients' demographic information not confirmed must run PDSA cycle to improve to 100% compliance.
Scheduling	CTQ	Monthly no show rate	USL = 5.5% LSL = 0%	Extract from database		Electronic Medical Record System	All patients for each month	Monthly	Black Belt	Spreadsheet	If no shows are greater than 5.5%, which is national best practice, the PDSA cycle will be run continuously until the no show rate is less than 5.5% or the CEO is satisfied with the results.

Table 14.10 Financial Impact for No Shows Project from Define Phase

Type of Appointment	Number of No Shows	Net Revenue per Slot	No Shows x Net Revenue/Slot
First appointment	2,435	$250	$608,750
Regular patient appointment	3,636	$150	$545,400
	Total financial impact of no shows 2014-2015 =		**$1,154,150**

The no show rate was reduced to roughly 8% in Improve phase pilot test. Assuming a no show rate of 8%, the financial impact would have been that shown in Table 14.11.

Table 14.11 Financial Impact Recalculated after Improve Phase Pilot Test

Type of Appointment	Number of No Shows	Net Revenue per Slot	No Shows x Net Revenue/Slot
First appointment	786	$250	$196,500
Regular patient appointment	1,180	$150	$177,000
	Total financial impact of no shows 2014-2015 =		**$373,500**

Therefore, assuming the number of patients stays the same, the financial impact in terms of net revenue per slot of the decreased no show rate would be $1,154,150 (the missed net revenue before) - $373,500 (the missed net revenue after) = $780,650

Input the Project into the Six Sigma Database

The Black Belt inputs the project into the project database.

Diffuse the Improvements throughout the Organization

A communication plan is created to spread the word regarding the success of the project. Given that there are other clinics in the hospital with problematic no shows rates, the next step is to set up a meeting with all the managers of those clinics to see which ones may be able to replicate what was done in the psychiatric clinic.

Champion, Process Owner, and Black Belt Review the Project

The team conducts the Control phase tollgate review using the Control phase checklist seen in Table 14.12 with the Project Champion, the Black Belt, the Process Owner, and the rest of the team on hand. The hospital CEO wants to make sure that the Control plan is followed and wants to receive a memo each month with a control chart on each of the three metrics in the control plan to ensure the gains are sustained and improved using the PDSA cycle. The Champion, Process Owner, and Black Belt sign off on the project; see Table 14.13. The team disbands and most importantly throws a party to celebrate the success of the project!

Table 14.12 Control Phase Tollgate Review Checklist for No Shows Project

Control Phase Component	Champion/Process Owner Sign-Off	Date	Comments
Reduce the effects of collateral damage to related processes.	CEO of Hospital XYZ	May 25/15	
Standardize improvements (ISO).	CEO of Hospital XYZ	May 25/15	
Develop a control plan for the Process Owner.	CEO of Hospital XYZ	May 25/15	
Identify and document the benefits and costs of the project.	CEO of Hospital XYZ	May 25/15	
Input the project into Six Sigma database.	CEO of Hospital XYZ	May 25/15	
Diffuse the improvements throughout the organization.	CEO of Hospital XYZ	May 25/15	
Champion, Process Owner, and Black Belt review project.	CEO of Hospital XYZ	May 25/15	

Table 14.13 Sign-Off Sheet for No Shows Project

	Signature	Date
Champion	CEO of Hospital XYZ	May 25/15
Process Owner	AVP of Behavioral Health, XYZ Hospital	May 25/15
Black Belt	Black Belt of Hospital XYZ	May 25/15

Takeaways from This Chapter

- The Control phase is the last of the five phases in the DMAIC model.
- The steps of the Control phase are
    - Identify and mitigate risk elements to the new process.
    - Standardize improvements using ISO protocols.
    - Develop a control plan to sustain improvements made.
    - Confirm the benefits and costs of the project.
    - Diffuse the improvements throughout the organization.
    - Have the Champion, Project Owner, and Black Belt review and sign off on the project.
    - Disband the team and celebrate success!

References

Cool, Karen, D. Igemar, and G. Szulanski, "Diffusion of Innovations Within Organizations: Electronic Switching in the Bell System, 1971-1982," *Organization Science*, vol. 8, no. 5, September/October 1997, pp. 543-559.

Gitlow, H., A. Oppenheim, R. Oppenheim, and D. Levine (2015), *Quality Management: Tools and Methods for Improvement*, 4th ed. (Naperville, IL: Hercher Publishing Company). This book is free online at hercherpublishing.com.

Gitlow, H. and D. Levine (2004), *Six Sigma for Green Belts and Champions: Foundations, DMAIC, Tools and Methods, Cases and Certification*, Prentice-Hall Publishers (Upper Saddle River, NJ).

Rogers, E. (2003), *Diffusion of Innovations*, 5th ed. (New York, NY: The Free Press).

Additional Readings

Rasis, D., H. Gitlow, and E. Popovich, "Paper Organizers International: A Fictitious Six Sigma Green Belt Case Study – Part 1," *Quality Engineering*, vol. 15, no. 1, 2002, pp. 127-145.

Rasis, D., H. Gitlow, and E. Popovich, "Paper Organizers International: A Fictitious Six Sigma Green Belt Case Study – Part 2," *Quality Engineering*, vol. 15, no. 2, pp. 259-274.

15

Maintaining Improvements in Processes, Products-Services, Policies, and Management Style

What Is the Objective of This Chapter?

The objective of this chapter is to clarify the critical importance of preventing backsliding in process improvement efforts in an organization over time. Backsliding is what happens when the Act phase of the PDSA cycle or the Control phase of the DMAIC model are not followed over time. Yesterday's gains are tomorrow's losses. The prevention of backsliding (maintaining the current process's performance) is just as important to process improvement efforts as the process improvement efforts themselves.

Improving Processes, Products-Services, and Processes: Revisited

Recall, the PDSA cycle aids management in improving and innovating processes; that is, in helping to reduce the difference between customers' needs and process performance. It consists of four stages: Plan, Do, Study, and Act. This chapter focuses on the consequences of failing to adhere to the process, product/service, or policy changes formalized in the Act step between turns of the PDSA cycle. Additionally, recall that the Six Sigma DMAIC model is also used to improve processes, products/services, and policies. The DMAIC model consists of five phases; they are Define, Measure, Analyze, Improve, and Control. Again, this chapter focuses on the consequences of failing to adhere to the process, product/service, or policy changes formalized in the Control phase between turns of the PDSA cycle.

Case Study 1: Failure in the Act Phase of the PDSA Cycle in Manufacturing

One of the authors was consulting for a large, international paper company in the Southeastern United States. The company conducted training for top managers and employees on Dr. Deming's theory of management; the training included heavy coverage of the PDSA cycle, as well as control charts and Pareto diagrams.

A first line supervisor (Randy C.) who attended the training sessions ran a crew at the end of the paper production process. He determined that the percentage of defective rolls of paper made per week was a stable and predictable process using a p chart. However, the process exhibited too much common variation, with an unacceptably high average proportion of defective rolls of paper. Consequently, Randy C. made a Pareto diagram of the number of defective rolls of paper by type of defective. He found more than 90 types of defectives; however, three types of defectives accounted for more than 80% of all the defective rolls.

Randy C. made several simple (and free) process changes to eliminate the three types of defectives, and consequently, saved the company $6,000,000 per year in reduced rework. Everyone at the company was ecstatic with the results. After about a year, Randy C. was promoted to a new position with more compensation; he was happy. A new supervisor took over his position, but he did not know about the process improvements Randy C. had made. So, after about six months, the paper production process was performing at its old level of defective rolls of paper. The $6,000,000 in reduced rework was lost to the company. This is a classic example of the failure to follow the Act phase of the PDSA cycle over time.

Case Study 2: Failure in the Act Phase of the PDSA Cycle in Accounts Receivable

One of the authors was contacted by the manager of the student Accounts Receivable department at a large, urban, research university. The A/R department was suddenly plagued by a huge increase in the percentage of abandoned telephone calls by students trying to resolve problems with their accounts. The author asked the manager to send to him as much data as was convenient. He sent 41 months of data. The author made a p chart of the 41 monthly percentage of abandoned call; see Figure 15.1. The data showed that something dramatic happened to the abandoned call process in January 2011.

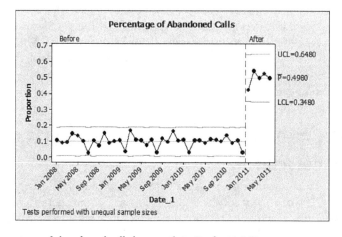

Figure 15.1 Percentage of abandoned calls by month in Students A/R

The author asked the manager what happened in January 2011. The author was told that the A/R department lost three of its seven employees due to university politics, and that the central administration was all over him due to student complaints about their telephone calls not being answered in a timely fashion. He didn't know what to do; he wanted help. The author told the manager of the A/R department about the 70 change concepts discussed in Chapter 6, "Non-Quantitative Techniques: Tools and Methods," and the manager and his team agreed to try to use them to decrease the percentage of abandoned calls.

The first change concept the A/R team identified that might help them was triage. They used triage by reassigning three work study students working in the A/R department to answer phone calls and take messages; the application of the triage change concept decreased the abandoned call rate by about 10%; in Figure 15.2, see the column Abandoned Calls after PDSA 1. The second change concept the A/R team identified was shift demand. The A/R team was able to use the shift demand change concept when the author asked them two questions:

Question 1: What is the average number of times per year a student calls the A/R department?

Answer 1: About ten times per year.

Question 2: Do all the students call about ten times per year, or do most students call once, and a few call 100 times per year?

Answer 2: Most students call once, and a few call 100 times per year.

Given the answers to these questions, the author suggested that the A/R team keep track of the number of times each student calls each year. They accomplished this by recording each student's identification number each time he called the A/R department. After a few months, they identified the frequent callers. Next, team members shifted demand by calling the frequent callers at 7:30 a.m. and telling them they had been identified as frequent callers and they could set up an appointment before or after regular hours to resolve their problem(s), once and for all. This change concept was amazingly successful; it decreased the percentage of abandoned calls to 8.2%; see Figure 15.2. So, after two turns of the PDSA cycle, using the change concept literature, the percentage of abandoned calls was lower with three employees (8.2%) than it had been with seven employees—originally 10% and then 49% after losing three employees, also shown in Figure 15.2.

The author went back to the Accounts Receivable department three months after his involvement with the project. He made a new p chart of the abandoned call data, shown in Figure 15.3. He discovered that the percentage of abandoned calls had climbed to 18.87%, up from its previous 8.2%. Looking at the p chart, the author realized that the reason for the increase was likely a failure to maintain the gains made from the triage change concept. The reason for this backsliding was simple: failure to successfully implement the Act phase of the PDSA cycle over time. Simply put, they discovered the triage change concept, used it for a short period of time, but never institutionalized it through training and revised job descriptions. Consequently, when the new school year began in August 2013, the work study students were not performing triage, and the abandoned call rate increased by 10%; Yesterday's improvement was today's loss; see Figure 15.3.

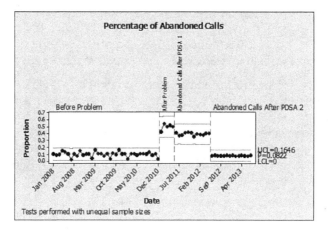

Figure 15.2 Percentage of abandoned calls by month in Students A/R after two turns of the PDSA cycle

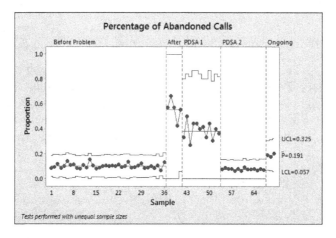

Figure 15.3 Percentage of abandoned calls by month three months later

A Method for Promoting Improvement and Maintainability

Dashboards

Dashboards are discussed in Chapter 8, "Project Identification and Prioritization: Building a Project Pipeline." Recall that dashboards are a cascading and interlocking system of mission statements, strategies, objectives, and indicators (metrics) throughout the entire organization. We propose that existing objectives and indicators are slightly modified to focus managerial attention on both improvement and maintainability of processes, products/services, and policies. Generic examples of such an objective and indicator are the following:

Objective: Improve *and maintain* stable outcomes of the process at their current level of process capability (before any process changes, improve and maintain them at the new capability)

Indicator: Process outcomes over time (for example, week, month, etc.)

Examples of the new type of objective and indicator are the following:

Objective: Improve *and maintain* stable cycle times for the selling process (time from initial sales call to time of signed sales contract)

Indicator: Cycle times for the selling process by contract over time

The point is, the preceding revised objective and indicator clearly include the word *maintain*. This means that when employees examine their dashboard, they are not just focused on turning the PDSA cycle to get better outputs; rather, they are also concerned with preventing backsliding of outputs. This requires that the indicators on a dashboard be periodically examined for maintainability until the next round of improvements creates a new process capability that must be maintained; this is repeated over and over again. This is all accomplished in the Act step of the PDSA cycle, or the Control phase of the DMAIC model. Figure 15.3 indicates what happens when maintainability is not a focus of employee attention.

Presidential Review of Maintainability Indicators

Presidential reviews are conducted to determine the state of the organization and to develop a plan of action for the promotion of corporate policy. Presidential reviews are high level studies of an organization's departments by the President or Chief Executive Officer.

During presidential reviews, the leaders of the departments explain to the president their mission and the status of projects emanating from the strategic and improvement plans. Normally, this information is conveyed through presentations. Much attention is devoted to the linkage between corporate and department strategies and the progress toward the achievement of these strategies. Problems in planning and executing these strategies are discussed, and attempts are made to identify the causes of these problems. Through the presidential review, the President is able to evaluate the state of quality and management in the organization

We suggest that presidential reviews also consider maintainability. Maintainability is most easily accomplished in the construction of the agendas for presidential reviews. The reviews consider both reports on improvements and innovations of processes, products/services, and policies, and reports on the maintainability innovations of processes, products/services, and policies.

The Funnel Experiment and Successful Management Style

Recall from Chapter 2, "Process and Quality Fundamentals," that W. Edwards Deming stated "If anyone adjusts a stable process (one exhibiting only common causes of variation) to try to compensate for a result that is undesirable, or for a result that is extra good, the output that follows will be worse than if he had left the process alone." The lessons of the Funnel Experiment hold especially true for the continuity of management style in an organization over the long term. If one tries to change the management style of an organization without profound knowledge of the effects of tampering with it, the management style of the organization will become increasingly variable, and unworkable, over time. In this section, we discuss rules 1 and 4 of the Funnel Experiment, as they relate to the continuity of management style in an organization. This is another form of maintainability.

Rule 1 Revisited

Recall management's use of rule 1 of the Funnel Experiment demonstrates an understanding of the distinction between special and common variation, and the different types of managerial action required for each type of variation. Rule 1 implies that the process is being managed by people who know how to reduce variation. In the case of continuity of management style, the organization's management style is established by the wisdom of top management that develops and improves their management style over the long term; for example, Dr. Deming's theory of management (Gitlow et al., 2015; Deming, 1994). Once set, it stands over time, with modifications coming only from advancements in management style. Rule 4 explains what happens to an organization as its management style changes with changes in top management over time.

Rule 4 Revisited

Again recall, management's use of rule 4 is an attempt to make things better in a process. Rule 4 implies that the process is being tampered with by people with inadequate knowledge of how to manage the process to reduce its variation. It implies acting on common variation as if it were special variation. Rule 4 explains what will happen to the culture of an organization over time, with changes in the top management, and the subsequent changes in management style. This scenario is a particularly scary and dangerous application of rule 4. If the new management style is a reaction to the last management style, the management style of the organization will become increasingly inconsistent (more variable) over time. Eventually, it will explode into a chaotic management style in which everyone is operating under her own conception of management style, and there is no consistency within areas, and between areas, in an organization with respect to management style over time. The organization is like a ship adrift in an ocean without a compass to guide it or a rudder to keep it on course.

The question remains: How does the top management create an environment conducive to improvement and maintainability without falling subject to rule 4 of the Funnel Experiment? The authors believe that this is accomplished through continuity of a management style that is process oriented and committed to improvement and maintainability. This type of management style requires succession plans for key managerial positions, methods to prevent the egos of top managers from changing management style to suit their particular needs, and plans to educate members of the board of directors in the critical role of management style in the functioning of the organization.

Succession Planning for the Maintainability of Management Style

The purpose of a succession plan is to increase the pool of available, experienced, and capable employees for leadership positions in an organization that can seamlessly pursue the organizational mission and strategy over time. It is based on the view that leaders in an organization must emerge from the organization itself; this is called *bench strength*.

Succession plans (Wikipedia) usually exhibit the following practices:

- Identify employees with potential for promotion within the organization.
- Provide developmental experiences for employees with potential for promotion.
- Engage the leadership in supporting the development of high potential leaders.
- Build a data base to make better staffing decisions for key jobs.
- Provide maintainability of management style over time. *(This bulleted item was added by the authors.)*
- Improve employee commitment and retention.
- Meet the career development expectations of existing employees.
- Counter the increasing difficulty and costs of recruiting employees externally.

There are three popular models for succession planning that promote maintainability of management style, as well as the other benefits just listed; they are discussed in the following sections (adapted from EOWA and Carter, 1986).

Succession Planning by Incumbent Model

The incumbent succession planning model has the employee currently holding a position identify one or more employees best qualified for his position within one year, within two years, and within three to five years. Additionally, the incumbent model identifies the developmental needs of each candidate for the position. This allows for a benefit/cost comparison of each candidate. Sometimes the organization decides that the succession plan is a strictly confidential document; consequently, only the developers of the succession plan are aware of it.

The incumbent model has four benefits. The first benefit is that it is simple. However, this assumes that the best person to identify the future holder of the position is the person currently holding the position. The second benefit is that many CEOs are most comfortable with the incumbent model because she is able to scan the list of candidates over all positions to see whether there are any positions that have no identified successors. The third benefit is that this approach is the least costly, and quickest, and does not require a high level of organizational commitment. It serves the purpose of ensuring at a minimum that managers are thinking about succession issues and are aware that succession planning is partly their responsibility. Finally, this approach hopefully transfers management style from the old generation to the new generation through mentoring over time, although this is not likely to happen without a strong guiding hand by the top management.

There are three costs of the incumbent model. The first cost is the high risk of passing on bad habits over time and/or denying fresh ideas into the organization. The second cost of the incumbent model is that the incumbent may not know employees across the organization, and hence, misses excellent candidates. The third cost of the incumbent model is that an employee identified for a position may not aspire to the promotion position.

Succession Planning by Creating Talent Pools Model

The succession planning by creating a talent pool model identifies a group of high potential candidates for positions that will open up in the future. The potential candidates are usually selected by a task force of senior managers, often with the assistance of the Human Resources department. The task force develops and uses a list of desirable characteristics for the position, including the ability to carry on the management style of the organization into the distant future. Once the high potential pool has been identified, those who make the list generally receive some special attention. Special attention frequently takes the form of a "fast track" program that provides training and mentoring by the incumbent or some other qualified person.

The talent pool model has two benefits. The first benefit is that it tends to be somewhat fairer than the incumbent model, because more employees are involved in the selection of the candidates, thus offsetting potential selection bias. The second benefit is that this model is more likely than the incumbent model to recognize the value of a broad background, rather than a single functional stream of experience.

The talent pool model has two costs. The first cost is that in large organizations, the majority of employees may not be well known to the task force members, and their view of the person may be influenced by the candidate's visibility in the organization. As a result, talented employees who do not have a high profile may be overlooked all together. The second cost is the demoralization of employees not selected by the succession planning process; this cost can be substantial. Be wary of this cost!

Succession Planning Using the Top-Down/Bottom-Up Model

The succession planning model using the top-down/bottom-up approach is based on determining the current and expected future needs of the organization. This model identifies the future needs of the organization and personnel interested in the promotion job. This is accomplished by surveying all stakeholders of the promotion job using an annual 360° feedback survey to identify the personal characteristics and domain expertise required by the promotion job.

This model has five benefits. The first benefit is that it conforms to Equal Opportunity principles by allowing a broad group of employees to participate though self-selection. The second benefit is that it empowers employees by giving them some control over their careers. The third benefit is that there is less chance to "work the system" (e.g., to wire jobs for favored applicants). The fourth benefit is that it allows employees to demonstrate their understanding of the organization's management style over time; this benefit was added by the authors. The last benefit is that the succession planning process is transparent. There are no secrets or hidden agenda items.

This model has two costs. The first cost is that this model requires serious commitment from the most senior levels of management. The second cost of this model is that current managers must not believe that they can run the organization through their successor. The focus of the succession plan must be on the maintainability of a healthy management style, not on strategic and tactical decision making.

Process Oriented Top-Down/Bottom-Up Succession Planning Model

The authors strongly favor a modified version of the top-down/bottom-up succession planning model. Our suggested model incorporates a process orientation into the model (Gitlow et al., 2015; Deming, 1994). In the following paragraphs, we repeat the top-down/bottom-up model with modifications and additions that tailor it to a process oriented theory of management.

The process oriented succession planning model using a top-down/bottom-up approach is based on determining the current and expected future needs of the organization, given the boundaries provided by the organizational mission, strategy, and values and beliefs, within the context of its management style. This model identifies the future needs of the organization and personnel interested in the promotion job, assuming a deep understanding of the management style. This is accomplished by surveying all stakeholders of the promotion job using a 360° feedback survey to identify the personal characteristics and domain expertise required by the promotion job.

We suggest that your list of personal characteristics and domain expertise for the promotion job holder, especially if the promotion job is for a top level executive, include the following:

- The holder of the promotion job has a deep understanding of the organizational culture and management style, and is committed to continuing the best practice management style until something better comes along that has strong theoretical and experiential roots.
- The holder of the promotion job sees the organization as *a system of interrelated components*, each with an aim, but all focused collectively to support the aim of the organization. The holder of the promotion job does not see the organization as *a system of independent components*, each with an aim, but all focused collectively to support the aim of the organization.

- The holder of the promotion job understands common and special causes of variation, and knows that the causes of common variations are the responsibility of management. Workers should not be blamed for common causes of variation. The holder of the promotion job does not use guesswork, opinion, and gut feel to judge, reward, and punish the performance of his employees.

- The holder of the promotion job understands that knowledge is developed by stating a theory, using the theory to predict a future outcome, comparing the observed outcome with the predicted outcome, and supporting, revising, or even abandoning the theory. This relates directly to any changes in the organizational management style.

- The holder of the promotion job understands that experience is of no value without the aid of theory. Theory allows people to understand and interpret experience. It allows people to ask questions and to learn. Management style is a theory.

- The holder of the promotion job understands that all plans are based on assumptions, and assumptions are the future outputs of processes. If the process underlying an assumption is not stable with an acceptable degree of predictability, then the assumption required for the plan cannot be relied upon with any degree of comfort. Consequently, the plan must be changed or abandoned, or the process must be improved to enhance the likelihood of the assumption being valid when called for by the plan. Management style is a plan that has assumptions.

- The holder of the promotion job understands that communication is possible when people share operational definitions. Operational definitions are statistical clarifications of the terms people use to communicate with each other. A term is operationally defined if the users of the term agree on a common definition.

- The holder of the promotion job understands that success cannot be copied from system to system. The reason for success in one system may not be present in another system. The theory underlying a success in one system can be used as a basis for learning in another system; this is called benchmarking. Management style cannot just be copied from one organization to another.

- The holder of the promotion job understands that helpful prediction of the near future is only possible by continuously working to create stable processes with low variation. Management style is a process that should be stable with low variation.

- The holder of the promotion job understands extrinsic and intrinsic motivation, and creates an environment in which employees can experience intrinsic motivation. Extrinsic motivation is stimulated by punishments and rewards. Intrinsic motivation is stimulated when management redefines work to be doing work and improving work using an appropriate process improvement method. This applies to management style.

- The holder of the promotion job understands the PDSA cycle and the Taguchi Loss function.

In the anecdotal experience of the authors, if an executive understands the preceding characteristics, it is likely that she believes in the characteristics and will manage the organization accordingly. She will not change the organization's management style without the benefit of profound knowledge. In other words, the executive will endeavor to maintain the management style of the organization and change it only based on strong theoretical grounds.

Additionally, the process oriented succession planning model using the top-down/bottom-up approach identifies candidates for promotion by providing special assignments that contain elements of the promotion job; utilizes an assessment center to observe candidates exercising the skills needed in the promotion job under realistic conditions (if available); and develops an organizational culture in which promotion is not the only vehicle for people to exercise leadership and influence, to get rewards and recognition, or to stretch and challenge themselves in their jobs and careers.

Further, the process oriented succession planning model using the top-down/bottom-up approach believes in advancing people from within the organization whenever possible. This requires that able people be hired into entry level positions and be offered the opportunity to be all they can be in the organization.

Egotism of Top Management as a Threat to the Maintainability of Management Style

Egotism is a perception that creates an extremely favorable view of your features and importance. The egotist is extremely concerned with the "Me"—of their personal qualities and puts him or her self at the center of the world with little concern for other people, even loved ones.

Egotism is a major challenge to the maintainability of a healthy management style. If every top executive who enters an organization decides to change the management style/culture of the organization, then rule 4 of the Funnel Experiment will cause a chaotic management style in the long term.

Six Indicators of Egotism That Threaten the Maintainability of Management Style

There are six indicators that a manager suffers from egotism; they are

1. **Extremely extrinsically motivated**—If you are an extremely extrinsically motivated person, you have an external locus of control and are subject to the opinions of other people. You must free yourself from the need for external approval to succeed at maintaining the management style/culture of a healthy organization.

2. **Overly concerned about appearing weak to other people**—If you are scared of asking others for help and demand getting all the credit for a management style, then you are an egotist.

3. **Compulsion to compete**—Egotists are compelled to compete when cooperation is a better course of action. A manager who is an egoist will change a perfectly healthy management culture to make it a reflection of his or her personality.

4. **Overly developed sense of greed**—Greed will cause a manager to change a perfectly healthy management culture to one that fits his or her personality just to make it his or her own. This rationale for change frequently causes a culture of fear, which is totally unacceptable to a healthy management culture.

5. **Failure to live in the moment**—Egotists do not live in the now; they live in the future, even if it doesn't make sense. This causes a manager to plan to change the current management style to a future management style for no other reason than his or her need to leave a mark on the future organization.

6. **Need to be right**—Egoists frequently lead successful lives because of their persistence and aggressiveness. However, this persistence and aggressiveness when focused on the need to leave their imprint on an organization's management style can destroy a healthy management style in an organization. Egotists are so demanding of the people who work for them that frequently they don't get support when they need it the most.

Summary

When a manager learns to let go of his ego, his level of success and fulfillment will increase dramatically. Only with your ego in check will you have the ability to reach your full potential and maintain a healthy organizational culture and management style.

The Board of Directors Fails to Understand the Need for Maintainability in the Organization's Culture and Management Style

This section is adapted from Dailey (2011). The culture of a board of directors is fundamental to maintaining a healthy management style in an organization over the long term. Without a long-term, consistent, healthy culture at the board of directors level, an organization is not likely to perform up to its potential. Why? Board culture dictates focus, behavior, risk appetite, and decision-making processes within an organization. Without board focus, the management style of the organization will follow rule 4 of the Funnel Experiment.

Definition of Culture/Management Style

Culture sets the foundation of an organization. It includes values and beliefs, assumptions, and attitudes, which in turn defines management style. Culture, and its ensuing management style, exerts enormous influence over individual behavior.

The members of the board of directors must understand culture and management style because they are critical to the success of an organization. First, culture is self-reinforcing; that is, it conveys how one is to respond to critical situations by providing common values and beliefs, rules for making decisions, behavioral norms, work styles and practices, as well as rules for dealing with all manner of informal matters. New directors become part of the existing board culture by identifying with the existing board's values and beliefs, assumptions, and attitudes. New directors "learn the ropes" of acceptable board behavior, and frequently, abide by the culture, practice its norms, and model the culture. And so, the culture of the board of directors continues over time. This characteristic of boards is critical to the maintainability of management style. Second, culture, and its ensuing management style, are self-defending and punish violators. The inner circle of a board frequently promotes the board culture. It punishes violations of acceptable board behavior by removing violators from the channel of communications; it marginalizes them and cuts them off from decision making. Third, culture is difficult to change; culture follows the Laws of Inertia. That is, a culture in place tends to stay in place unless enormous energy is exerted by the leaders of the board to change it. The kind of change required to create a process oriented management style requires that the inner circle of the board have a profound understanding of a process orientation.

Components of Board Culture

There are five components of board culture:

1. Shared mission and shared values/beliefs
2. Allocation of work
3. Reducing variability
4. Engagement
5. Trust

Shared Mission and Shared Values/Beliefs

A shared mission, as well as shared values and beliefs, by all board members creates a focus and maintainable management style for an organization over the long term. It dictates acceptable behavior, and creates and sustains a management style that directs decision making by all members of the organization. One example of a process orientation value/belief that dictates decision making in the long term is that jobs have two components: doing work (having the personal discipline to follow the best practice method) and improving work (having the team discipline to improve the best practice method) using the PDSA cycle or the DMAIC model.

Allocation of Work

Culture determines how work gets allocated among individual directors, as well as among board committees. The decisions that are the output of this work strongly influence organizational management style. These decisions include answers to the following questions (Gitlow and Levine, 2004; Deming, 1994):

- Is the organization internally cooperative or competitive?
- Does the organization create win-win decisions, or win-lose decisions?

- Does the organization improve processes to get results, or does it just demand results that are enforced through Management by Objectives (MBO) and performance appraisal systems?
- Does the organization balance intrinsic and extrinsic motivators, or does it rely only on extrinsic motivators?

The answers to these questions determine the management style of an organization. Without answers to these questions, the stakeholders of an organization have no rallying point for a maintainable management style.

Reducing Variability

Directors are constantly looking for changes in the environment (variability) that affect the organization. Effective boards are comfortable with variability, although they strive to minimize it by promoting a healthy, functional, and maintainable management style. A process oriented management is such a management style.

Engagement

Board members must be fully engaged with their responsibilities. They are expected to be competent, to be prepared, and to contribute. Culture conveys the management style needed for directors to do their jobs. Management style requires commitment, not just support, from the Board of Directors.

Trust

Board level decision making demands that directors trust their peers. Without trust, mutual influence does not emerge, reliability is a chronic concern, and the board never matures beyond individual contributors. Culture conveys the standards for trust in the form of a maintainable management style. A process oriented management style promotes trust and respect for all.

Takeaways from This Chapter

- The prevention of backsliding (maintaining the current process's performance) is just as important to process improvement efforts as the process improvement efforts themselves.
- Utilization of dashboards and presidential reviews are two methods that can maintain improvements in an organization.
- Without continuity of management style, the management style of an organization will deteriorate over time into a chaotic nightmare that shifts with every change in management.

- Succession planning can be used to increase the pool of available, experienced, and capable employees for leadership positions in an organization that can seamlessly pursue the organizational mission and strategy over time
- One of the main drivers of variability in management style is the ego of top managers to imprint their personal mark on the organizational culture, and hence, its management style.
- Without strong commitment from the board of directors about maintainability of management style, the ego of top managers will likely cause management style to change with every change in top management. This results in a chaotic culture.

References

Carter, N. (1986), "Guaranteeing Management's Future through Succession Planning," *Journal of Information Systems Management*, pp. 13-14.

Dailey, Patrick R. (2011), "The Anatomy of Board of Director Culture," *European Business Review*, http://www.europeanbusinessreview.com/?p=3080.

Deming, W. E. (1994), *The New Economics for Industry, Government, Education*, 2nd ed. (Cambridge, MA: Massachusetts Institute of Technology).

EOWA, http://www.eowa.gov.au/Developing_a_Workplace_Program/Six_Steps_to_a_Workplace_Program/Step_4/Women_in_Management_Tools/Develop_And_Implement_Succession_Plans/Three_Models_Of_Succession_Planning.asp

Gitlow, H. G., A. Oppenheim, R. Oppenheim, and D. M. Levine (2015), *Quality Management: Tools and Methods for Improvement*, 4th ed. (Naperville, IL: Hercher Publishing Company). This book is free online at hercherpublishing.com.

Gitlow, H. and D. Levine (2004), *Six Sigma for Green Belts and Champions: Foundations, DMAIC, Tools and Methods, Cases and Certification,* Prentice-Hall Publishers (Upper Saddle River, NJ).

Wikipedia, "Succession planning." http://en.wikipedia.org/wiki/Succession_planning.

16

W. Edwards Deming's Theory of Management: A Model for Cultural Transformation of an Organization

Background on W. Edwards Deming

W. Edwards Deming was born in Sioux City, Iowa, on October 14, 1900, and died in Washington, DC, on October 20, 1993. He developed a theory of management called the System of Profound Knowledge that promotes Six Sigma management and is described in the remainder of this chapter.

Deming's System of Profound Knowledge

Deming developed a theory of management that promotes "joy in work" through the acquisition of process knowledge (learning) gained from experience and coordinated by theory. This theory is called the System of Profound Knowledge (Deming, 1993; Deming, 1994). Joy in work doesn't imply that every day is Hawaiian shirt day and lets out at 3:30 p.m.

The System of Profound Knowledge is appropriate for leadership in any culture. However, applying this theory in a particular culture requires a focus on issues unique to that culture. For example, in the Western world, managers have frequently operated using the following assumptions (often without realizing it):

- Rewards and punishments are the most effective motivators for people.
- Optimization of every area in an organization leads to optimization of the entire organization.
- Results are achieved by setting objectives.
- Quality is inversely related to quantity.
- Rational decisions can be made based on guesswork and opinion.
- Organizations can be improved by fighting fires.
- Competition is a necessary aspect of life.

Leaders who manage in the context of the preceding assumptions are lost in the twenty-first century. They have no idea how to manage their organizations because they do not know the assumptions required for success in tomorrow's marketplace. Such leaders need a theory from which they can understand the assumptions of Six Sigma management.

Purpose of Deming's Theory of Management

Deming's theory of management promotes joy in work for all the stakeholders of an organization. Deming believed that joy in work will "unleash the power of human resource contained in intrinsic motivation. Intrinsic motivation is the motivation an individual experiences from the sheer joy of an endeavor" (Deming, 1986; Deming, 1993; Deming, 1994).

Paradigms of Deming's Theory of Management

Deming's theory of management is based on four paradigms, or belief systems that an individual or group uses to interpret data about conditions and circumstances. You can think of each of Deming's paradigms as a shift in assumptions for the practice of management, designed to create the management style and culture required to promote joy in work, and hence, release the power contained in intrinsic motivation.

Paradigm 1. People are best inspired by a mix of intrinsic and extrinsic motivation, not only by extrinsic motivation.

Intrinsic motivation comes from the sheer joy of performing an act. It releases human energy that can be focused into improvement and innovation of a system. It is management's responsibility to create a management style and culture that fosters intrinsic motivation. This atmosphere is a basic element of Deming's theory of management. Extrinsic motivation comes from the desire for reward or the fear of punishment. It restricts the release of energy from intrinsic motivation by judging, policing, and destroying the individual. Management based on extrinsic motivation will "squeeze out from an individual, over his lifetime, his innate intrinsic motivation, self-esteem, dignity, and build into him fear, self-defense" (Deming, 1993; Deming, 1994).

Paradigm 2. Manage using both a process and results orientation, not only a results orientation.

Management's job is to improve and innovate the processes that create results, not just to manage results; this includes management style. This paradigm shift allows management to define the capabilities of processes, and consequently, to predict and plan the future of a system to achieve organizational optimization. This type of optimization requires that managers make decisions based on facts, not on guesswork and opinion. It is critical that top management change the culture of their organization from "management by guts" (called KKD in Japan) to "management by data." It is easy to refute an argument based on guesswork or opinion, but it is difficult to refute an argument based on solid, scientific data. Managers must consider visible figures, as well as unknown and unknowable figures (for example, the cost of an unhappy customer or the benefit of a prideful employee).

Paradigm 3. Management's function is to optimize the entire system so that everyone wins, not to maximize only their component of the system.

Managers must understand that individuals, organizations, and systems of organizations are interdependent. Optimization of one component may cause suboptimization of another component. Management's job is to optimize the entire system towards its aim, or mission. This may require the managers of one or more components of a system to knowingly suboptimize their component of the system to optimize the entire system. Management style plays a significant role in the optimization of the entire organization so that everyone wins.

Paradigm 4. Cooperation works better than competition.
In a cooperative environment, everybody wins.

Customers win products and services they can brag about. The firm wins returns for investors and secure jobs for employees. Suppliers win long-term customers for their products. The community wins an excellent corporate citizen.

In a competitive environment, most people lose. The costs resulting from competition are huge. They include the costs of rework, waste, and redundancy, as well as the costs for warranty, retesting, re-inspection, customer dissatisfaction, schedule disruptions, and destruction of the individual's joy in work. Individuals and organizations cannot reap the benefits of a win-win point of view when they are forced to compete.

Is competition ever the preferred paradigm? The answer is yes, if and only if the aim of the system is to win. If the aim of the system is anything other than to win, for example, to improve or have fun, then competition is not the preferred paradigm. Cooperation is the preferred paradigm in all systems with non-competitive aims.

If leaders practice these four paradigms, they will reap enormous benefits. They will begin to understand the ground on which the organization's management style must be built.

Components of Deming's Theory of Management

Deming's theory of management consists of four components: appreciation of a system, theory of variation, theory of knowledge, and psychology (Deming, 1993 and Deming, 1994). All four components are interdependent. This discussion presents some of the highlights of Deming's theory of management.

Appreciation of a System

A system is a collection of components that interact and have a common purpose or aim (a mission). It is the job of top management to optimize the entire system toward its mission. It is the responsibility of the managers of the components of the system to promote the mission of the entire system; this may require that they suboptimize some components.

Theory of Variation

Variation is inherent in all processes. Recall, there are two types of variation, special and common. Special causes of variation are external to the system. It is the responsibility of local people and engineers to determine and resolve special causes of variation. Common causes of variation are due to the inherent design and structure of the system; they define the system. It is the responsibility of management to isolate and reduce common causes of variation. A system that does not exhibit special causes of variation is stable; that is, it is a predictable system of variation. Its output is predictable in the near future.

Two types of mistakes can be made in the management of a system. The first mistake is to treat a common cause of variation as a special cause of variation; this is by far the more frequent of the two mistakes. It is called tampering and invariably increases the variability of a system; recall the Funnel Experiment. The second mistake is to treat a special cause of variation as a common cause of variation. Shewhart (1931, 1939) developed the control chart to provide an economic rule for minimizing the loss from both types of mistakes.

Management requires knowledge about the interactions between the components of a system and its environment. Interactions can be positive or negative; they must be managed.

Theory of Knowledge

Information, no matter how speedy or complete, is not knowledge. Knowledge is indicated by the ability to predict future events with a quantifiable risk of being wrong and the ability to explain past events without fail. Knowledge is developed by stating a theory, using the theory to predict a future outcome, comparing the observed outcome with the predicted outcome, and supporting, revising, or even abandoning the theory.

Experience is of no value without the aid of theory. Theory allows people to understand and interpret experience. It allows people to ask questions and to learn.

All plans are based on assumptions. An assumption is the future output of a process. If the process underlying an assumption is not stable with an acceptable degree of predictability, the assumption required for the plan cannot be relied upon with any degree of comfort. Consequently, the plan must be changed or abandoned, or the process must be improved to enhance the likelihood of the assumption being valid when called for by the plan.

Communication is possible when people share *operational definitions*. Operational definitions are statistical clarifications of the terms people use to communicate with each other. A term is operationally defined if the users of the term agree on a common definition.

Success cannot be copied from system to system. The reason for success in one system may not be present in another system. The theory underlying a success in one system can be used as a basis for learning in another system.

Psychology

Psychology helps us understand people, the interactions between people, and the interactions between people and the system of which they are part. Management must understand the difference between intrinsic motivation and extrinsic motivation. All people require different amounts of intrinsic and extrinsic motivation. It is the job of a manager to learn the proper mix of the two types of motivation for each person.

Overjustification occurs when an extrinsic motivator is used to reward a person who did something for the sheer joy of it. The result of overjustification is to throttle future desire to act.

People are different. They learn in different ways and at different speeds. A manager of people must use these differences to optimize the system of interdependent stakeholders of an organization.

Deming's 14 Points for Management

The System of Profound Knowledge generates an interrelated set of 14 Points for leadership in the Western world (Deming, 1982; Gabor, 1990; and Gitlow and Gitlow, 1987). These 14 Points provide guidelines, or a road map, for the shifts in thinking required for organizational success. They form a highly interactive system of management; no one point can be studied in isolation.

Point 1: Create constancy of purpose toward improvement of product and service with a plan to become competitive, stay in business, and provide jobs.

Leaders must state their organization's values and beliefs. They must create a mission statement for their organizations based on these values and beliefs. Values and beliefs are the fundamental operating principles that provide guidelines for organizational behavior and decision making. A *mission statement* serves to inform stakeholders of the reason for the existence of the organization. The values and beliefs plus the mission statement provide a frame of reference for focused, consistent behavior and decision making by all stakeholders of an organization. This framework permits stakeholders to feel more secure because they understand where they fit into the organization.

Point 2: Adopt the new philosophy. We are in a new economic age. We can no longer live with commonly accepted levels of delays, mistakes, defective material, and defective workmanship.

Point 2 encompasses the paradigm shifts that leaders must accept as a consequence of Deming's System of Profound Knowledge.

Point 3: Cease dependence on mass inspection. Require, instead, statistical evidence that quality is built in to eliminate the need for inspection on a mass basis.

There is a hierarchy of views on how to pursue predictable dependability and uniformity at low cost:

1. **Defect detection**—Involves dependence upon mass inspection to sort conforming material from defective material. Mass inspection does not make a clean separation of good from bad. It involves checking products with no consideration of how to make them better. Management must eliminate the need for inspection on a mass basis and build quality into the processes that generate goods and services. Mass inspection does nothing to decrease the variability of the quality characteristics of products and services. Dependence on mass inspection to achieve quality forces quality to become a separate subsystem (called Quality Assurance), whose aim is to police defects without the authority to eliminate the defects. As the Quality Assurance department optimizes its efforts, it causes other departments to view quality as someone else's responsibility.

2. **Defect prevention**—Involves improving processes so that all output is predictably within specification limits; this is often referred to as *zero defects*. Defect prevention leaves employees with the impression that their job (with respect to reducing variation) is accomplished if they achieve zero defects. Unfortunately, zero defects will be eroded by a force similar to the concept of entropy in thermodynamics, or the natural tendency of a system to move toward disorder or chaos. This force makes a stable and capable process eventually stray out of specification limits. Further, when people are rewarded for zero defects, they may attempt to widen specification limits, rather than improve the process's ability to predictably create output within specification limits. Defect prevention is illustrated by the goal post view of quality; see Figure 2.6.

3. **Continuous improvement**—Is the ongoing reduction of process (unit-to-unit) variation, even within specification limits. Products, services, and processes are improved in a relentless and continuous manner. It is always economical to reduce unit-to-unit variation around the nominal value, even when a process is producing output within specification limits, absent capital investment. Continuous improvement is illustrated by the Taguchi Loss Function in Figure 2.7.

kp Rule. Deming advocated a plan that minimizes the total cost of incoming materials and final product. Simply stated, the rule is an "inspect all-or-none" rule. Its logical foundation has statistical evidence of quality as its base. The rule for minimizing the total cost of incoming materials and final product is referred to as the *kp rule* (see Deming, 1994). It specifies when mass inspection of all items should be performed and when only routine monitoring of a sample of items should be done. This method facilitates the collection of process or product data such that variation can be continually reduced; this means progressing from defect detection to continuous improvement.

Point 4: End the practice of awarding business on the basis of price tag. Instead, minimize total cost. Move toward a single supplier for any one item on a long-term relationship of loyalty and trust.

Buyers and vendors form a system. If each individual player in this system attempts to optimize his own position, the system will be suboptimized. Optimization requires that policy makers understand the three scenarios in which purchasing can take place. Deming called these three scenarios World 1, World 2, and World 3.

World 1 is characterized by a purchasing situation in which the customer knows what she wants and can convey this information to a supplier. In this scenario, purchase price is the total cost of buying and using the product; no supplier provides better service than any other supplier. Several suppliers can precisely meet the customer's requirements, and the only difference between suppliers is the price. In this world, purchasing on lowest price is the most rational decision.

In World 2, the customer knows what she wants and can convey this information to a supplier. The purchase price is not simply the total cost of buying and using the product; one supplier may provide better service than any other supplier. Several suppliers can precisely meet the customer's requirements, and all suppliers quote identical prices. In this world, purchasing based on best service is the most rational decision. World 2 frequently includes the purchasing of commodities.

In World 3, the customer thinks she knows what she wants and can convey this information to a supplier. However, she will listen to advice from the supplier and make changes based on that advice. Purchase price is not the total cost of buying and using the product; there is also a cost to use the purchased goods. Several suppliers tender their proposals (all of which are different in many ways), and all suppliers quote different prices. In this world, selecting a supplier will be difficult.

Some purchasing agents buy as if all purchases were World 1 scenarios; that is, they purchase solely on the basis of price, without adequate measures of quality and service.

In World 3, after careful and extensive research, it makes sense for customers and suppliers to enter into long-term relationships based on trust (that is, relationships without the fear caused by threat of alternative sources of supply) and statistical evidence of quality. Such long-term relationships promote continuous improvement in the predictability of uniformity and reliability of products and services, and, hence, lower costs. The ultimate extension of reducing the supply base is moving to a single supplier and purchasing agent for a given item (Gitlow and Gitlow, 1987). Single supplier relationships should include contingencies on the part of the supplier and customer for disasters.

The concept of single supplier extends beyond the purchasing function. For example, employees should focus on improvement of existing information channels, rather than create additional information channels when the main channel does not yield the desired information (Gitlow and Gitlow, 1987).

Point 5: Improve constantly and forever the system of production and service to improve quality and productivity, and thus constantly decrease costs.

Improvement of a system requires statistical and behavioral methods.

Management should understand the difference between special and common causes of variation and the capability of a system. They must realize that only when a system is stable (that is, when it exhibits only common causes of variation) can management use process knowledge to predict the output of the system in the near future. This allows management to plan the future state of the system. Further, management of a system requires knowledge of the interrelationships between all functions and activities in the system; this includes the interactions between people and the system, as well as between people.

Operational Definitions. Any two people may have different ideas about what constitutes knowledge of an event. This leads to the need for people to agree on the definitions of characteristics that are important about a system. Operational definitions increase communication between people and help to optimize a system; they require statistical and process knowledge. Operational definitions are fully discussed in Chapter 6, "Non-Quantitative Techniques: Tools and Methods."

SDSA Cycle. The *Standardize–Do–Study–Act (SDSA) cycle* is a technique that helps employees to standardize a process. It includes four steps:

1. **Standardize**—Employees study the process and develop best practice methods with key indicators of process performance. The best practice method is characterized by a flowchart. It is important for all employees doing a job to agree on (operationally define) a best practice method. If multiple employees perform the same job differently, there will be increased variation in output and problems will result for the customer(s) of those outputs. For example, the medical records department in a hospital receives, processes, and files patients' medical records. The director of the department decided to standardize the medical records process. First, she trained all of her personnel on how to construct a flowchart. Second, she asked each employee to create a detailed flowchart of the medical records process. Third, she reviewed all the flowcharts with her entire staff and created a best practice flowchart. The best practice flowchart incorporated all the strengths and eliminated all the weaknesses of each employee's flowchart. Fourth, she identified the key objectives and indicators for the medical records department. The key objective is file more than 80% of all medical records within 30 days of a patient's checking out of the hospital. This is a state-mandated objective. The key indicator is the percentage of medical records filed within 30 days of a patient checking out of the hospital.

2. **Do**—Employees conduct planned experiments using the best practice methods on a trial basis. In the case of the medical records department, the director collected baseline data on the key indicator for a period of months.

3. **Study**—Employees collect and analyze data on the key indicators to determine the effectiveness of the best practice methods. Again, in the case of the medical records department, the director studied the key indicator data and determined that the percentage of medical records filed within 30 days of a patient leaving the hospital was a predictable process with an average of 35% per month, and would rarely go above 45% per month or below 25% per month. She knew that this was woefully inadequate given her state-mandated key objective.

4. **Act**—Managers establish standardized best practice methods and formalize them through training. In the case of the medical records department, the director formalized the best practice method by training all employees in the method and putting it in the department's training manual for the training of all future employees. Finally, she prepared to move onto the PDSA cycle to improve the best practice method.

The Japanese developed a method to promote good housekeeping practices; it is called the *5S movement* (Hirano, 1996). "In a 5S environment there is a place for everything, and everything in its place. Time spent searching for items is essentially eliminated, and out of place or missing items are immediately obvious in a properly functioning 5S facility" (Bullington, 2003, p. 56).

The name 5S movement is derived from five Japanese words that begin with the letter "S." The words are *seiri* (sort), *seiton* (systematize), *seiso* (spic and span), *seiketsu* (standardize), and *shitsuke* (self-discipline). The five words are part of a basic management system that focuses employees' attention on the following:

- **Sort**—Simplify a process by omitting unnecessary work-in-progress, unnecessary tools, unused machinery, defective products, and unnecessary documents and papers.
- **Systematize**—Label things so they are easy to identify. For example, label storage locations with tape on the floor so that one glance identifies missing or improperly stored items. Keep things organized and ready for their next use by putting them in their proper place; for example, put tools and materials in their assigned places.
- **Spic and span**—Maintain a clean workplace; promote a proactive system for maintenance.
- **Standardize**—Be a clean and neat person; assist in the development of best practice methods for your area. The first 3Ss (seiri, seiton, and seiso) prevent backsliding in seiketsu.
- **Self-discipline**—Spread the 5Ss throughout the entire organization.

Some employees at Motorola Corporation added a sixth "S" to the list: *shitake*—be well-mannered.

Deming (PDSA) Cycle. The Deming cycle (Deming, 1982, pp. 86-89) can aid management in improving and innovating processes; that is, in helping to reduce the difference between customers' needs and process performance. The Deming cycle consists of four stages: *Plan-Do-Study-Act (PDSA)*. Often, the Deming cycle is referred to as the PDSA cycle. Initially, a plan is developed to improve or innovate the standardized best practice method developed using the SDSA cycle. The revised best practice method is characterized by a revised flowchart. Hence, a process improvement team plans to modify a process from operating under the existing best practice flowchart to operating under a revised and improved best practice flowchart, as shown in Figure 16.1.

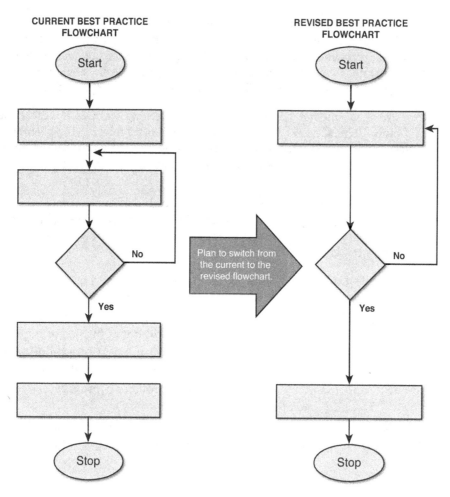

Figure 16.1 Plan portion of the PDSA cycle

The revised best practice method is identified using five possible methods:

- Process improvers can statistically analyze key indicator data on the components of the process under study (using the existing flowchart) to identify an effective change concept to allow for the construction of a revised and improved flowchart. This is the Plan portion of the PDSA cycle.
- Process improvers can *benchmark* their process using a flowchart against another organization's process (using the other organization's flowchart), which is considered excellent to identify an effective change concept. The other organization should be one that is known for the quality of the process under study. Benchmarking is accomplished by comparing your flowchart with another organization's flowchart to determine whether anything in their flowchart makes sense in your organization.

If it does, utilize the new information to improve your flowchart, with appropriate modifications. This is the Plan portion of the PDSA cycle.

- Process improvers can utilize a list of 70 tried and proven ideas to identify an effective *change concept* and determine whether it makes sense within the context of the process under study. This list of change concepts is discussed in Chapter 10, "DMAIC Model: 'D' Is for Define." The change concepts are used to move from the current flowchart to a revised and improved flowchart that uses one or more of the change concepts. Again, this is the Plan portion of the PDSA cycle. Process improvers can talk with experts to identify a change concept that promotes a turn of the PDSA cycle. Frequently, experts have valuable insights into what change concepts are most appropriate for a given situation. Again, this the Plan section of the PDSA cycle.

- Process improvers can use a search engine, such as Google, to identify other people's solutions to their process problem(s). It would be unusual for an individual to have a problem that someone else has not experienced, studied, and solved. This is an excellent method for finding a change concept that promotes a turn of the PDSA cycle.

As an example of statistically analyzing key indicator data to find a change concept, in a medical records department, cycle time data was collected for the length of time from when a physician ordered a medical report until the medical records department received (in the inbox) the patient's medical report, from each of 16 departments, such as EEG, EKG, and laboratory. Statistical analysis showed that 15 of the 16 departments' cycle times were stable and predictable processes with cycle times measured in hours. However, the laboratory department had cycle times measured in weeks, with an average of six weeks. From this analysis it was obvious that a huge proportion of medical records could not be filed within 30 days if one of the component reports took an average of six weeks to get to the medical records department.

The director of the medical records department went to the laboratory department and was greeted by the director with the comment: "We grow cultures and they can't be rushed." The director of the medical records department asked if she could visit the laboratory department anyway. The laboratory director agreed. After poking around the lab, the medical records director noticed that each lab report required three signatures before it could be released. She asked the first signer how often he refused to sign a lab report. He replied never. She asked the second signer how often he refused to sign a lab report. The second signer had seen the director's interaction with the first signer and said: "It happens." She asked: "Does it happen every day?" He said: "No." She asked: "Every week?" He said: "No." She asked: "Every month?" He said: "No." She asked: "Every quarter?" He said: "No." She asked: "Every year?" He said: "No." The director of medical records asked the director of the laboratory department if he would eliminate the need for the two signatures since they were no screen for quality. The laboratory director agreed with a modicum of irritation. The average cycle time for the laboratory reports fell from six weeks to 3.75 weeks. The percentage of medical records filed on time rose from an average of 35% to an average of 60%. This was better but still woefully inadequate for the state-mandated goal of 80% per month.

The plan is then tested using an experiment on a small scale or trial basis (Do), the effects of the plan are studied using measurements from key indicators (Study), and appropriate corrective actions are taken (Act). These corrective actions can lead to a new or modified plan and are formalized through training. The PDSA cycle continues forever in an uphill progression of continuous improvement.

One method for validating the effectiveness of a change concept is to conduct a series of tests alternating between the flowchart before the change concept and the flowchart after the change concept; this is a repetitive cycle between the Do and Study phases of the PDSA cycle. If failure appears and disappears every time you switch between the before and after flowcharts, your degree of confidence grows in the effectiveness of the change concept.

Empowerment. *Empowerment* is a term commonly used by managers in today's organizational environment (Pietenpol and Gitlow, 1996). However, empowerment has not been operationally defined, and its definition varies from application to application. Currently the prevailing definition of empowerment relies loosely on the notion of dropping decision making down to the lowest appropriate level in an organization. Empowerment's basic premise is that if people are given the authority to make decisions, they will take pride in their work, be willing to take risks, and work harder to make things happen. While this sounds ideal, frequently employees are empowered until they make a mistake and then the hatchet falls. Most employees know this and treat the popular definition of empowerment without too much respect. Consequently, empowerment in its current form is destructive to Quality Management.

Empowerment in a Quality Management sense has a dramatically different aim and definition. The aim of empowerment in Quality Management is to increase joy in work for all employees. Empowerment can be defined so as to translate the preceding aim into a realistic objective. Empowerment is a process that provides employees with (1) the opportunity to define and document their key systems, (2) the opportunity to learn about systems through training and development, (3) the opportunity to improve and innovate the best practice methods that make up systems, (4) the latitude to use their own judgment to make decisions within the context of best known methods, and (5) an environment of trust in which superiors do not react negatively to the latitude taken by people in decision making within the context of a best practice method.

Empowerment starts with leadership but requires the commitment of all employees. Leaders need to provide employees with all five of the preceding conditions. Item (5) requires that the negative results emanating from employees using their judgment within the context of a best practice method lead to improvement or innovation of best practice methods, not to judgment and punishment of employees. Employees need to accept responsibility for (1) increasing their training and knowledge of the system, (2) participating in the development, standardization, improvement, and innovation of best known methods that make up the system, and (3) increasing their latitude in decision making within the context of best known methods.

Individual workers must be educated to understand that increased variability in output will result if each worker follows his own best practice method. They must be educated about the need to reach consensus on one best practice method. Management should understand the differences between workers and channel these differences into the development of the best practice method in a constructive, or team-building, manner.

The best practice method will consist of generalized procedures and individualized procedures. Generalized procedures are standardized procedures that all workers must follow. The generalized procedures can be improved or innovated through team activities. Individualized procedures are procedures that afford each worker the opportunity to utilize his individual differences. However, the outputs of individualized procedures must be standardized across individuals. The individualized procedures can be improved through individual efforts. In the beginning of a quality improvement effort, management may not have the knowledge to allow for individualized procedures.

Note that latitude to make decisions within the context of a best practice method refers to the options an employee has in resolving problems within the confines of a best practice method, not to modification of the best practice method. Differentiating between the need to change the best practice methods and latitude within the context of the best practice methods must take place at the operational level.

Teams must work to improve or innovate best practice methods. Individuals can also work to improve or innovate best practice methods; however, the efforts of individuals must be shared with and approved by the team. Empowerment can only exist in an environment of trust that supports planned experimentation concerning ideas to improve and innovate best practice methods. Ideas for improvement and innovation can come from individuals or from the team, but tests of ideas' worthiness must be conducted through planned experiments under the auspices of the team. Anything else will result in chaos because everybody will "do his own thing."

Empowerment is operationalized at two levels. First, employees are empowered to develop and document best practice methods using the SDSA cycle. Second, employees are empowered to improve or innovate best practice methods through application of the PDSA cycle.

Point 6: Institute training on the job.

Employees are an organization's most important asset. Organizations must make long-term commitments to employees that include the opportunity to take joy in their work. This requires training in job skills.

Training in job skills is a system. Effective training changes the distribution for a job skill, as shown in Figure 16.2. Management must understand the capability of the training process and the current distribution of job skills to improve the future distribution of job skills. Data, not guesswork or opinion, should be used to guide the training plans for employees.

Training is a part of everyone's job and should include formal classwork, experiential work, and instructional materials. Training courseware must take into consideration how the trainee learns and the speed at which she learns. It should utilize statistical methods that indicate when an employee reaches a state of statistical control; that is, only common causes of variation are present in the key indicator(s) used to measure the employee's output. If an employee is not in statistical control with respect to a job characteristic, more training of the type she is receiving will be beneficial. However, if an employee is in a state of statistical control with respect to a job characteristic, more training of that type will not be beneficial; the employee has learned all that is possible from the training program.

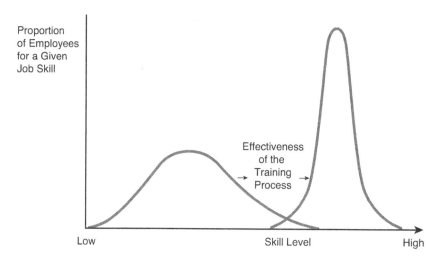

Figure 16.2 Distribution of job skills

Point 7: Institute leadership. The aim of leadership should be to help people and machines and gadgets to do a better job. Leadership of management is in need of overhaul, as well as leadership of production workers.

A leader (Deming, 1993, pp. 125-128) must see the organization as a system of interrelated components, each with a mission, but all focused collectively to support the aim of the organization. This type of focus may require suboptimization of some system components.

According to Deming, "A leader uses plots of points and statistical calculations, with knowledge of variation, to try to understand both his performance and that of his people" (Deming, 1993, p. 127).

Leaders know when employees are experiencing problems that make their performance fall outside the system, and leaders treat the problems as special causes of variation. These problems could be common causes to the individual (e.g., long-term alcoholism), but special causes to the system (an alcoholic works differently from his peers).

A leader must understand that experience without theory does not facilitate prediction of future events. For example, a leader cannot predict how a person will perform in a new job based solely on experience in the old job. A leader must have a theory to predict how an individual will perform in a new job.

A leader must be able to predict the future to plan the actions necessary to pursue the organization's aim. Prediction of future events requires that the leader continuously work to create stable processes with low variation to facilitate rational prediction.

Point 8: Drive out fear so that everyone may work effectively for the company.

There are two kinds of negative reactive behaviors: fear and anxiety. Fear is a reaction to a situation in which the person experiencing the fear can identify its source. Anxiety is a reaction to a situation in which the person experiencing the anxiety cannot identify its source. We can remove the source of fear because it is known; this is not the case with anxiety. Thus, Point 8 focuses on driving out fear.

Fear has a profound impact on those working in an organization, and consequently, on the functioning of the organization. On an individual level, fear can cause physical and physiological disorders such as a rise in blood pressure or an increase in heart rate. Behavioral changes, emotional problems, and physical ailments often result from fear and stress generated in work situations, as do drug and alcohol abuse, absenteeism, and burnout. These maladies impact heavily on any organization. An employee subjected to a climate dominated by fear experiences poor morale, poor productivity, stifling of creativity, reluctance to take risks, poor interpersonal relationships, and reduced motivation to optimize the system of interdependent stakeholders. The economic loss to an organization from fear is immeasurable, but huge.

A statistically based system of management will not work in a fear-filled environment. This is because people in the system will view statistics as a vehicle for policing, judging, and punishing, rather than a method that provides improvement opportunities.

Fear emanates from lack of job security, possibility of physical harm, ignorance of company goals, shortcomings in hiring and training, poor supervision, lack of operational definitions, failure to meet quotas, blame for the problems of the system (fear of being below average and being punished), and faulty inspection procedures, to name a few causes. Management is responsible for changing the organization to eliminate the causes of fear. Generally, fear creates variability in the behavior of employees within an organization; fear creates common causes of variation. Managers may have unknowingly designed fear into the structure of their organization through the construction and deployment of policies and procedures such as management by objectives and traditional performance appraisal systems, discussed later in this chapter.

Point 9: Break down barriers between departments. People in research, design, sales, and production must work as a team to foresee problems of production and in use that may be encountered with the product or service.

Management's job is to optimize the system of interdependent stakeholders of an organization. This may require suboptimization of some parts of the system. An example of suboptimization of a part, which leads to optimization of the whole, is a supermarket's "loss leader" product (a product carrying an extremely low price). The aim of a loss leader is to entice buyers into a store. Once in the store, buyers purchase other products, thereby creating a greater profit for the store. Profit from the loss leader is suboptimized to optimize store profit. Managers must remove disincentives for suboptimization of areas if they want to optimize the organization. For example, rating departments or divisions with respect to profit alone usually fosters suboptimization of the organization.

Barriers between the areas of an organization thwart communication and cooperation. The greater the interdependence between the components of a system, the greater is the need for communication and cooperation between them.

Point 10: Eliminate arbitrary numerical goals, posters, and slogans for the work force that seek new levels of productivity without providing methods.

Slogans, exhortations, and targets do not help to form a plan or method to improve or innovate a process, product, or service. They do not operationally define process variables in need of improvement or innovation. They do not motivate individuals or clarify expectations. Slogans, exhortations, and targets are meaningless without methods to achieve them.

Generally, targets are set arbitrarily by someone for someone else. If a target does not provide a method to achieve it, it is a meaningless plea. Examples of slogans, exhortations, and targets that do not help anyone do a better job are

>Do it right the first time.
>
>Safety is job number 1.
>
>Zero Defects.
>
>Just Say No.

These kinds of statement do not represent action items for employees; rather, they show management's wishes for a desired result. How, for example, can an employee "do it right the first time" without a method? People's motivation can be destroyed by slogans.

Slogans, exhortations, and targets shift responsibility for improvement and innovation of the system from management to the worker. The worker is powerless to make improvements to the system. This causes resentment, mistrust, and other negative emotions.

Point 11a: Eliminate work standards (quotas) on the factory floor. Substitute leadership.

Work standards, *measured day work*, and *piecework* are names given to a practice that can have devastating effects on quality and productivity. A work standard is a specified level of performance determined by someone other than the worker who is actually performing the task.

The effects of work standards are, in general, negative. They do not provide a road map for improvement, and they prohibit good supervision and training. In a system of work standards, workers are blamed for problems beyond their control. In some cases, work standards actually encourage workers to produce defectives to meet a production quota. This robs workers of their pride and denies them the opportunity to produce high-quality goods and thus to contribute to the stability of their employment.

Work standards are negotiated values that have no relationship to the capability of a process. When work standards are set too high or too low, there are additional devastating effects. Setting work standards too high increases pressure on workers and results in the production of more defectives. Worker morale and motivation are diminished because the system encourages the production of defectives. Setting work standards too low also has negative effects. Workers who have met their quota spend the end of the day doing nothing; their morale is also destroyed.

Work standards are frequently used for budgeting, planning, and scheduling, and provide management with invalid information on which to base decisions. Planning, budgeting, and scheduling would improve greatly if they were based on process capability studies as determined by statistical methods.

Point 11b: Eliminate management by objective. Eliminate management by numbers and numerical goals. Substitute leadership.

The Old Way. Setting arbitrary goals and targets is a dysfunctional form of management. Numerical goals are frequently set without understanding a system's capability. They do not include methods, and hence, do not provide a mechanism for improvement of a process. In a stable system, the proportion of the time an individual is above or below a specified quota/goal is a random lottery. This causes people below the quota to copy the actions of those above the quota even though they are both part of the same common cause system. This increases the variability of the entire system due to inappropriate copying of actions.

Deploying arbitrary goals and targets causes problems in most organizations. Managers use *management by objectives (MBO)* to systematically break down a "plan" into smaller and smaller subsections. Next, managers assign the subsections to individuals or groups who are accountable for achieving results. This is considered fair because subsection goals emerge out of a negotiation between supervisor and supervisee. For example, an employee may negotiate a 3% increase in output instead of a 3.5% increase as long as the subsection's goals yield the goals of the plan. Note that employees are not being given any new tools, resources, or methods to achieve the 3% increase. Consequently, they must abuse the existing system to meet the goal. This type of behavior may allow an employee to meet a goal, or to work a lot of uncompensated overtime. The result of either option creates system failure due to a lack of resources. Arbitrary numerical goals hold people accountable for the problems of the system, and consequently, steal their pride of workmanship.

The New Way. The types of relationships that managers establish between the aim (or mission) of a system, methods, and goals (or targets) can define a functional style of management. A group of components come together to form a system with an aim. The aim requires that the components organize in such a way that they create subsystems. The subsystems are complex combinations of the components. The subsystems require certain methods to accomplish the aim. Resources are allocated between the methods by setting goals or targets that may be numerical and that optimize the overall system, not the subsystems, with respect to the aim. For example, a group of individuals form a team with an aim. The individuals must combine their efforts to form subsystems. These combinations may require complex interactions between the individuals. The subsystems require methods, and the methods require resources. Resources are allocated between the methods, and ultimately, the subsystems and individuals by setting goals that optimize the team's aim. The aim, methods, and goals are all part of the same system; they cannot be broken into three separate entities. Separation of the aim, methods, and goals destroys them because they are defined by their interactions.

Variation can cause a good method to yield undesirable results. Therefore, one should not overreact (tamper) and change methods by considering negative results in the absence of theory.

Point 12: Remove barriers that rob the hourly worker of his right to pride of workmanship. The responsibility of supervisors must be changed from stressing sheer numbers to quality. Remove barriers that rob people in management and engineering of their right to pride of workmanship. This means abolishment of the annual merit rating and of management by objective.

People are born with the right to find joy in their work; it provides the impetus to perform better and to improve quality for the worker's self-esteem, for the company, and ultimately for the customer. People enjoy taking joy in their work, but few are able to do so because of poor management. Management must remove the barriers that prevent employees from finding joy in their work.

In the current system of management there are many such barriers: (1) employees not understanding their company's mission and what is expected of them with respect to the mission, (2) employees being forced to act as automatons who are not allowed to think or use their skills, (3) employees being blamed for problems of the system, (4) hastily designed products and inadequately tested prototypes, (5) inadequate supervision and training, (6) faulty equipment, materials, and methods, (7) management by objective systems that focus only on results, such as daily production reports, and (8) the traditional performance appraisal process. Organizations reap tremendous benefits when management removes barriers to joy in work.

The Case Against Management by Objectives. Traditional management creates and enforces decision making through an interrelated pair of systems; they are management by objectives (MBO) and performance appraisal. Figure 16.3 shows the results for a physician who sees patients on a daily basis. The physician attends to about 20 patients on average, with a standard deviation of 2 patients per day. This distribution of patients has been stable and predictable for some period of time.

Assume the chief medical officer and the chief financial officer are under extreme pressure to increase physician productivity. The question is: How will they do it? Traditional management offers three basic options.

The first option sets a stretch goal of 24 patients per day. The rationale is that a stretch goal provides a strong incentive to do better. Unfortunately, it does not indicate how to do better. In such a case, doing better usually means working more uncompensated overtime, or cutting corners to meet the goal. Given the current operational system used by the physician, from Figure 16.3 we see that he can expect to exceed 24 patients per day 2.28% of the days, based on statistical calculation.

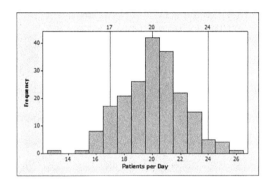

Figure 16.3 Histogram and patients per day by physician X

The second option is to set the goal at 20 patients per day, the average number. The rationale assumes that, if you try a little harder, you can always do better than the average. This option may seem logical, but it is not. Given the symmetry of the distribution of patients seen per day by the physician, as shown in Figure 16.3, you can expect 50% of the days to be above the average, and 50% of the days to be below the average.

Using the average as a goal encourages the physician to look for reasons why he is above or below average on any given day, encouraging him to replicate the above average experience, or prevent the below average experience. Unfortunately, the hospital's operational system determines the distribution of patients seen per day. Consequently, the physician's search for a special cause in a process exhibiting only common causes of variation is fruitless and will likely result in overreaction to random noise, which serves to increase the variability in the system.

The third option is to set an easy goal of 17 patients per day. The rationale is that an easy goal provides a strong incentive to do better, thereby enhancing the physician's perception of his performance. Given the operational process currently used, from Figure 16.3 we see that he can expect, based on statistical calculation, to exceed 17 patients per day about 93.32% of the days.

The options described do not achieve an improvement in physician productivity because the distribution of patients seen per day derives from the operational system (common cause of variation) and is not responsive, over time, to the efforts of the individual physician.

The Case Against Traditional Performance Appraisal Systems. Performance appraisal systems are used by managers to enforce MBO within their organization. If a worker, Dr. A, does very well with respect to a particular goal, deadline, or other mandate, she gets a high performance score, say 5, on a 1 to 5 scale, where 1 is unacceptable, 3 is average, and 5 is excellent. If she does poorly with respect to a particular goal, deadline, or mandate, she gets a low performance score, say a 1. This all seems rational and fair. But it rests on the underlying assumption that an individual's performance score is due solely to her efforts. That may not be the case. Let us restate this in the form of an equation:

$$\text{Dr. A's Performance Score} = \text{Dr. A}_{\text{Individual Effort}} = 4.8$$

Actually, the individual's performance score reflects both her individual efforts and the effect of the system in which she performs her job. The system may treat everyone equally and fairly, or not. The problem is, you don't know how much of Dr. A's 4.8 to attribute to her individual effort, and how much to attribute to the system in which she works. Again, restate this in the form of an equation:

$$\text{Dr. A's Performance Score} = \text{Dr. A}_{\text{Individual Effort}} + \text{Dr. A}_{\text{System Effect}}$$

The equation has two variables; hence there are many solutions to it, for example:

Dr. A's Performance Score = 4.8 + 0

Dr. A's Performance Score = 0 + 4.8

or any values in between that sum to 4.8. Managers lose the ability to use traditional performance appraisals to score an individual's performance. They cannot separate the individual's effort from the effect of the system on the individual.

Comparing two workers:

$$\text{Dr. A's Performance Score} = \text{Dr. A}_{\text{Individual Effort}} + \text{Dr. A}_{\text{System Effect}}$$

$$\text{Dr. B's Performance Score} = \text{Dr. B}_{\text{Individual Effort}} + \text{Dr. B}_{\text{System Effect}}$$

Traditional managers assume that both Dr. A and Dr. B are equally affected by the system in which they work. The assumption is not likely to be the reality. Managers cannot make this distinction given the information contained in traditional performance appraisal systems.

The Case Against Forced Ranking. Traditional performance appraisal systems are frequently used to force the ranking of employees. Figure 16.4 shows the distribution of performance appraisal scores for 100 employees, in this case physicians, in a hospital. They all see the same patient mix. Figure 16.4 shows the top 10%, the middle 80%, and the bottom 10%. In economically challenging times, management may decide to terminate the bottom 10% of employees. This seems rational, but it causes several types of potentially serious collateral damage.

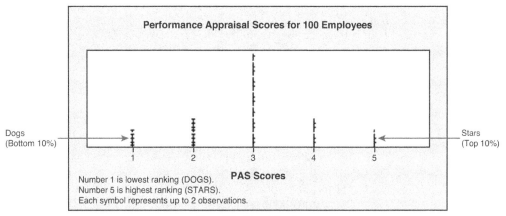

Figure 16.4 Forced ranking from performance appraisal scores

First, performance appraisal scores used in the forced ranking distribution are flawed because they do not take into account the effect of the operational system on the individual employee. So decisions based on those scores will be flawed as well.

Second, when you fire the bottom 10% of employees, you automatically have a new bottom 10%. The new bottom 10% are different from the top 90% likely due only to common causes of variation, so your average performance for the group is not improving. You cannot fire your way out of having a bottom 10% of employees! Further, a typical result of firing the bottom 10% of employees is that many remaining employees obsess about whether they will be in the next round of firings. Consequently, morale is adversely impacted, and performance suffers.

Third, there is a reduction of organizational cohesion and inclination to cooperate. Warring factions, with each attempting to protect its eroding turf, surface, at the expense of the organization.

In short, traditional management, using management by objectives coupled with performance appraisal and forced ranking of employees, does not achieve solutions to escalating organizational costs. The why is clear: They do not deal with the operational system of the organization. And the operational system is the source of the common causes of variation that produce escalating costs, and only management can improve a stable system.

Point 13: Encourage education and self-improvement for everyone.

Education and self-improvement are important vehicles for continuously improving employees, both professionally and personally. Leaders are obligated to educate and improve themselves and their people to optimize the system of interdependent stakeholders. Education for leaders may have to come from outside the system.

Remember, training (Point 6) is to improve job skills, while education (Point 13) is to improve the individual, regardless of his job. So, if one of the authors takes a course in Advanced Statistical Theory, it is an exercise in training for job skills. However, if one of us takes a course in floral arrangement or cooking, it is an educational endeavor. It is impossible to predict the benefits of education to the organization.

Point 14: Take action to accomplish the transformation.

The transformation of an organization from its current paradigm of management to the System of Profound Knowledge cannot occur without the expenditure of energy by its stakeholders. Top management expend this energy due to a variety of causes: for example, if they are confronted with a crisis or if they have a mission that they want to pursue. Other stakeholders expend this energy if stimulated by top management. The transformation cannot take place without a critical mass of stakeholders. The critical mass must include some policy makers.

Individuals have different reasons for wanting, or not wanting, to accomplish the transformation. Individuals have different interpretations of what is involved in the transformation. To be able to plan, control, and improve the transformation, a leader must know (1) each person's reasons for wanting (or not wanting) the transformation and (2) how each of those different reasons interact with each other and with the aim of the transformation.

Deming's 14 Points and the Reduction of Variation

In this section, each of the 14 points is repeated with a brief discussion of how it is related to the reduction of variation in a process.

Point 1: Create constancy of purpose toward improvement of product and service with a plan to become competitive, stay in business, and provide jobs.

Establishing a mission statement is synonymous with setting a process's nominal or target level. Getting all employees (management, salaried, and hourly), members of the board of directors, and shareholders to behave in accordance to the common interpretation of a mission statement is a problem of reducing variation (Scherkenbach, 1986, pp. 133-134).

Point 2: Adopt the new philosophy. We are in a new economic age. We can no longer live with commonly accepted levels of delays, mistakes, defective material, and defective workmanship.

All people in an organization should embrace the System of Profound Knowledge as the focus of all action. As everyone uniformly embraces the System of Profound

Knowledge, variation in how people view the organization—and in how they interpret their job responsibilities—decreases.

Point 3: Cease dependence on mass inspection. Require, instead, statistical evidence that quality is built in to eliminate the need for inspection on a mass basis.

Dependence on mass inspection does nothing to decrease variation. Moreover, inspection does not create a uniform product within specification limits—rather, product is bunched around specification limits, or, at best, product is distributed within specification limits with large variance and tails truncated at the specification limits. Instead, eliminate defectives and defects, using attribute statistical control charts and eliminate unit-to-unit variation within specification limits using measurement statistical control charts, absent capital investment.

Point 4: End the practice of awarding business on the basis of price tag. Instead, minimize total cost. Move toward a single supplier for any one item on a long-term relationship of loyalty and trust.

Multiple supplier processes, each of which has small variations, combine to create a process with large variation. This means an increase in the variability of inputs to the organization, which is counter to the reduction of variation. Consequently, reducing the supply base from many suppliers to one supplier is a rational action. This idea applies to both external and internal suppliers.

Point 5: Improve constantly and forever the system of production and service to improve quality and productivity, and thus constantly decrease costs.

The Taguchi Loss Function explains the need for the continuous reduction of variation in a system, as discussed in Chapter 2, "Process and Quality Fundamentals." Management must realize that when a system is stable, or exhibits only common causes of variation, it is able to predict the system's future condition. This allows management to plan the future state of the system and use the PDSA cycle to decrease the difference, or variation, between customer needs and process performance. The PDSA cycle is a procedure for improving a process by reducing variation.

Point 6: Institute training on the job.

Statistical methods should be used to determine when training is complete. In chaos, more training of the same type is effective. In stability, more training of the same type is not effective; management may have to find the trainee a new job for which he is trainable.

Point 7: Institute leadership. The aim of leadership should be to help people and machines and gadgets to do a better job. Leadership of management is in need of overhaul, as well as leadership of production workers.

A leader must understand that variation in a system can come from the individual, the system, or the interaction between the system and the individual. A leader must not rank the people who perform within the limits of a system's capability.

Point 8: Drive out fear, so that everyone may work effectively for the company.

Managers who do not understand variation rank individuals within a system; that is, they hold individuals accountable for system problems. This causes fear, which stifles the desire to change and improve a process. Fear creates variability between an individual's or team's actions and the actions required to surpass customer needs and wants.

Point 9: Break down barriers between departments. People in research, design, sales, and production must work as a team to foresee problems of production and in use that may be encountered with the product or service.

Barriers between departments result in multiple interpretations of a given message. This increases variability in the actions taken with respect to a given message.

Point 10: Eliminate arbitrary numerical goals, posters, and slogans for the work force that seek new levels of productivity without providing methods.

Slogans and posters try to shift the responsibility for common causes of variation to the worker. This is sure to increase fear and variability in employees' behavior.

Point 11a: Eliminate work standards (quotas) on the factory floor. Substitute leadership.

Point 11b: Eliminate management by objective. Eliminate management by numbers and numerical goals. Substitute leadership.

If a work standard is between a system's upper capability and lower capability, there's a possibility that the standard can be met, but meeting the standard this way is simply a random lottery. If a work standard is above the system's capability, there's little chance that the standard will be met unless management changes the system. Rather than focusing on the standard as a means to productivity, management should focus on stabilizing and improving the process to increase productivity by empowering employees to do their work and improve their work using the PDSA cycle.

Point 12: Remove barriers that rob the hourly worker of his right to pride of workmanship.

Performance appraisal systems can increase variability in employee performance, resulting from actions such as rewarding everyone who is above average and penalizing everyone who is below average. In such a situation, below-average employees try to emulate above-average employees. However, as the employees who are above average and those who are below average are part of the same system (only common variation is present), the below-average ones are adjusting their behavior based on common variation.

Point 13: Encourage education and self-improvement for everyone.

The education of employees will lower variability in processes, products, and jobs, continuing the never-ending cycle of improvement (Scherkenbach, 1986, p. 126).

Point 14: Take action to accomplish the transformation.

The current paradigm of Western management is shaped by reactive forces. Therefore, it has an explosive and high degree of variation in its application. The transformation must emanate out of a new paradigm shaped by the System of Profound Knowledge, not reactive forces. This new paradigm will have a stable, reducible degree of variation in its application.

Transformation or Paradigm Shift

The issues involved in understanding the transformation of people and organizations from management's prevailing style to the System of Profound Knowledge is presented in Figure 16.5, which displays: (1) the prevailing paradigm of leadership and the business and education systems it creates, (2) the System of Profound Knowledge and the business and education systems it creates, and (3) the 14 Points' role in the transformation process from the prevailing style of management to the System of Profound Knowledge.

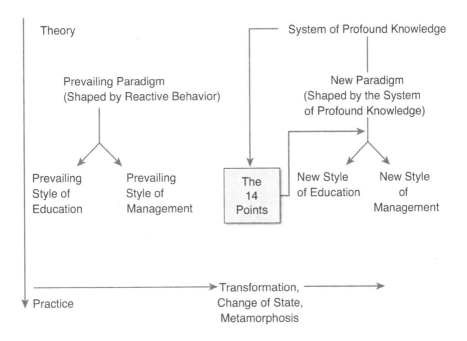

Figure 16.5 Issues involved in transformation

The Prevailing Paradigm of Leadership

According to Deming, "The prevailing style of management was not born with evil intent. It grew up little by little by reactive behavior, unsuited to any world, and especially unsuited to the new kind of world of dependence and interdependence that we are in now" (Deming, 1994). The prevailing paradigm of management, shown on the left side of Figure 16.5, is not based on any holistic or comprehensive theory; it is just the cumulative result of assorted theories and experiences.

The New Paradigm of Leadership

The System of Profound Knowledge allows leadership to change and to develop a new basis for understanding the interrelationships between themselves and their environment. The environment includes people, systems, and organizations. It is based on a holistic and comprehensive theory of management.

Transformation

It is not easy to move from the prevailing style of leadership to the new style of leadership. The 14 Points provide a framework that helps explain the relationship between the prevailing style and the System of Profound Knowledge. They provide a window for managers operating under the prevailing techniques to compare and contrast their business practices with business practices in the System of Profound Knowledge. The real work of transformation comes from understanding the System of Profound Knowledge. According to Deming, "Transformation of American style of management is not a job of reconstruction, nor is it revision. It requires a whole new structure, from foundation upward" (Deming, 1994).

Managers in one organization should not use the experiences of managers in another organization to focus their transformation efforts. This is because organizations are unique, having their own idiosyncrasies and nuances. Conditions that led to the experiences of managers in one organization may not exist for managers of the other organization. However, this is not to say that managers' experiences in one organization cannot stimulate development of theories for improvement and innovation on the part of another organization's managers.

Quality in Service, Government, and Education

The U.S. Census shows that the overwhelming majority of Americans works in service, government, or educational organizations, or performs service functions in manufacturing organizations. Thus, improvement in our standard of living is highly dependent on better quality and productivity in these sectors of the economy.

A denominator common to all organizations is that mistakes and defects are costly. The further a mistake goes without correction, the greater the cost to correct it. A defect that reaches the consumer or recipient may be costliest of all (Deming, 1982). The principles and methods for process improvement are the same in all organizations. The System of Profound Knowledge and 14 Points apply equally to all sectors of the economy.

Quotes from Deming

We thought you would enjoy some quotes from Dr. Deming on different aspects of management:

On who is responsible for quality:

"Quality is everyone's responsibility."

On importance of documenting processes:

"If you can't describe what you are doing as a process, you don't know what you're doing."

On the importance of identifying who owns a process:

"To manage one must lead. To lead, one must understand the work that he and his people are responsible for."

On the importance of measuring the right thing:

"Just because you can measure everything doesn't mean that you should."

On the importance of asking the right question:

"If you do not know how to ask the right question, you discover nothing."

On the importance of knowing the aim of a system or process:

"A system is a network of interdependent components that work together to try to accomplish the aim of the system. A system must have an aim. Without the aim, there is no system."

On the importance of management being responsible for processes:

"Divide responsibility and nobody is responsible."

On the importance of understanding variation:

"Understanding variation is the key to success in quality and business."

On the importance of flowcharting:

"You can see from a flow diagram who depends on you and whom you can depend on. You can now take joy in your work."

On the importance of managing with data:

"In God we trust; all others must bring data." (This quote has been attributed to a couple of individuals, Deming being one of them.)

Summary

W. Edwards Deming developed a theory of management, called the System of Profound Knowledge, which promotes joy in work through the acquisition of process knowledge gained from experience coordinated by theory. Deming's theory can be used to create the organizational culture needed for Six Sigma management.

Deming's theory of management is based on four paradigms, which create the environment required to promote joy in work. They are: (1) People are best inspired by a mix of intrinsic and extrinsic motivation, not only by extrinsic motivation; (2) Manage using both a process and results orientation, not only a results orientation; (3) Management's function is to optimize the entire system so that everyone wins, not to maximize only one component of the system; and (4) Cooperation works better than competition, if the aim of the system is not to win.

Deming's theory of management comprises four components: appreciation of a system, theory of variation, theory of knowledge, and psychology. All four components are interdependent and do not stand alone. Fortunately, it is not necessary to be expert in any of the components to understand and apply the System of Profound Knowledge.

The System of Profound Knowledge generates an interrelated set of 14 Points for leadership in the Western world. These 14 Points provide a road map for the shifts in thinking required for organizational success. They form a highly interactive system of management; no one point should be studied in isolation.

The System of Profound Knowledge allows leadership to change and to develop a new basis for understanding the interrelationships between themselves and their environment. The environment includes people, systems, and organizations. It is based on a holistic and comprehensive theory of management.

The U.S. Census Bureau shows that the overwhelming majority of U.S. citizens are employed in service, government, or educational organizations, or performs service functions in manufacturing organizations. Hence, improvement in our standard of living is highly dependent on better quality and productivity in these sectors of the economy.

References and Additional Readings

Bullington, K. E., "5S for Suppliers," *Quality Progress*, January 2003, pp. 56–59.

Deming, W. E. (1982), *Quality, Productivity, and Competitive Position* (Cambridge, MA: Massachusetts Institute of Technology).

Deming, W. E. (1986), *Out of the Crisis*, (Cambridge, MA: Massachusetts Institute of Technology, Center for Advanced Engineering Studies).

Deming, W. E. (1994), *The New Economics for Industry, Government, Education*, 2nd ed. (Cambridge, MA: Massachusetts Institute for Technology Center for Advanced Engineering Study).

Deming, W. E., "Foundation for Management of Quality in the Western World," revised April 1, 1990, delivered at a meeting of The Institute of Management Sciences in Osaka, July 24, 1989.

Deming, W. E. (1993), *The New Economics for Industry, Government, Education* (Cambridge, MA: Massachusetts Institute of Technology).

Gabor, A. (1990), *The Man Who Discovered Quality* (New York: Time Books).

Gitlow, H. and S. Gitlow (1987), *The Deming Guide to Quality and Competitive Position* (Englewood Cliffs, NJ: Prentice-Hall).

Gitlow, H., "Total Quality Management in the United States and Japan," *APO Productivity Journal*, Asian Productivity Organization (Tokyo, Japan) Winter 1993-1994, pp. 3-27.

Gitlow, H., "A Comparison of Japanese Total Quality Control and Dr. Deming's Theory of Management," *The American Statistician*, vol. 48, no. 3, August 1994, pp. 197-203.

Gitlow, H., "Understanding Total Quality Creation (TQC): The Japanese School of Thought," *Quality Engineering*, vol. 7, no. 3, 1995, pp. 523-542.

Gitlow, H. (2000), *Quality Management Systems* (Boca Raton, FL: St. Lucie Press).

Hirano, H. (1996), *5S for Operators: 5 Pillars of the Visual Workplace* (Portland, OR: Productivity Press).

Pietenpol, D. and H. Gitlow (1996), "Empowerment and the System of Profound Knowledge," *International Journal of Quality Science*, vol. 1, no. 3.

Scherkenbach, W. (1986), *The Deming Route to Quality and Productivity: Road Maps and Roadblocks* (Washington, DC: CeePress).

Shewhart, W. A. (1931), *Economic Control of Quality of Manufactured Products* (New York: Van Nostrand and Company; reprinted by the American Society for Quality Control, Milwaukee, 1980).

Shewhart, W. A. and W. E. Deming (1939), *Statistical Methods from the Viewpoint of Quality Control* (Washington, DC: Graduate School, Department of Agriculture; Dover Press, 1986).

Index

A

accounts receivable, 394-396
activating Six Sigma team
    case study, 292-293
    overview, 274-276
Act phase
    PDSA cycle, 207
    SDSA cycle, 204
actual versus ideal, 6
adopting new philosophy, 413
affinity diagrams
    creating, 181-182
    defined, 156
    example, 156
    purpose of, 156
affordances, 197
aim, 18
Allied Signal, 253
allocating
    resources, 250-251
    work, 405-406
alternative methods
    generating
        case study, 366-367
        explained, 358-359
    selecting
        case study, 367-368
        explained, 360-361
American National Standards Institute (ANSI) standard flowchart symbols, 43-44
Analyze phase (DMADV cycle), 218-219
Analyze phase (DMAIC model)
    case study: reducing patient no shows in outpatient psychiatric clinic, 344-355
    current state flowchart, creating, 334, 344-345
    data collection plan for Xs, 339-340, 348
    FMEA (failure modes and effects analysis), 338
    go-no go decision point, 342-343
    hypotheses about relationship between critical Xs and CTQs, 342, 354
    identification of Xs for CTQs, 335-338, 344-346
    measurement system for Xs, validating, 340, 348
    operational definitions for Xs, 338, 346-347
    overview, 214-215
    pitfalls to avoid, 343-344
    purpose of, 333-334
    test of theories to determine critical Xs, 340-342, 348-353
    tips for success, 343-344
    tollgate reviews, 354-355
ANSI (American National Standards Institute) standard flowchart symbols, 43-44
anxiety, 423
appreciation of a system, 411
arbitrary goals, posters, and slogans, eliminating, 424
Arena, 359
assumptions in planning, 13-14
attribute check sheets, 177-178
attribute classification data
    explained, 47-48
    measures of central tendency, 61-62
attribute control charts, 90
attribute count data
    explained, 48-49
    graphing
        bar charts, 50-51
        line graphs, 52-54
        Pareto diagrams, 51-52

439

attribute data, 47
    c charts, 104-106
    p charts, 98-104
    u charts, 106-108
automation, 189
autonomous maintenance, 223
average, 266

B

background check process, 19-20
backsliding, preventing, 393
    board of directors culture, 404-406
        allocation of work, 405-406
        components of, 405
        engagement, 406
        reducing variability, 406
        shared values/beliefs, 405
        trust, 406
    dashboards, 396-397
    egotism, 403-404
    failure in Act phase of PDSA cycle
        in accounts receivable, 394-396
        in manufacturing, 393-394
    Funnel Experiment, 398-399
    presidential review of maintainability indicators, 397
    succession planning
        creating talent pools model, 400
        explained, 399
        incumbent model, 399-400
        process oriented top-down/bottom-up model, 401-403
        top-down/bottom-up model, 400-401
balanced scorecards, 232
Bar Chart: Data Options dialog box (Minitab), 74-75
bar charts
    explained, 50-51
    obtaining in Minitab, 74-76
Bar Charts dialog box (Minitab), 74-76
barriers, removing
    barrier between departments, 423-424
    barriers to pride of workmanship, 426-429

baseline data analysis for CTQs
    case study, 328-329
    explained, 317-321
benchmarking, 359, 367, 418
benefits of project
    Define phase (DMAIC model), 281-282
    documenting
        case study, 387
        explained, 383
best alternative method, selecting, 367-368
Black Belts, 258-260
board of directors, 404-406
Bossidy, Larry, 253
bottlenecks, 189
boundaries (process), 37-38
brainstorming, 358-359, 366
    conducting brainstorming sessions, 179-181
    defined, 155
    example, 155-156
    purpose of, 155
business case
    case study, 293-294
    explained, 276-277

C

Calculator dialog box (Minitab), 78, 84
cause and effect (C&E) diagrams
    creating, 182
    defined, 157
    examples, 157-159
    purpose of, 157
causes of variation, 6-8
C Chart dialog box (Minitab), 134
c charts, 104-106
    creating in Minitab, 134
    example, 105-106
    explained, 104
    when to use, 105
ceasing dependence on mass inspection, 413-414
C&E diagrams. See cause and effect (C&E) diagrams
Centers for Medicare and Medicaid Services (CMS), 237

central tendency, measures of
  mean, 59
  median, 60
  mode, 60-61
  proportion, 61-62
Champions, 256-257
change concepts, 2, 419
  changing work environment, 191-192
  defined, 160
  designing systems to avoid mistakes, 196-197
  eliminating waste, 187-188
  enhancing producer/customer relationship, 193-194
  example, 162-163
  focusing on product/service, 197-198
  generating
    case study, 366-367
    explained, 358-359
  improving work flow, 188-190
  managing time, 194-195
  managing variation, 195-196
  optimizing inventory, 190-191
  overview, 185-187
  purpose of, 162
  selecting, 360-361, 367-368
  70 change concepts, 359
changing
  set points, 188
  targets, 188
  work environment, 191-192
charter. *See* **project charter**
charts
  affinity diagrams
    creating, 181-182
    defined, 156
    example, 156
    purpose of, 156
  bar charts
    explained, 50-51
    obtaining in Minitab, 74-76
  cause and effect (C&E) diagrams
    creating, 182
    defined, 157
    examples, 157-159
    purpose of, 157

control charts, 7
  attribute control charts, 90
  case study (defective surgical screws), 119-125
  c charts, 104-106, 134
  choosing, 119
  control limits, 93
  explained, 90
  I-MR (Individuals and Moving Range) charts, 109-112, 136-137
  p charts, 98-104, 131-133
  rules for determining out of control points, 93-98
  three-sigma limits, 93
  type one errors, 92
  type two errors, 92
  u charts, 106-108, 134-136
  variables control charts, 91
  X Bar and R charts, 112-115, 137-139
  X Bar and S charts, 115-119, 139-142
dot plots
  explained, 55-56
  obtaining in Minitab, 82-83
flowcharts
  advantages of, 39-40
  analyzing, 44-45
  ANSI standard flowchart symbols, 43-44
  creating, 165-166
  current state flowchart, 334, 344-345
  defined, 146
  deployment flowcharts, 42-43, 148, 334
  future state flowchart, 361, 368-369
  process flowcharts, 40-41, 147, 334
  purpose of, 146
  simple generic flowchart, 39
  symbols and functions, 165
Gage run charts, 313-317, 327
Gantt charts, 279-280
  case study, 296-297
  creating, 185
  defined, 159
  example, 161
  purpose of, 160
histograms
  explained, 54-55
  obtaining in Minitab, 79-82

line graphs
 explained, 52-54
 obtaining in Minitab, 78-79
Pareto diagrams
 creating, 182-185
 definition of, 159
 example, 159
 explained, 51-52
 obtaining in Minitab, 76-77
 purpose of, 159
run charts
 explained, 56-58
 obtaining in Minitab, 84-85
checklist, measurement system analysis checklist, 126-127
check sheets
 attribute check sheets, 177-178
 defect location check sheets, 179
 defined, 153
 example, 153-155
 measurement check sheets, 178
 purpose of, 153
classification data, 47
classifications, 188
clean workplace, 417
CMS (Centers for Medicare and Medicaid Services), 237
collateral damage to related processes, reducing
 case study, 386
 explained, 376-378
collecting baseline data for CTQs
 case study, 328-329
 explained, 317-321
common cause feedback loops, 23-24
common variation, 6-8, 89
 explained, 25-26
 Funnel Experiment, 27-29
 Red Bead Experiment, 30-31
communication
 communication plans
  case study, 299, 300
  creating, 198-200
  defined, 163
  example, 164
  explained, 282-283

 pilot tests, 362
 purpose of, 163
 operational definitions, 11-13
competition, 411
complaints
 customer feedback, 237
 employee feedback, 234-235
compliance, regulatory, 237-238
constancy of purpose, 413
constraints, 197
contingency plans, 195
continuous data, 49-50
continuous improvement, 9-11, 414
continuous improvement definition of quality, 32-33
control charts, 7
 attribute control charts, 90
 case study (defective surgical screws), 119-125
 c charts, 104-106
  creating in Minitab, 134
  example, 105-106
  explained, 104
  when to use, 105
 choosing, 119
 control limits, 93
 explained, 90
 I-MR (Individuals and Moving Range) charts, 109-112
  creating in Minitab, 136-137
  example, 109-112
  explained, 109
 p charts, 98-104
  creating in Minitab, 131-133
  explained, 98
  p chart with equal subgroup size, 99-102
  p chart with unequal subgroup size, 102-104
  when to use, 98-99
 rules for determining out of control points, 93-98
 three-sigma limits, 93
 type one errors, 92
 type two errors, 92
 u charts, 106-108
  creating in Minitab, 134-136
  example, 107-108

442   Index

explained, 106
when to use, 107
variables control charts, 91
X Bar and R charts, 112-115
example, 113-115
explained, 112
obtaining from Minitab, 137-139
X Bar and S charts, 115-119
example, 115-119
explained, 115
obtaining from Minitab, 139-142
control limits, 93
Control phase (DMAIC model)
case study: reducing patient no shows in outpatient psychiatric clinic, 386-391
collateral damage to related processes, reducing, 376-378
control plans, developing, 380-381
costs/benefits, documenting, 383
diffusion of improvements, 383-384
overview, 216
pitfalls to avoid, 385
projects, inputting into Six Sigma database, 383
purpose of, 375-376
standardization, 379-380
tips for success, 385
tollgate reviews, 384-385
control plans, developing
case study, 387-389
explained, 380-381
cooperation, 411
coordinators, 193
costs
benefits and, 281-282
cost avoidance, 242, 281
cost reduction, 242, 281
costs/benefits, documenting
case study, 387
explained, 383
intangible costs, 242, 281
reduction, 242
tangible costs, 242, 281
count data, 47
Critical-to-Quality characteristics. *See* **CTQs (Critical-to-Quality characteristics)**

cross-training, 192
CTQs (Critical-to-Quality characteristics), 145, 260
baseline data analysis
case study, 328-329
explained, 317-321
data collection plan, 312-313, 325-326
defined, 145, 260, 333
definition of, 288-289, 308
measurement system, validating, 313-317
operational definitions, 312
operation definitions, 325-326
process capability estimation, 321-323
culture
board of directors, 404-406
defined, 404-405
current state flowchart, creating
case study, 344-345
explained, 334
customer feedback, 237
customers
customer focus groups, 236
customer segments, 285
customer surveys, 236
identifying, 285
producer/customer relationship, enhancing, 193-194
Voice of the Customer analysis. *See* VoC (Voice of the Customer) analysis

D

dashboards, 232 238-240, 396-397
data
attribute classification data
explained, 47-48
measures of central tendency, 61-62
attribute count data
explained, 48-49
graphing, 50-54
attribute data, 47
central tendency, measures of
mean, 59
median, 60
mode, 60-61
proportion, 61-62

Index   443

defined, 47
measurement data, 49-50
   graphing, 54-58
   measures of central tendency, 59-61
shape, measures of, 66-68
skewness
   defined, 66
   negative or left skewness, 67-68
   positive or right skewness, 66-67
   symmetrical distribution, 66
variables, 47
variation, measures of
   range, 62-63
   standard deviation, 63-66, 68-69
   variance, 63-66
database, inputting projects into, 383
data collection plan
   for CTQs, 312-313, 325-326
   VoC (Voice of the Customer) analysis, 167, 287
   for Xs, 339-340, 348
data interpretation, VoC (Voice of the Customer) analysis, 167, 288
data redundancy, 187
decision matrix, 361, 367
decision symbol, 43
defects, 289, 322
   defined, 261, 289
   defect location check sheets, 179
   defect opportunities, 261, 289, 322
   defective surgical screws case study, 119-125
   detection, 414
   DPMO (defects per million opportunities), 262
   DPOs (defects per opportunity), 261
   DPUs (defects per unit), 261
   latent defects, 253
   prevention, 414
Define phase (DMADV cycle), 218
Define phase (DMAIC model)
   case study: reducing patient no shows in outpatient psychiatric clinic, 292-309
   definition of CTQs, 288-289, 308
   go-no go decision point, 290-291, 308-309
   initial draft of project objective, 289-290, 308
   overview, 213
   pitfalls to avoid, 291-292
   project charter, 276-283, 293-299
   purpose of, 273-274
   SIPOC analysis, 283-286, 299
   Six Sigma team, activating, 274-276, 292-293
   tips for success, 291-292
   tollgate review, 290-291, 308-309
   VoC (Voice of the Customer) analysis, 286-288, 299-308
defining processes
   boundaries, 37-38
   flowcharts, 40-41
   importance of, 35-36
   objectives, 38
   ownership, 36
Deming cycle, 417-419
Deming, W. Edwards, 27, 30. *See also* System of Profound Knowledge
   biographical information, 409
   quotations, 434-435
dependence on mass inspection, ceasing, 413-414
deployment flowcharts, 42-43, 148, 334
descriptive statistics, obtaining in Minitab, 85-87
desensitization, 196
Design for Six Sigma for Green Belts and Champions: Foundations, DMADV, Tools and Methods, Cases and Certification (Gitlow et al.), 359
designing systems to avoid mistakes, 196-197
Design of Experiments (DoE), 359
Design phase (DMADV cycle), 219
determining out of control points, 93-98
developing
   control plans
      case study, 387-389
      explained, 380-381
   hypotheses, 342
diagrams. *See also* charts
   affinity diagrams
      creating, 181-182
      defined, 156
      example, 156
      purpose of, 156

cause and effect (C&E) diagrams
  creating, 182
  defined, 157
  examples, 157-159
  purpose of, 157
  Pareto diagrams, 51-52
    adding to Minitab worksheets, 76-77
    creating, 182-185
    definition of, 159
    example, 159
    purpose of, 159
differentiation, 196
diffusing improvements throughout organization
  case study, 390
  explained, 383-384
Display Descriptive Statistics dialog box (Minitab), 86
diversity, 5
DMADV model
  example, 219-221
  explained, 218
DMAIC model
  Analyze phase
    case study: reducing patient no shows in outpatient psychiatric clinic, 344-355
    current state flowchart, creating, 334, 344-345
    data collection plan for Xs, 339-340, 348
    FMEA (failure modes and effects analysis), 338
    go-no go decision point, 342-343
    hypotheses about relationship between critical Xs and CTQs, 342, 354
    identification of Xs for CTQs, 335-338, 344-346
    measurement system for Xs, validating, 340, 348
    operational definitions for Xs, 346-347, 338
    overview, 214-215
    pitfalls to avoid, 343-344
    purpose of, 333-334
    test of theories to determine critical Xs, 340-342, 348-353
    tips for success, 343-344
    tollgate reviews, 354-355
  Control phase
    case study: reducing patient no shows in outpatient psychiatric clinic, 386-391
    collateral damage to related processes, reducing, 376-378
    control plans, developing, 380-381
    costs/benefits, documenting, 383
    diffusion of improvements, 383-384
    overview, 216
    pitfalls to avoid, 385
    projects, inputting into Six Sigma database, 383
    purpose of, 375-376
    standardization, 379-380
    tips for success, 385
    tollgate reviews, 384-385
  Define phase
    case study: reducing patient no shows in outpatient psychiatric clinic, 292-309
    definition of CTQs, 288-289, 308
    go-no go decision point, 290-291, 308-309
    initial draft of project objective, 289-290, 308
    overview, 213
    pitfalls to avoid, 291-292
    project charter, 276-283, 293-299
    purpose of, 273-274
    SIPOC analysis, 283-286, 299
    Six Sigma team, activating, 274-276, 292-293
    tips for success, 291-292
    tollgate review, 290-291, 308-309
    VoC (Voice of the Customer) analysis, 286-288, 299-308
  example, 216-217
  Improve phase
    alternative methods, generating, 358-359
    best alternative method, selecting, 360-361
    case study: reducing patient no shows in outpatient psychiatric clinic, 366-373
    future state flowchart, creating, 361
    go-no go decision point, 364-365
    overview, 216
    pilot testing, 362-364
    pitfalls to avoid, 365-366
    purpose of, 357-358
    risk mitigation, 362

    tips for success, 365-366
    tollgate reviews, 365
  Measure phase
    baseline data analysis for CTQs, 317-321, 328-329
    data collection plan for CTQs, 312-313, 325-326
    go-no go decision point, 323-324, 330
    operational definitions for CTQs, 312, 325-326
    overview, 213-214
    pitfalls to avoid, 324
    process capability estimation for CTQs, 321-323
    purpose of, 311-312
    tips for success, 324
    tollgate reviews, 323-324, 330
    validation of measurement system for CTQs, 326-327
  Measure phase (DMAIC model)
    validation of measurement system for CTQs, 313-317
    overview, 212-213
documenting
  costs/benefits
    case study, 387
    explained, 383
  processes
    flowcharts, 39-44
    importance of, 35-36
DoE (Design of Experiments), 359
Do phase
  PDSA cycle, 207
  SDSA cycle, 203
dot plots
  explained, 55-56
  obtaining in Minitab, 82-83
Dotplots dialog box (Minitab), 82
DPMO (defects per million opportunities), 262
DPOs (defects per opportunity), 261
DPUs (defects per unit), 261
driving out fear, 423

E

education
  encouraging, 430
  quality in, 434
egotism, 403-404
80-20 rule, 8
eliminating
  arbitrary goals, posters, and slogans, 424
  management by objective, 425-426
  waste, 187-188
  work standards (quotas), 424-425
employee feedback, 234-235
employee focus groups, 233
employee forums, 233-234
employee surveys, 234
empowerment, 420-421
ending practice of awarding business on basis of price, 414-415
engagement, board of directors, 406
enhancing producer/customer relationship, 193-194
errors, 92
estimating
  process capability for CTQs, 321-323
  project benefits, 242
  time to complete project, 245-246
executing projects, 250-251
executive steering committee, 256
expansion of knowledge through theory, 13
expectations, 193
experimental design, 359
external proactive sources, 235-236
external reactive sources, 236-238
extrinsic motivation, 410

F

failure in Act phase of PDSA cycle
  in accounts receivable, 394-396
  in manufacturing, 393-394
Failure Modes and Effects Analysis. *See* FMEA (Failure Modes and Effects Analysis)
fear, 423

feedback
    customer feedback, 237
    employee feedback, 234-235
    loops
        common cause feedback loops, 23-24
        defined, 19
        lack of, 23
fishbone diagrams. See cause and effect (C&E) diagrams
5S methods, 221-223, 417
flowcharts
    advantages of, 39-40
    analyzing, 44-45
    ANSI standard flowchart symbols, 43-44
    creating, 165-166
    current state flowchart, creating, 334, 344-345
    defined, 146
    deployment flowcharts, 42-43, 148, 334
    future state flowchart, 361, 368-369
    process flowcharts, 40-41, 147, 334
    purpose of, 146
    simple generic flowchart, 39
    symbols and functions, 165
flowline symbol, 43
FMEA (Failure Modes and Effects Analysis), 283
    case study, 299-301, 346
    conducting, 174-177
    defined, 153
    example of, 175
    explained, 338
    purpose of, 153
focus groups
    customer focus groups, 236
    employee focus groups, 233
focus on product/service, 197-198
focus points, 288
forced ranking of employees, 428-429
form value, 4
forums, employee, 233-234
14 Points (Deming), 413-430
    adopting new philosophy, 413
    breaking down barriers between departments, 423-424
    ceasing dependence on mass inspection, 413-414
    creating constancy of purpose, 413
    driving out fear, 423
    eliminating arbitrary goals, posters, and slogans, 424
    eliminating management by objective, 425-426
    eliminating work standards (quotas), 424-425
    encouraging education and self-improvement, 430
    ending practice of awarding business on basis of price, 414-415
    improving constantly the system of production and service, 415-421
    instituting leadership, 422-423
    reduction of variation and, 430-433
    removing barriers to pride of workmanship, 426-429
    taking action to accomplish transformation, 430
    training on the job, 421-422
Funnel Experiment, 27-29, 398-399
future state flowchart, 361, 368-369

G

Gage R&R studies, 127-131
Gage run charts, 313-317, 327
Gantt charts, 279-280
    case study, 296-297
    creating, 185
    defined, 159
    example, 161
    purpose of, 160
General Electric, 253
generating alternative methods
    case study, 366-367
    explained, 358-359
goal post view of quality, 31-32
goal statement, 278, 294
go-no go decision point
    Analyze phase, 342-343, 354-355
    Define phase, 290-291, 308-309
    Improve phase, 364-365, 373
    Measure phase, 323-324, 330

government, quality in, 434
graphing
    attribute count data
        bar charts, 50-51
        line graphs, 52-54
        Pareto diagrams, 51-52
    measurement data
        dot plots, 55-56
        histograms, 54-55
        run charts, 56-58
    in Minitab
        bar charts, 74-76
        dot plots, 82-83
        histograms, 79-82
        line graphs, 78-79
        Pareto diagrams, 76-77
        run charts, 84-85
Green Belts, 259-260

H

handoffs, minimizing, 189
hard benefits, 281-282
Health Insurance Portability and Accountability Act (HIPAA), 237
health maintenance, 224
high-level project charters, 246-247, 275
HIPAA (Health Insurance Portability and Accountability Act), 237
histograms
    explained, 54-55
    obtaining in Minitab, 79-82
Histogram: Scale dialog box (Minitab), 81
Histograms dialog box (Minitab), 80
history of Six Sigma, 253-254
hypothesis development, 342

I

ideal versus actual, 6
identifying
    customers, 285
    inputs, 285
    outputs, 285
    potential Xs, 308

projects
    customer feedback, 237
    customer focus groups, 236
    customer surveys, 236
    employee feedback, 234-235
    employee focus groups, 233
    employee forums, 233-234
    employee surveys, 234
    managerial dashboards, 238-240
    project identification matrix, 231
    regulatory compliance issues, 237-238
    strategic/tactical plans, 232
    VoC (Voice of the Customer) interviews, 235-236
    VoE (Voice of the Employee) interviews, 232-233
suppliers, 285
impact/effort matrix, 360
improvements, maintaining, 393
    board of directors culture, 404-406
    dashboards, 396-397
    diffusing throughout organization
        case study, 390
        explained, 383-384
    egotism, 403-404
    failure in Act phase of PDSA cycle
        in accounts receivable, 394-396
        in manufacturing, 393-394
    Funnel Experiment, 398-399
    presidential review of maintainability indicators, 397
    standardization
        case study, 386-387
        explained, 379-380
    succession planning
        creating talent pools model, 400
        explained, 399
        incumbent model, 399-400
        process oriented top-down/bottom-up model, 401-403
        top-down/bottom-up model, 400-401
Improve phase (DMAIC model)
    alternative methods, generating, 358-359
    best alternative method, selecting, 360-361
    case study: reducing patient no shows in outpatient psychiatric clinic, 366-373
    future state flowchart, creating, 361

go-no go decision point, 364-365
overview, 216
pilot testing, 362-364
pitfalls to avoid, 365-366
purpose of, 357-358
risk mitigation, 362
tips for success, 365-366
tollgate reviews, 365
I-MR Chart: Options dialog box (Minitab), 136
I-MR (Individuals and Moving Range) charts, 109-112
creating in Minitab, 136-137
example, 109-112
explained, 109
incumbent succession planning model, 399-400
independent components, system of, 401
Individuals and Moving Range (I-MR) charts, 109-112
creating in Minitab, 136-137
example, 109-112
explained, 109
initial draft of project objective
case study, 308
explained, 289-290
inputs, 18, 285
inputting projects into Six Sigma database, 383
inspect all-or-none rule, 414
inspection, 194
intangible costs, 242, 281
intermediaries, 188
internal proactive sources
employee focus groups, 233
employee forums, 233-234
employee surveys, 234
explained, 232
strategic/tactical plans, 232
VoE (Voice of the Employee) interviews, 232-233
internal reactive sources, 234-235
International Standards Organization (ISO), 379-380
interrelated components, system of, 401

interviews
VoC (Voice of Customer) interviews, 235-236
VoE (Voice of the Employee) interviews, 232-233
intrinsic motivation, 1, 410
inventory optimization, 190-191
ISO (International Standards Organization), 379-380

J-K

JCAHO (Joint Commission on Accreditation of Healthcare Organizations), 237

Kaizen
example, 210-212
explained, 209-210
key performance indicators (KPIs), 232
knowledge
expanding through theory, 13
theory of (Deming), 412
KPIs (key performance indicators), 232
kp rule, 414

L

latent defects, 253
LCL (lower control limit), 7, 92
leadership, 422-423
lean thinking, 341
5S methods, 221-223
overview, 221
poka-yoke, 224-225
SMED (Single Minute Exchange of Dies), 224-225
tools and methods, 359
TPM (Total Productive Maintenance), 223-224
value streams, 226-227
left skewness, 67-68
life as a process, 3-4
line graphs
explained, 52-54
obtaining in Minitab, 78-79

lower control limit (LCL), 7, 92
lower specification limit (LSL), 31
LSL (lower specification limit), 31

M

maintainability indicators, presidential review of, 397
maintaining improvements. *See* improvements, maintaining
maintenance
    autonomous maintenance, 223
    health maintenance, 224
    planned maintenance, 223-224
    Total Productive Maintenance (TPM), 223-224
"management by data," 410
"management by guts," 410
management by objectives (MBO), 425-426
management terminology, 260-264
    CTQs. *See* CTQs (Critical-to-Quality characteristics)
    defective, 261
    defects, 261
    DPMO (defects per million opportunities), 262
    DPOs (defects per opportunity), 261
    DPUs (defects per unit), 261
    process sigma, 262-264
    RTY (rolled throughput yield), 262
    units, 261
    yield, 262
management theory (Deming). *See* System of Profound Knowledge
managerial dashboards, 238-240
managing
    time, 194-195
    variation, 195-196
manufacturing, failure in Act phase of PDSA cycle, 393-394
market segmentation, 166-167
mass inspection, 413-414
Master Black Belts, 257-258

matrices
    decision matrix, 361, 367
    impact/effort matrix, 360
    project identification matrix, 231
    project prioritization matrix, 248-250
MBO (management by objectives), 425-426
mean, 59, 266, 341
measured day work, 424
measurement check sheets, 178
measurement data, 49-50
    control charts, 108-109
        I-MR *(Individuals and Moving Range) charts, 109-112*
        X Bar and R *charts, 112-115*
        X Bar and S *charts, 115-119*
    graphing
        dot plots, 55-56
        histograms, 54-55
        run charts, 56-58
    measures of central tendency
        mean, 59
        median, 60
        mode, 60-61
measurement system analysis checklist, 126-127
measurement systems analysis
    for CTQs, 313-317, 326-327
    explained, 126
    Gage R&R studies, 127-131
    measurement system analysis checklist, 126-127
    for Xs, 340, 348
Measure phase (DMADV cycle), 218
Measure phase (DMAIC model)
    baseline data analysis for CTQs, 317-321, 328-329
    data collection plan for CTQs, 312-313, 325-326
    go-no go decision point, 323-324, 330
    operational definitions for CTQs, 312, 325-326
    overview, 213-214
    pitfalls to avoid, 324
    process capability estimation for CTQs, 321-323
    purpose of, 311-312

tips for success, 324
tollgate reviews, 323-324, 330
validation of measurement system for CTQs, 313-317, 326-327
measures. *See* statistical analysis
median, 60, 341
methodology selection, 243-245
milestones, 279-280, 295
minimizing handoffs, 189
Minitab, 70
   bar charts, 74-76
   control charts, 131
      c charts, 134
      I-MR (Individuals and Moving Range) charts, 136-137
      p charts, 131-133
      u charts, 134-136
      X Bar and R charts, 137-139
      X Bar and S charts, 139-142
      zone limits, plotting, 131
   descriptive statistics, 85-87
   dot plots, 82-83
   histograms, 79-82
   line graphs, 78-79
   Pareto diagrams, 76-77
   run charts, 84-85
   worksheets, 70-74
mission statements, 18
mistakes, designing systems to avoid, 196-197
mitigating risk
   case study, 369
   explained, 362
mode, 60-61
monthly steering committee reviews, 251
motivation, 1, 410
Motorola Corporation, 417
multiple processing units, 190

N

negative reactive behaviors, 423
negative skewness, 67-68
new paradigm of leadership, 434
nominal value, 31

non-quantitative tools
   affinity diagrams, 156, 181-182
   brainstorming, 155-156, 179-181
   cause and effect (C&E) diagrams, 157-159, 182
   change concepts
      changing work environment, 191-192
      defined, 160
      designing systems to avoid mistakes, 196-197
      eliminating waste, 187-188
      enhancing producer/customer relationship, 193-194
      example, 162-163
      focusing on product/service, 197-198
      improving work flow, 188-190
      managing time, 194-195
      managing variation, 195-196
      optimizing inventory, 190-191
      overview, 185-187
      purpose of, 162
   check sheets, 153-155, 177-178
   communication plans, 163-164, 198-200
   flowcharts, 146-148, 165-166
   FMEA (Failure Modes and Effects Analysis), 153, 174-177
   Gantt charts, 159-161, 185
   operational definitions, 151-153
   overview, 145
   Pareto diagrams, 159, 182-185
   SIPOC analysis, 149-151, 172-173
   VoC (Voice of the Customer) analysis
      case study: reducing patient no shows at outpatient psychiatric clinic, 168-172
      data collection, 167
      data interpretation, 167
      defined, 146
      market segmentation, 166-167
      planning, 167
      purpose of, 149
non-technical definition of Six Sigma, 253
normal distribution, 68, 266
numeric data
   attribute classification data
      explained, 47-48
      measures of central tendency, 61-62

attribute count data
  explained, 48-49
  graphing, 50-54
attribute data, 47, 61-62
central tendency, measures of
  mean, 59
  median, 60
  mode, 60-61
  proportion, 61-62
measurement data, 49-50
  graphing, 54-58
  measures of central tendency, 59-61
shape, measures of, 66-68
skewness
  defined, 66
  negative or left skewness, 67-68
  positive or right skewness, 66-67
  symmetrical distribution, 66
variation, measures of
  range, 62-63
  standard deviation, 63-66, 68-69
  variance, 63-66

O

objectives of processes, 38
opening Minitab worksheets, 71-74
Open Worksheet dialog box (Minitab), 72
operational definitions, 11-13, 412, 416
  creating, 173-174
  defined, 151
  example, 151-153
  importance of, 153
  of CTQs
    case study, 325-326
    explained, 312
  purpose of, 151
  for Xs
    case study, 346-347
    explained, 338
optimization, 190-191, 411
out of control points, determining, 93-98
outpatient psychiatric clinic case study. *See* patient no shows at outpatient psychiatric clinic (case study)

outputs, 18, 285
overjustification, 413
ownership of processes, 36

P

paradigm shift, 433-434
Pareto Chart dialog box (Minitab), 76-77
Pareto diagrams
  adding to Minitab worksheets, 76-77
  creating, 182-185
  definition of, 159
  example, 159
  explained, 51-52
  purpose of, 159
passive baseline data, 318
patient no shows at outpatient psychiatric clinic (case study)
  Analyze phase
    current state process flowchart, 344-345
    data collection plan for Xs, 348
    FMEA (failure modes and effects analysis), 346
    go-no go decision point, 354-355
    hypotheses about relationship between critical Xs and CTQs, 354
    identification of Xs for CTQs, 344-346
    measurement system for Xs, validating, 348
    operational definitions of Xs, 346-347
    test of theories to determine critical Xs, 348-353
    tollgate reviews, 354-355
  Control phase, 386-391
    collateral damage to related processes, reducing, 386
    control plan, developing, 387-389
    diffusion of improvements, 390
    financial impact, 387, 390
    project, inputting into Six Sigma database, 390
    standardized improvements, 386-387
    tollgate review, 390-391
  Define phase
    definition of CTQs, 308
    go-no go decision point, 308-309
    initial draft of project objective, 308

project charter, 293-299
SIPOC analysis, 299
Six Sigma team, activating, 292-293
tollgate reviews, 308-309
VoC (Voice of the Customer) analysis, 299-308
Improve phase, 366-373
alternative methods, generating, 366-367
best alternative method, selecting, 367-368
future state flowchart, creating, 368-369
go-no go decision point, 373
pilot testing, 369-372
risk mitigation, 369
tollgate review, 373
Measure phase
baseline data analysis, 328-329
data collection plan for CTQs, 325-326
go-no go decision point, 330
operational definitions of CTQs, 325-326
tollgate review, 330
validation of measurement system for CTQs, 326-327
VoC (Voice of the Customer) analysis, 168-172
pay system, 191
P Chart dialog box (Minitab), 132
p charts, 98-104
creating in Minitab, 131-133
explained, 98
p chart with equal subgroup size, 99-102
p chart with unequal subgroup size, 102-104
when to use, 98-99
PDSA (Plan-Do-Study-Act) cycle, 417-419
example, 207-209
explained, 206-207
failure in Act phase, 393-396
performance appraisal systems, 428
piecework, 424
pilot tests, 362-364
case study, 369-372
communication plans, 362
data analysis, 362-364
employee training, 362
pilot test charter, 362-363
place value, 3
Plan-Do-Study-Act (PDSA) cycle, 417-419

planned maintenance, 223-224
Plan phase (PDSA cycle), 206
plans
assumptions, 13-14
communication plan
case study, 299-300
explained, 282-283
communication plans
creating, 198-200
defined, 163
example, 164
purpose of, 163
contingency plans, 195
control plans, 380-381
risk abatement plan, 283, 299
stability, 13-14
strategic/tactical plans, 232
succession planning, 399-403
VoC (Voice of the Customer) analysis, 167, 287
plotting zone limits, 131
poka-yoke, 224-225
positive skewness, 66-67
potential Xs, identifying, 308, 335-338
predictions, improving, 195
presidential review of maintainability indicators, 397
prevailing paradigm of leadership, 433
preventing backsliding. *See* backsliding, preventing
pride of workmanship, 426-429
prioritizing projects, 247-250
proactive data, 167
proactive sources. *See* external proactive sources; internal proactive sources
problem statements, 247, 277-278, 294
process capability estimation for CTQs, 321-323
processes, 17. *See also* quality; variation
analyzing, 44-45
defined, 18
defining, 35-38
DMAIC model. *See* DMAIC model
documenting
ANSI standard flowchart symbols, 43-44
benefits of flowcharts, 39-40

deployment flowcharts, 42-43
process flowcharts, 40-41
simple generic flowchart, 39
examples, 19-22
feedback loops, 19, 23-24
flowcharts, 39-41, 44-45, 147, 334
importance of, 19
maintaining improvements. *See* improvements, maintaining
orientation, 410
process flowcharts, 40-41
process sigma, 262-264
variation. *See* variation
where processes exist, 18-19
process improvement methodologies
    DMAIC model. *See* DMAIC model
    Kaizen/Rapid Improvement Events
        example, 210-212
        explained, 209-210
    lean thinking
        5S methods, 221-223
        overview, 221
        poka-yoke, 224-225
        SMED (Single Minute Exchange of Dies), 224-225
        TPM (Total Productive Maintenance), 223-224
        value streams, 226-227
    non-quantitative tools. *See* non-quantitative tools
    PDSA cycle
        example, 207-209
        explained, 206-207
    principles, 3-14
        continuous improvement, 9-11
        expansion of knowledge through theory, 13
        life as a process, 3-4
        operational definitions, 11-13
        special and common causes of variation, 6-8
        stability in planning, 13-14
        stable versus unstable processes, 8-9
        variation in processes, 5-6
        waste in processes, 11
    SDSA cycle, 203-204

processing symbol, 43
process oriented top-down/bottom-up succession planning, 401-403
Process Owners, 257
process sigma, 262-264
producer/customer relationship, enhancing, 193-194
production, improving, 415-421
products
    focus on, 197-198
    maintaining improvements. *See* improvements, maintaining
Profound Knowledge, System of. *See* System of Profound Knowledge
Project Champions, 256-257
project charter, 276-283
    benefits and costs, 281-282
    business case, 276-277
    case study, 293-299
    communication plan, 282-283
    goal statement, 278
    high level project charter, 275
    problem statement, 277-278
    project plan with milestones, 279-280
    project scope, 278-279
    risk abatement plan, 283
    roles and responsibilities, 282
project objective, initial draft of, 289-290, 308
project plan with milestones, 279-280, 295
project prioritization matrix, 248-250
projects
    benefits, 242
    executing, 250-251
    identifying
        customer feedback, 237
        customer focus groups, 236
        customer surveys, 236
        employee feedback, 234-235
        employee focus groups, 233
        employee forums, 233-234
        employee surveys, 234
        managerial dashboards, 238-240
        project identification matrix, 231
        regulatory compliance issues, 237-238
        strategic/tactical plans, 232

VoC (Voice of the Customer) interviews, 235-236
VoE (Voice of the Employee) interviews, 232-233
  prioritizing, 247-250
  screening and scoping, 278-279, 294-295
    estimation of project benefits, 242
    estimation of time to completion, 245-246
    high-level project charters, 246-247
    overview, 240-241
    problem statements, 247
    project methodology selection, 243-245
    questions to ask, 241
  selecting, 250
  tracking, 250-251
proportion, 61-62
psychiatric clinic case study. *See* patient no shows at outpatient psychiatric clinic (case study)
psychology, 412-413
pull systems, 190
push systems, 190

Q-R

quality, 31-33, 434
quick changeover, 224-225
quotas, 424-425
quotations from W. Edwards Deming, 434-435

range, 62-63
ranking of employees, 428-429
Rapid Improvement Events, 209-210
reactive data, 167
reactive sources. *See* external reactive sources; internal reactive sources
recycling, 188
Red Bead Experiment, 30-31
reduction of patient no shows at outpatient psychiatric clinic. *See* patient no shows at outpatient psychiatric clinic (case study)
regulatory compliance issues, 237-238
reminders, 196
resources, allocating, 250-251
responsibilities (team), 282, 298

reviews
  monthly steering committee reviews, 251
  presidential review of maintainability indicators, 397
  tollgate reviews
    Analyze phase, 342-344, 354-355
    Control phase, 384-385, 390-391
    Define phase, 290-291, 308-309
    Improve phase, 365, 373
    Measure phase, 323-324, 330
right skewness, 66-67
risk mitigation, 362, 369
risk abatement plan, 283, 299
risk sharing, 192
roles
  Black Belts, 258-259, 260
  executive steering committee, 256
  Green Belts, 259-260, 260
  Master Black Belts, 257-258
  Process Owners, 257
  Project Champions, 256-257
  Senior Executives, 255
RTY (rolled throughput yield), 262
Run Chart dialog box (Minitab), 85
run charts
  explained, 56-58
  obtaining in Minitab, 84-85

S

sampling, 188
Sarbanes Oxley Act (SOX), 237
Save Worksheet As dialog box (Minitab), 73
saving Minitab worksheets, 71-74
scope
  case study, 294-295
  project scope, 278-279
scoping projects
  estimation of project benefits, 242
  estimation of time to completion, 245-246
  high-level project charters, 246-247
  overview, 240-241
  problem statements, 247
  project methodology selection, 243-245
  questions to ask, 241

scorecards, balanced, 232
screening projects
    estimation of project benefits, 242
    estimation of time to completion, 245-246
    high-level project charters, 246-247
    overview, 240-241
    problem statements, 247
    project methodology selection, 243-245
    questions to ask, 241
SDSA (Standardize-Do-Study-Act) cycle, 203-204, 416-417
segmentation
    customer segments, 285
    market segmentation, 166-167
seiketsu, 417
seiri, 417
seiso, 417
seiton, 417
self-discipline, 222, 417
self-improvement, 430
Senior Executives, 255
service
    focus on, 197-198
    improving, 415-421
    quality in, 434
    maintaining improvements. *See* improvements, maintaining
set points, changing, 188
70 change concepts, 359
shape, measures of, 66-68
shared mission, 405
shared risks, 192
shared values/beliefs, 405
shitake, 417
shitsuke, 417
simulation, 359
Single Minute Exchange of Dies (SMED), 224-225
single suppliers, 414-415
SIPOC (Supplier-Input-Process-Output-Customer) analysis, 149
    case study, 299
    creating, 172-173
    defined, 149
    example, 149-151

    explained, 283-286
    purpose of, 149
Six Sigma
    benefits of, 254
    history of, 253-254
    importance of, 272
    management opportunities, 261
    management terminology, 260-264
        CTQs. See CTQs (Critical-to-Quality characteristics)
        defective, 261
        defects, 261
        DPMO (defects per million opportunities), 262
        DPOs (defects per opportunity), 261
        DPUs (defects per unit), 261
        process sigma, 262-264
        RTY (rolled throughput yield), 262
        units, 261
        yield, 262
    non-technical definition, 253
    roles
        Black Belts, 258-260
        executive steering committee, 256
        Green Belts, 259-260
        Master Black Belts, 257-258
        Process Owners, 257
        Project Champions, 256-257
        Senior Executives, 255
    teams, activating
        case study, 292-293
        overview, 274-276
    technical definition, 253, 266-272
        normal distribution, 266
        relationship between VoP and VoC, 266-271
    tips for success, 255
skewness, 66-68
slogans, 424
SMART (Specific, Measurable, Attainable, Relevant, and Time Bound), 290
SMED (Single Minute Exchange of Dies), 224-225
Smith, Bill, 253
soft benefits, 281-282
sorting, 222, 417

SOX (Sarbanes Oxley Act), 237
special variation, 6-8, 90
    explained, 25-26
    Funnel Experiment, 27-29
    Red Bead Experiment, 30-31
specification limits, 31
spic and span, 222, 417
stability in planning, 13-14
stable processes, 8-9
standard deviation, 68-69, 266, 341
standard flowchart symbols, 43-44
standardization, 195, 222, 417
    case study, 386-387
    explained, 379-380
    improvements
        case study, 386-387
        explained, 379-380
Standardize-Do-Study-Act (SDSA) cycle, 203-204, 416-417
Standardize phase (SDSA cycle), 203
start/stop symbol, 43
statistical analysis, 341
    central tendency
        mean, 59
        median, 60
        mode, 60-61
        proportion, 61-62
    Minitab, 70
        bar charts, 74-76
        descriptive statistics, 85-87
        dot plots, 82-83
        histograms, 79-82
        line graphs, 78-79
        Pareto diagrams, 76-77
        run charts, 84-85
        worksheets, 70-74
    shape (skewness), 66-68
    variation
        range, 62-63
        standard deviation, 63-69
        variance, 63-66
strategic/tactical plans, 232
Study phase (PDSA cycle), 207
Study phase (SDSA cycle), 203
subgroups, 7

substitution, 188
succession planning
    creating talent pools model, 400
    explained, 399
    incumbent model, 399-400
    process oriented top-down/bottom-up model, 401-403
    top-down/bottom-up model, 400-401
suggestions
    customer feedback, 237
    employee feedback, 234-235
Supplier-Input-Process-Output-Customer analysis. *See* SIPOC (Supplier-Input-Process-Output-Customer) analysis
suppliers
    identifying, 285
    single suppliers, 414-415
surveys
    customer surveys, 236
    employee surveys, 234
symbols (flowchart), 43-44
symmetrical distribution, 66
synchronization, 189
system, appreciation of, 411
systematization, 222, 417
system of independent components, 401
system of interrelated components, 401
System of Profound Knowledge
    appreciation of a system, 411
    14 Points, 413-430
        adopting new philosophy, 413
        breaking down barriers between departments, 423-424
        ceasing dependence on mass inspection, 413-414
        creating constancy of purpose, 413
        driving out fear, 423
        eliminating arbitrary goals, posters, and slogans, 424
        eliminating management by objective, 425-426
        eliminating work standards (quotas), 424-425
        encouraging education and self-improvement, 430

ending practice of awarding business on basis of price, 414-415
improving constantly the system of production and service, 415-421
instituting leadership, 422-423
reduction of variation and, 430-433
removing barriers to pride of workmanship, 426-429
taking action to accomplish transformation, 430
training on the job, 421-422
overview, 409-410
paradigms, 410-411
psychology, 412-413
purpose, 410
quality in service, government, and education, 434
quotations from W. Edwards Deming, 434-435
theory of knowledge, 412
theory of variation, 412
transformation, 433-434

T

tactical plans, 232
Taguchi Loss Function (TLF), 32-33
talent pool model of succession planning, 400
tampering, 195
tangible costs, 242, 281
targets, changing, 188
technical definition of Six Sigma, 253, 266-272
    normal distribution, 266
    relationship between VoP and VoC, 266-271
testing
    pilot tests, 362-364
        case study, 369-372
        communication plans, 362
        data analysis, 362-364
        employee training, 362
        pilot test charter, 362-363
    test of theories to determine critical Xs, 340-342, 348-353
theory of knowledge, 412
theory of management. *See* System of Profound Knowledge

theory of variation, 412
three-sigma limits, 93
time management, 194-195
time series plot, 78. *See also* line graphs
Time Series Plot: Simple dialog box (Minitab), 79
time to complete project, estimating, 245-246
time value, 3
TLF (Taguchi Loss Function), 32-33
tollgate reviews
    Analyze phase
        case study, 354-355
        explained, 342-343
    Control phase
        case study, 390-391
        explained, 384-385
    Define phase, 290-291, 308-309
    Improve phase
        case study, 373
        explained, 365
    Measure phase, 323-324, 330
top-down/bottom-up model of succession planning, 400-401
TPM (Total Productive Maintenance), 223-224
tracking projects, 250-251
training, 192, 224, 421-422
transformation, 430, 433-434
type one errors, 92
type two errors, 92

U

U Chart dialog box (Minitab), 135
u charts, 106-108
    creating in Minitab, 134-136
    example, 107-108
    explained, 106
    when to use, 107
UCL (upper control limit), 7, 92
units, 261, 289, 322
unit-to-unit variation, 5-6
unstable processes, 8-9
unwanted variation, 6
USL (upper specification limit), 31

V

validating
   measurement system for CTQs
      case study, 326-327
      explained, 313-317
   measurement system for Xs
      case, 348
      explained, 340

value engineering, 188

values
   form value, 4
   nominal value, 31
   place value, 3
   time value, 3
   value streams, 226-227

variability, reducing, 406

variables, 47
   variables control charts, 91
   variables data, 49-50

variation
   common variation, 6-8, 25-26, 89
   control charts
      attribute control charts, 90
      case study (defective surgical screws), 119-125
      c charts, 104-106, 134
      choosing, 119
      control limits, 93
      explained, 90
      I-MR (Individuals and Moving Range) charts, 109-112, 136-137
      p charts, 98-104, 131-133
      rules for determining out of control points, 93-98
      three-sigma limits, 93
      type one errors, 92
      type two errors, 92
      u charts, 134-136
      variables control charts, 91
      X Bar and R charts, 112-115, 137-139
      X Bar and S charts, 115-119, 139-142
   defined, 5-6, 24-25, 89
   Funnel Experiment, 27-29
   importance of, 25
   measurement systems analysis, 126-131
   measures of
      range, 62-63
      standard deviation, 63-66, 68-69
      variance, 63-66
   Minitab, 131
      c charts, 134
      I-MR (Individuals and Moving Range) charts, 136-137
      p charts, 131-133
      u charts, 134-136
      X Bar and R charts, 137-139
      X Bar and S charts, 139-142
      zone limits, plotting, 131
   Red Bead Experiment, 30-31
   reducing, 430-433
   special variation, 6-8, 25-26, 90
   theory of variation (Deming), 412
   variation management, 195-196

Verify/Validate phase (DMADV cycle), 219

VoC (Voice of the Customer) analysis, 286-288
   case study, 168-172, 299-308
   data collection, 167, 287
   data interpretation, 167, 288
   defined, 146
   interviews, 235-236
   market segmentation, 166-167, 287
   planning, 167, 287
   purpose of, 149
   Six Sigma, 266-271

VoE (Voice of the Employee) interviews, 232-233

VoP (Voice of the Process), 266-271

W

wait time, reducing, 195

waste
   eliminating, 187-188
   in processes, 11

Welch, Jack, 253

work environment, changing, 191-192

work flow, improving, 188-190

worksheets (Minitab)
   bar charts, 74-76
   descriptive statistics, 85-87

    dot plots, 82-83
    explained, 70-71
    histograms, 79-82
    line graphs, 78-79
    opening, 71-74
    Pareto diagrams, 76-77
    run charts, 84-85
    saving, 71-74
work standards (quotas), 424-425

X

X Bar and R charts, 112-115, 137-139
X Bar and S charts, 115-119, 139-142
Xbar-R Chart dialog box (Minitab), 138-140
Xbar-S Chart dialog box (Minitab), 141
Xs
    data collection plan, 339-340
    defined, 333
    hypotheses about relationship between critical Xs and CTQs, 354
    identification of potential Xs, 308, 335-338
    operational definitions, 346-348
    test of theories to determine critical Xs, 340-342, 348-353

Y-Z

yield, 262

zone limits, plotting, 131
ZQC (Zero Quality Control), 225